SEISMIC DESIGN AIDS
for Nonlinear Pushover Analysis of
REINFORCED CONCRETE
and STEEL BRIDGES

Advances in Earthquake Engineering Series

Series Editor: Franklin Y. Cheng
Missouri University of Science and Technology

Seismic Design Aids for Nonlinear Pushover Analysis of Reinforced Concrete and Steel Bridges

Jeffrey Ger and Franklin Y. Cheng

Seismic Design Aids for Nonlinear Analysis of Reinforced Concrete Structures

Srinivasan Chandrasekaran, Luciano Nunziante, Giorgio Serino, and Federico Carannante

SEISMIC DESIGN AIDS
for Nonlinear Pushover Analysis of
REINFORCED CONCRETE
and STEEL BRIDGES

Jeffrey Ger
Franklin Y. Cheng

CRC Press
Taylor & Francis Group
Boca Raton London New York

CRC Press is an imprint of the
Taylor & Francis Group, an **informa** business

CRC Press
Taylor & Francis Group
6000 Broken Sound Parkway NW, Suite 300
Boca Raton, FL 33487-2742

First issued in paperback 2017

© 2012 by Taylor & Francis Group, LLC
CRC Press is an imprint of Taylor & Francis Group, an Informa business

No claim to original U.S. Government works

Version Date: 20110608

ISBN-13: 978-1-4398-3763-4 (hbk)
ISBN-13: 978-1-138-11462-3 (pbk)

Library of Congress Cataloging-in-Publication Data

Ger, Jeffrey.
  Seismic design aids for nonlinear pushover analysis of reinforced concrete and steel bridges / Jeffrey Ger and Franklin Y. Cheng.
    p. cm. -- (Advances in earthquake engineering)
    Includes bibliographical references and index.
    ISBN 978-1-4398-3763-4 (alk. paper)
    1. Concrete bridges--Design and construction. 2. Iron and steel bridges--Design and construction. 3. Structural engineering--Mathematics. 4. Reinforced concrete. 5. Lateral loads. I. Cheng, Franklin Y. II. Title. III. Series.

TG340.G46 2011
624.2'5--dc23   2011014658

Visit the Taylor & Francis Web site at
http://www.taylorandfrancis.com

and the CRC Press Web site at
http://www.crcpress.com

Disclaimer

In using the computer coding given in *Seismic Design Aids*, the reader accepts and understands that no warranty is expressed or implied by the authors on the accuracy or the reliability of the programs. The examples presented are only introductory guidelines to explain the applications of proposed methodology. The reader must independently verify the results and is responsible for the results.

To our families for their immense love and support

Jeffrey Ger

Father—Chia Chian

Mother—Mei Yu

Wife—Jenny

Son—Max

Daughter—Christie

Franklin Y. Cheng

Wife—Pi Yu (Beatrice)

Son and daughter-in-law—George and Annie

Daughter and son-in-law—Deborah and Craig

Grandchildren—Alex, Camille, and Natalie

Contents

Series Preface

The new *2009 AASHTO Guide Specifications for LRFD Seismic Bridge Design* requires pushover analysis for Seismic Design Category D bridges. The pushover analysis can identify the failure modes with collapse sequence of damaged bridges for the limit state design of the system. This is a benchmark book that provides readers with an executable file for a computer program, INSTRUCT, to serve the engineering community's needs. The book includes step-by-step numerical procedures with five different nonlinear element stiffness formulation methods that vary from the most sophisticated to the simplest and are suitable for users with varying levels of experience in nonlinear analysis. Most of the numerical examples provided with the demonstration of the accuracy of analytical prediction conformed well with the full- or large-scale test results. The key features of this book are as follows:

1. A complete handbook for pushover analysis of reinforced concrete and steel bridges with confined and nonconfined concrete column members of either circular or rectangular cross sections as well as steel members of standard shapes
2. New technology for displacement-based seismic analysis with various in-depth, nonlinear member stiffness formulations
3. Step-by-step pushover analysis procedures and applications in bridge engineering
4. A computer execute file for readers to perform pushover analysis
5. Real engineering examples with performance-based bridge design
6. Detailed figures/illustrations as well as detailed input and output descriptions

This book is a useful reference for researchers and practitioners working in the field of structural engineering. It is also a key resource for senior undergraduates and all postgraduates that provides an organized collection of nonlinear pushover analysis applications.

Preface

Nonlinear static monotonic analysis, or pushover analysis, has become a common practice for performance-based bridge seismic design. The *2009 AASHTO Guide Specifications for LRFD Seismic Bridge Design* (AASHTO, 2009) explicitly requires pushover analysis for Seismic Design Category D bridges. The *2006 FHWA Seismic Retrofitting Manual for Highway Structures: Part I—Bridges* (FHWA, 2006) adopted pushover analysis for bridges in Seismic Retrofit Categories C and D to assess bridge seismic capacity. The popularity of pushover analysis is mainly due to its ability to identify failure modes and design limit states of bridge piers and provide the progressive collapse sequence of damaged bridges when subjected to major earthquakes. Unfortunately, there is no complete technical reference in this field to give the practical engineer step-by-step procedures for pushover analyses and various nonlinear member stiffness formulations. This book includes step-by-step procedures for pushover analysis and provides readers an executable file for a computer program, INSTRUCT (INelastic STRUCTural Analysis of Reinforced-Concrete and Steel Structures) to perform pushover analysis. The readers can download the INSTRUCT executable file from the website at http://www.crcpress.com/product/isbn/9781439837634. Many examples are provided to demonstrate the accuracy of analytical prediction by comparing numerical results with full- or large-scale test results.

The computer program INSTRUCT was developed based on a microcomputer program INRESB-3D-SUPII (Cheng et al., 1996a and b) and mainframe program INRESB_3D-SUP (Cheng and Mertz, 1989a). INRESB-3D-SUPII was a modular computer program consisting of six primary blocks. The first block (STRUCT) defines the structural model. The remaining five blocks (SOL01, SOL02, SOL03, SOL04, and SOL05) are independent solutions for static loading, seismic loading, natural frequency and buckling loading, static cyclic or pushover loading, and response spectrum analysis, respectively. Since the purpose of INSTRUCT is mainly to perform nonlinear pushover analysis of reinforced concrete and steel bridge bents, it includes only SOL01 and SOL04. During the development of INSTRUCT, SOL04 was enhanced significantly, and it includes five different nonlinear element stiffness formulation methods for pushover analysis. They are finite segment–finite string (FSFS), finite segment–moment curvature (FSMC), axial load–moment interaction (PM), constant moment ratio (CMR), and plastic hinge length (PHL) methods. These range from the most sophisticated to the simplest and are suitable for engineers with varying levels of experience in nonlinear structural analysis. The results from these methods have been compared during the development of the program. They generally exhibit reasonable differences due to the different numerical operation of individual methods, but are consistent in general. SOL04 is capable of performing not only unidirectional pushover analysis but also cyclic pushover analysis. Depending on future needs, SOL02, SOL03, and SOL05 can be incorporated into future versions of INSTRUCT.

Chapter 1 describes the evolution of seismic bridge design codes in the United States over the past 70 years and includes a comparison between force-based and displacement-based design approaches. Regardless of the design approach being used, it demonstrates the importance of using pushover analysis for seismic bridge design and retrofitting evaluation.

Chapter 2 summarizes the application of pushover analysis in force-based bridge design as well as in displacement-based seismic bridge design. Other applications such as capacity/demand analysis for the evaluation of existing bridges, quantitative bridge redundancy evaluation, moment–curvature analysis, and estimation of inelastic response demand for buildings are also described in this chapter.

Nonlinear pushover analysis procedure is described in Chapter 3. The flowchart for structural modeling and the procedures for solutions SOL01 and SOL04 are described. Material and element libraries are provided, including 12 material and 7 element types. The material library covers elastic material and hysteresis models of bilinear, Takeda, gap/restrainer, hinge, interaction axial load–moment, finite-segment (steel), finite-segment (reinforced concrete), FSMC, plate, point, and brace materials. The element library includes elastic three-dimensional (3D) beam, spring, inelastic 3D beam, finite-segment, plate, point, and brace elements.

The nonlinear bending stiffness matrix formulations for reinforced concrete members are described in Chapter 4, including the above-mentioned FSFS, FSMC, PM, CMR, and PHL methods. Since most bridge columns in the United States are reinforced concrete columns, it is necessary to check all the possible concrete column failure modes in the pushover analysis. Possible concrete column failure modes include

1. Compression failure of unconfined concrete due to fracture of transverse reinforcement
2. Compression failure of confined concrete due to fracture of transverse reinforcement
3. Compression failure due to buckling of the longitudinal reinforcement
4. Longitudinal tensile fracture of reinforcing bars
5. Low cycle fatigue of the longitudinal reinforcement
6. Failure in the lap-splice zone
7. Shear failure of the member that limits ductile behavior
8. Failure of the beam–column connection joint

INSTRUCT is capable of checking all the possible concrete column failure modes. The approaches used to check individual failure modes are also described in this chapter.

Chapter 5 describes how to combine bending, shear, axial, and torsional stiffnesses to form the 3D element stiffness matrices for bridge columns and cap beams. The stiffness matrix formulation for other elements such as brace and plate elements is introduced in this chapter. Once all the element stiffness matrices are formulated, a 3D structural system subjected to both static and nonlinear push-over loadings can be analyzed. The definitions of structural joints and degrees of freedom (dofs), including free, restrained, condensed, or constrained dofs, are also described in detail.

Chapter 6 contains detailed input data instructions. The modular form of INSTRUCT allows the addition of new materials and/or new elements into the program depending on future needs. The structural analysis adopted in the program is based on the matrix method. The system formulation in INSTRUCT has the following attributes: (1) joint-based degrees of freedom, (2) rigid body and planar constraints, (3) material and geometric stiffness matrix formulation, and (4) unbalanced load correction. INSTRUCT has been developed to achieve efficiency in both computation and data preparation. The output solutions include the results of joint forces and displacements, member forces and deformations, member ductility factors, and structural displacement capacities corresponding to different performance-based limit states.

Chapter 7 provides 13 numerical examples to illustrate the preparation of input data and the output solutions for the bridge pushover analysis of reinforced concrete and steel bridge bents. Most examples provide a comparison between the numerical results and available experimental test results. Many existing steel diaphragms (cross frames) in steel or prestressed concrete girder bridges were not designed for high seismic loads, and the inelastic buckling of brace members could occur when subjected to lateral loads. For steel pile cap bents, the steel piles may develop plastic hinges and the diagonal brace members may buckle due to lateral seismic load. As shown in some of the examples, INSTRUCT is capable of performing pushover analysis for steel pile cap bents and steel diaphragms, with consideration of post-buckling effects of steel members.

The majority of the mathematic derivations for the nonlinear stiffness matrices of various structural elements, nonlinear member cross-sectional properties, and different numerical analyses described in this book are included in Appendices A through E, I, and J. Although this book is mainly for readers who have fundamental earthquake engineering and structural dynamics background, Appendices F through H provide structural engineers with basic knowledge of dynamic analysis of structures, including elastic and inelastic time history analyses, damped free vibration, damped vibration with dynamic force, the development elastic and inelastic response spectra, equivalent viscous damping, and the response spectrum analysis of the multiple-degrees-of-freedom system.

The photo shown on the book cover is of the Tanana River Bridge near Tok, Alaska, which was one of the first bridges in Alaska designed using the *AASHTO Guide Specifications for LRFD Seismic Bridge Design* (AASHTO, 2009) and pushover analysis to ensure that the displacement capacities of individual piers are greater than the corresponding seismic displacement demands. The authors wish to thank Derek Soden, the former Alaska DOT structural designer who designed this bridge, for providing this photo cover and proofreading a majority of the book manuscripts.

Series Editor

Franklin Y. Cheng, PhD, PE, is a distinguished member (formerly honorary member) of ASCE; member of the Academy of Civil Engineers, Missouri University of Science and Technology (MST); and curators' professor emeritus of civil engineering, MST. He is one of pioneers in allying computing expertise to large, complex, seismic-resistant structures with major impact as follows: (1) development of computer algorithms and programs for supercomputers, PCs, and executable files on the web available worldwide; response results that conform well with field observations such as the 22-story Pino-Suraez steel buildings in Mexico City and River Crossing reinforced concrete bridges in California; (2) optimum design of 2D and 3D tall buildings in the United States and abroad for both practice and parametric investigations such as design logic procedures and criteria of the Tentative Provisions of ATC-3 for improvement recommendations; (3) development of a smart HDABC (hybrid damper actuator braced control) system considering soil–structure interaction effective for various earthquake magnitudes with shaking-table verification; (4) leadership in integrating frontier design and retrofitting techniques through international workshops. Dr. Cheng has numerous publications to his credit, the most recent being *Structural Optimization—Dynamic and Seismic Applications*, *Smart Structures—Innovative Systems for Seismic Response Control*, and *Matrix Analysis of Structural Dynamic—Applications and Earthquake Engineering*.

Authors

Jeffrey Ger is the Federal Highway Administration (FHWA) division bridge engineer in Florida, Puerto Rico, and U.S. Virgin Islands. Before joining FHWA, he worked with the Missouri Department of Transportation where he supervised 10 bridge designers and was extensively involved with projects for the seismic retrofitting and strengthening of bridges in St. Louis, Missouri. He received his PhD in civil engineering from the University of Missouri-Rolla. His research experience has been in the field of earthquake engineering, nonlinear structural response, and building and highway bridge design. He has published more than 40 technical papers in structural engineering. Dr. Ger is a professional engineer and is a member of FHWA's National Seismic Virtual Team. He is also the FHWA Ex-Officio for the Technical Committee T-4 (Construction) of the American Association of State Highway and Transportation Officials (AASHTO) Subcommittee on Bridges and Structures. He is a panel member of current NCHRP 12-86 and 20-07 research projects on "Bridge System Safety and Redundancy," and "Update AASHTO Guide Spec. for Bridge Temporary Works," respectively. Dr. Ger received the U.S. Secretary of Transportation's Team Award in 2004 "for providing extraordinary transportation services to move food, water and shelter materials to relieve the pain and suffering by millions of victims of the 2004 Hurricanes." He provided critical support in the wake of Florida's 2004 hurricanes, completing an emergency interstate bridge repair project 26 days ahead of schedule. In 2006, he received the FHWA Bridge Leadership Council's Excellent Award, recognizing his outstanding customer service in carrying out the bridge program in Florida. In 2007, he received the FHWA Engineer of the Year Award as well as an award from the National Society of Professional Engineers, as one of the top 10 federal engineers of the year among the 26 federal agencies. In 2008, he received the Civil Engineering Academy Award from the Department of Civil Engineering at the University of Missouri-Rolla. Recently, Dr. Ger was appointed as one of the seven members of the U.S. Transportation Infrastructure Reconnaissance Team traveling to Chile in April 2010 to assess the bridge damage condition due to the February 27, 2010 Chile earthquake.

Franklin Y. Cheng was appointed Curators' Professor of Civil Engineering at the University of Missouri-Rolla (now Missouri University of Science and Technology, MST) in 1987, the highest professorial position in Missouri's university system, and is the senior investigator, Intelligent Systems Center, University of Missouri-Rolla. Dr. Cheng has received 4 honorary professorships abroad and chaired 7 of his 24 National Science Foundation (NSF) delegations to various countries for research and development cooperation. He has served as either chairman or member of 37 professional societies and committees, 12 of which are American Society of Civil Engineers (ASCE) groups. He was the first chair of the Technical Administrative Committee on Analysis and Computation and initiated the Emerging Computing Technology Committee and Structural Control Committee. He also initiated and chaired the Stability Under Seismic Loading Task Group of the Structural Stability Research Council. Dr. Cheng has served as a consultant for Martin Marietta Energy Systems Inc., Los Alamos National Laboratory, and Martin & Huang International, among others. The author, coauthor, or editor of 26 books and over 250 publications, Dr. Cheng is the recipient of numerous honors, including the MSM-UMR Alumni Merit, ASCE State-of-the-Art (twice), the Faculty Excellence, and the Haliburtan Excellence awards. In 2007, he was elected as the 565th honorary member of ASCE since 1852. Dr. Cheng gained industrial experience with C.F. Murphy and Sargent & Lundy in Chicago, Illinois. He received his PhD (1966) in civil engineering from the University of Wisconsin-Madison.

1 Overview of Seismic Design of Highway Bridges in the United States

1.1 INTRODUCTION

The nonlinear static monotonic analysis, or pushover analysis, has become a common procedure in current structural engineering practice (ATC-40, 1996; FEMA-273, 1997; FEMA-356, 2000). The American Association of State Highway and Transportation Officials (AASHTO) Guide Specifications for load and resistance factors design (LRFD) Seismic Bridge Design explicitly require pushover analysis for seismic design category D (SDC D) bridges. The *2006 FHWA Seismic Retrofitting Manual for Highway Structures: Part I—Bridges* (FHWA, 2006) adopted pushover analysis in evaluation method D2 for bridges of seismic retrofit categories C and D (SRC C and SRC D) to assess bridge seismic performance.

This chapter describes the evolution of seismic bridge design codes in the United States. The intent is not to introduce the seismic design codes in detail, but to illustrate the differences among these codes and discuss major code improvements over the past 70 years. The history of code development can explain why the current *AASHTO Guide Specifications for LRFD Seismic Bridge Design* and the *FHWA Seismic Retrofitting Manual* require using nonlinear pushover analysis for bridge design and retrofit, respectively. This chapter also provides a discussion of possible future code improvement.

1.2 AASHTO BRIDGE SEISMIC DESIGN PHILOSOPHY

The highway bridge design code in the United States has evolved several times over the past 70 years. The first highway bridge design code was published in 1931 by the American Association of State Highway Officials (AASHO), later by the AASHTO. From 1931 through 1940, AASHO codes did not address seismic design. The 1941 edition of the AASHO code required that bridges be designed for earthquake load; however, it did not specify how to estimate that load. In 1943, the California Department of Transportation (Caltrans) developed various levels of equivalent static lateral forces for the seismic design of bridges with different foundation types, with individual members designed using the working stress design (WSD) method (Moehle et al., 1995).

Following Caltrans' criteria, the 1961 edition of the AASHO specifications for the first time specified an earthquake loading for use with the WSD design approach. This seismic provision, used until 1975, did not include a national seismic map. The AASHO design code provisions from this period are briefly described as follows.

1.2.1 AASHO ELASTIC DESIGN PROCEDURES (1961–1974)

In regions where earthquakes may be anticipated, the equivalent earthquake static lateral force was calculated (AASHO, 1969) as follows:

$$EQ = CD \qquad (1.1)$$

where
   EQ is the lateral force applied horizontally at the center of gravity of the structure
   D is the dead load of structure
   $C = 0.02$ for structures founded on spread footings on material rated as 4 t or more per square foot
   $C = 0.04$ for structures founded on spread footings on material rated as less than 4 t per square foot
   $C = 0.06$ for structures founded on piles

The earthquake force, EQ, calculated from Equation 1.1 was part of the Group VII loading combination given by

$$\text{Group VII} = D + E + B + SF + EQ \qquad (1.2)$$

in which D, E, B, and SF are dead load, earth pressure, buoyancy, and stream flow, respectively. With WSD, the code allowed a $33\frac{1}{3}\%$ increase in the allowable stress for member design due to earthquake consideration. For reinforced concrete columns subjected to bending, the allowable compression stress at the extreme fiber was $0.4 f_c'$, and tension stress at the extreme fiber of the member was not permitted.

Despite the Caltrans design criteria, many highway bridges were severely damaged or collapsed during the 1971 San Fernando earthquake. The post-earthquake damage assessment indicated that the elastic WSD provisions for bridges subjected to earthquake were not adequate. This event illustrated the drawbacks of elastic design, such as (1) the seismic lateral force levels of 2%, 4%, and 6% of the total structural dead load were too low in California, (2) the actual column moment demand reached the column moment capacity, (3) columns were not designed for ductility, which resulted in brittle failure during the earthquake, and (4) energy dissipation was very small.

Following the San Fernando earthquake, Caltrans developed a new force-based seismic design procedure for highway bridges. The new design criteria included soil effects on seismic load and the dynamic response characteristics of bridges. It increased the amount of column transverse reinforcement for ductility, and beam seat lengths were increased to minimize the risk of unseating of the superstructure. In 1975, AASHTO adopted an interim seismic design specification, which was based on the Caltrans' design criteria. The same design criteria were used in the 1977, 1983, 1989, and 1992 AASHTO Standard Specifications. The following describes the design criteria during this time period.

1.2.2   AASHTO Force-Based Design Procedures (1975–1992)

The equivalent static force method was used to calculate the design earthquake loading. The design earthquake load is given as follows:

$$EQ = CFW \qquad (1.3)$$

where
   EQ is the equivalent static horizontal force applied at the center of gravity of the structure
   F is the framing factor
   $F = 1.0$ for structures where single columns or piers resist the horizontal forces
   $F = 0.8$ for structures where continuous frames resist the horizontal forces applied along the frame
   W is the total dead weight of the structure
   C is the combined response coefficient, expressed as

$$C = A \times R \times \frac{S}{Z} \qquad (1.4)$$

where
   A is the maximum expected peak ground acceleration (PGA) as shown in the seismic risk map of the United States in Figure 1.1
   R is the normalized acceleration response (PGA = 1 g) spectral value for a rock site
   S is the soil amplification factor
   Z is the force-reduction factor, which accounts for the ductility of various structural components

The first U.S. seismic map, as shown in Figure 1.1, was included in this version of the AASHTO code. Although the definitions of R, S, and Z were described in the code, the numerical values of R, S, and Z were not provided. Instead, four plots of C as a function of structural period were provided with each plot representing a certain depth range of alluvium to rocklike material. One of the combined response coefficient plots is shown in Figure 1.2. The PGA values corresponding to three seismic zones (zones 1, 2, and 3) in the seismic map are shown in Table 1.1.

The same Group VII load combination given by Equation 1.2 was used for WSD with a $33\frac{1}{3}\%$ increase in the allowable stress. From the lessons learned in the 1971 San Fernando earthquake, for the first time, AASHTO provided the option of using

FIGURE 1.1 National seismic risk map. (From American Association of State Highway Transportation Officials (AASHTO), *Standard Specifications for Highway Bridges*, 12th edn., Washington, DC, 1977.)

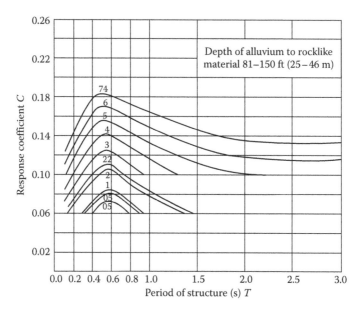

FIGURE 1.2 Combined response coefficient *C* for different rock acceleration *A*. (From American Association of State Highway Transportation Officials (AASHTO), *Standard Specifications for Highway Bridges*, 12th edn., Washington, DC, 1977.)

load factor strength design (LFD) and allowed inelastic deformations in ductile column members. For LFD, the Group VII load combination was

$$\text{Group VII} = \gamma[\beta_D D + \beta_E E + B + SF + EQ] \tag{1.5}$$

in which the load factor $\gamma = 1.3$, $\beta_D = 0.75$ for checking the column for minimum axial load and maximum moment, $\beta_D = 1.0$ for checking the column for maximum axial

TABLE 1.1

Maximum Expected

PGA for Different Zones

PGA Value (g)	Zone
0.09	1
0.22	2
0.5	3

load and minimum moment, $\beta_E = 1.3$ for lateral earth pressure and 0.5 for checking positive moments in rigid frames, and B and SF are the buoyancy and stream flow pressure, respectively.

Since values for Z were not provided in the specifications, a designer did not have a clear idea what column ductility demand was required. Without knowing the ductility demand, the ductility capacity of the design column was of questionable adequacy. This drawback was improved in the 1992 AASHTO specifications as described in the next section.

1.2.3   AASHTO FORCE-BASED DESIGN PROCEDURES (1992–2008)

The 1992 edition of the AASHTO Standard Specifications was based on the Applied Technology Council (ATC) publication entitled "Seismic Design Guidelines for Highway Bridges" (ATC-6, 1981). The primary departure from the previously mentioned AASHTO specification (1975–1992) is described as follows:

1. Instead of the equivalent static force method, structures were analyzed by elastic response spectrum analysis. The detailed description of response spectrum analysis is given in Appendix H.
2. The design acceleration spectrum included consideration of soil type at the bridge site, ranging from hard (S_1) to very soft (S_4).
3. The elastic member forces calculation considered two horizontal seismic components. The combination of structural responses due to multicomponent seismic input is described in Appendix H.
4. The elastic member forces from the response spectrum analysis were reduced by a response modification factor, R, which mainly represented the column ductility demand with consideration of the redundancy of the structure.
5. The specifications emphasized the ductile detailing of columns via a minimum transverse reinforcement requirement.

As mentioned above, the elastic force demand of the ductile member is divided by the code-provided response modification factor R (also called force-reduction factor or strength-reduction factor). The intent of R is to estimate the column ductility

TABLE 1.2

Response Modification Factors

Substructure	R
Wall-type pier	2
Reinforced concrete pile bents	
1. Vertical piles only	3
2. One or more battered piles	2
Single columns	3
Steel or composite and steel	
Concrete pile bents	
1. Vertical piles only	5
2. One or more battered piles	3
Multiple column bent	5

Source: American Association of State Highway Transportation Officials (AASHTO), *Standard Specifications for Highway Bridges*, 16th edn., Washington, DC, 1996.

demand. The response modification factors in the 1992 and 1996 editions of the AASHTO Standard Specifications are shown in Table 1.2.

Based on these specifications, the LFD Group VII load combination for seismic performance categories (SPCs) C and D was

$$\text{Group VII} = 1.0[D + E + B + SF + EQM] \tag{1.6}$$

in which

$$EQM = \frac{EQ}{R} \tag{1.7}$$

where

EQ is the elastic seismic member force calculated from the response spectrum analysis

EQM is the elastic seismic member force modified by the appropriate R-factor given in Table 1.2

In the response spectrum analysis, the design spectrum value corresponding to the mth mode shape is in terms of the elastic seismic response coefficient, C_{sm}, expressed by

$$C_{sm} = \frac{1.2AS}{T_m^{2/3}} \tag{1.8}$$

where

A is the acceleration coefficient from the seismic PGA map

S is the site coefficient having the values of 1.0, 1.2, 1.5, and 2.0 for soil types of S_1, S_2, S_3, and S_4 (or called soil types I, II, III, and IV), respectively

T_m is the structural period corresponding to the mth mode

Figure 1.3 shows the AASHTO 500-year return period seismic contour map, which is much more refined than the previous AASHTO map shown in Figure 1.1. The design spectrum with soil types of S_1, S_2, S_3, and S_4 is shown in Figure 1.4, which was determined from the generation of many response spectra based on many earthquake records, primarily from earthquakes in the western United States (Seed et al., 1976). A description of how to generate response spectra is given in Appendix G. The specifications defined four SPCs (A, B, C, and D) on the basis of the acceleration coefficient, A, for the site, and the importance classification (IC) of the bridge to be designed, as shown in Table 1.3, in which IC = I for essential bridges and IC = II for other bridges. An essential bridge is one that must be designed to function during and after an earthquake. The specifications provided different degrees of sophistication of seismic analysis and design for each of the four SPCs.

In 1994, AASHTO published the first edition of the AASHTO *LRFD Bridge Design Specifications*, with the second, third, and fourth editions published in 1998, 2004, and 2007, respectively. Similar to the previous 1992 and 1996 AASHTO standard specifications, the LRFD specifications account for column ductility

FIGURE 1.3 PGA acceleration coefficient *A*. (From American Association of State Highway Transportation Officials (AASHTO), *Standard Specifications for Highway Bridges*, 16th edn., Washington, DC, 1996.)

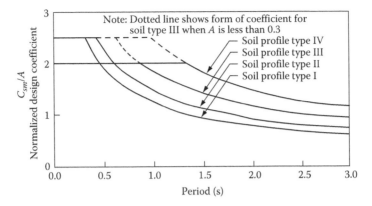

FIGURE 1.4 Normalized seismic response spectra for various soil types. (From American Association of State Highway Transportation Officials (AASHTO), *Standard Specifications for Highway Bridges*, 16th edn., Washington, DC, 1996; American Association of State Highway Transportation Officials (AASHTO), *LRFD Bridge Design Specifications*, 4th edn., Washington, DC, 2007.)

TABLE 1.3

Seismic Performance Category

Acceleration Coefficient (g)	IC	
A	I	II
$A \leq 0.09$	A	A
$0.09 < A \leq 0.19$	B	B
$0.19 < A \leq 0.29$	C	C
$0.29 < A$	D	C

using response modification R factors. The R factors in the LRFD specifications are shown in Table 1.4. The number of levels of bridge importance was increased from two levels ("essential" and "other") to three levels ("critical," "essential," and "other"). Critical bridges are those that must remain open to all traffic after the design earthquake. Essential bridges are those that should be open to emergency vehicles and for security/defense purposes immediately after the design earthquake.

Instead of using SPCs, the LRFD requires each bridge to be assigned to one of the four seismic zones in accordance with Table 1.5. Similar to the AASHTO Standard Specifications, the seismic zone reflects the different requirements for methods of analysis and bridge design details.

In LRFD design, load combinations are based on the following equation:

$$Q = \sum \eta_i \gamma_i Q_i \tag{1.9}$$

TABLE 1.4

Response Modification Factors

Substructure	IC		
	Critical	Essential	Other
Wall-type piers, larger dimension	1.5	1.5	2.0
Reinforced concrete pile bents			
1. Vertical piles only	1.5	2.0	3.0
2. With batter piles	1.5	1.5	2.0
Single columns	1.5	2.0	3.0
Steel or composite steel and concrete pile bents			
Vertical piles only	1.5	3.5	5.0
With batter piles	1.5	2.0	3.0
Multiple column bents	1.5	3.5	5.0

Source: American Association of State Highway Transportation Officials (AASHTO), *LRFD Bridge Design Specifications*, 4th edn., Washington, DC, 2007.

TABLE 1.5

Seismic Zones

Acceleration Coefficient (g)	Seismic Zone
$A \leq 0.09$	1
$0.09 < A \leq 0.19$	2
$0.19 < A \leq 0.29$	3
$0.29 < A$	4

where

Q_i is the force effect from loading type i

γ_i is the load factor for load Q_i

η_i is the load modifier relating to ductility, redundancy, and operational importance for load Q_i

In most cases, the value of each η_i is between 0.95 and 1.05, though normally, a constant η is used for all force effects, Q_i. The load combination including earthquake load is considered as the "EXTREME EVENT I" limit state in the code, given by

$$Q = \eta[\gamma_{DC}DC + \gamma_{DW}DW + \gamma_{EQ}LL + WA + FR + EQM] \qquad (1.10)$$

where
    DC is the dead load of structural components
    DW is the dead load of wearing surfaces and utilities
    LL is the vehicular live load
    WA is the water load
    FR is the friction load
    EQM is the elastic seismic member force, EQ, modified by the appropriate
      R-factor given in Table 1.4

The elastic seismic member force, EQ, is calculated via response spectrum analysis. The design spectrum value, C_{sm}, corresponding to the mth mode shape is expressed by Equation 1.8. Essentially, the same design spectrum shown in Figure 1.4 was used in the 1994–2007 LRFD specifications.

The 2008 AASHTO LRFD interim bridge design specifications use the same R factors shown in Table 1.4. However, they incorporate some major changes to the calculation of the elastic force demand, including (1) three 1000-year USGS seismic maps (PGA, 0.2 and 1.0 s) are provided in the interim specifications (Frankel et al., 1996) and (2) more realistic site effects are incorporated into the design acceleration spectrum. The revised site effects are the result of studies carried out following the 1989 Loma Prieta earthquake in California, which culminated in recommendations that have also been adopted by the Uniform Building Code (ICBO, 1997), NEHRP Building Provisions (BSSC, 1998), and the International Building Code (ICC, 2000).

The design response spectrum in the 2008 interim specifications as shown in Figure 1.5 is constructed using accelerations taken from three seismic maps mentioned above. The design earthquake response spectral acceleration coefficients, A_S, S_{DS} (the short period 0.2 s), and S_{D1} (the 1 s period acceleration coefficient) are determined using Equations 1.11 through 1.13, respectively:

$$A_S = F_{pga}PGA \tag{1.11}$$

$$S_{DS} = F_a S_S \tag{1.12}$$

$$S_{D1} = F_v S_1 \tag{1.13}$$

where
    PGA is the peak horizontal ground acceleration coefficient from the PGA seismic
      map
    F_{pga} is the site factor corresponding to the PGA coefficient
    $S_S = 0.2\,\mathrm{s}$ period spectral acceleration coefficient from 0.2 s seismic map
    F_a is the site factor for S_S
    $S_1 = 1.0\,\mathrm{s}$ period spectral acceleration coefficient from 1.0 s seismic map
    F_v is the site factor for S_1

The value of S_{D1} is used to determine the seismic zone level, as shown in Table 1.6.

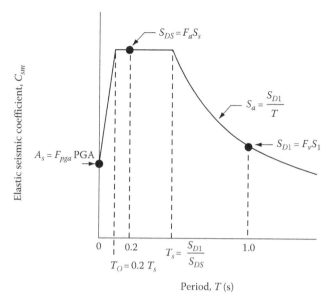

FIGURE 1.5 Design response spectrum.

TABLE 1.6
Seismic Zones

Acceleration Coefficient, $S_{D1} = F_v S_1$	Seismic Zone
$S_{D1} \leq 0.15$	1
$0.15 < S_{D1} \leq 0.30$	2
$0.30 < S_{D1} \leq 0.50$	3
$0.50 < S_{D1}$	4

The code recognizes that a well-designed structure should have enough ductility to be able to deform inelastically to the deformations imposed by the earthquake without loss of the post-yield strength. R-factors are used in the code to estimate the inelastic deformation demands on the resisting members when a bridge is subjected to the design earthquake.

The concept of R-factor is based on the equal-displacement approximation, as illustrated in Figure 1.6.

The equal-displacement approximation assumes that the maximum seismic displacement of an elastic system is the same as (or very close to) that of an inelastic system when subjected to the same design earthquake. Figure 1.6 shows two structures with the same lateral stiffness, K_e, but with different lateral yield strengths, F_{Y1} and F_{Y2}. Based on the equal-displacement approximation, the inelastic deformation,

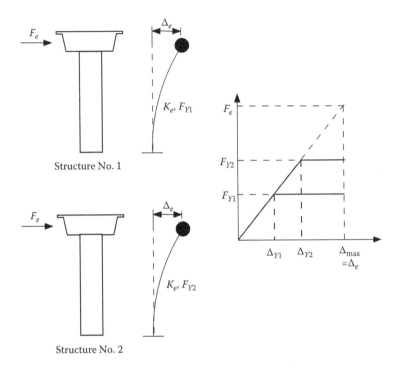

FIGURE 1.6   *R*-factor based on equal-displacement approximation.

Δ_{max}, is equal to the elastic deformation from the elastic lateral force, F_e. Therefore, the ductility demands of structures 1 and 2 can be expressed as follows:

$$\mu_1 = \frac{\Delta_{max}}{\Delta_{Y1}} = \frac{F_e}{F_{Y1}} = R_1 \tag{1.14}$$

and

$$\mu_2 = \frac{\Delta_{max}}{\Delta_{Y2}} = \frac{F_e}{F_{Y2}} = R_2 \tag{1.15}$$

From Equations 1.14 and 1.15, the force-reduction factor *R* represents the ratio of the elastic strength demand to the inelastic strength demand. Based on the equal-displacement approximation, the force-reduction factors R_1 and R_2 also represent the member ductility demands μ_1 and μ_2, respectively. Sound seismic design dictates that a structure should be designed for the ductility capacity greater than the seismic-induced ductility demand. However, the code-specified *R*-factor has its drawbacks, which will be discussed in the following section.

1.2.3.1 Force-Reduction *R*-Factor

The problems with the force-reduction factor are described as follows:

1. *Period independence*: As described in the previous section, AASHTO force-based design specifications define constant *R*-factors for different substructure types, independent of the period of the structure. In fact, the *R*-factor is a function of the period of vibration, *T*, of the structure, the structural damping, the hysteretic behavior of the structure, soil conditions at the site, and the level of inelastic deformation (i.e., ductility demand). Figure 1.7 shows the mean force-reduction factor spectrum for a single-degree-of-freedom system, using a large number of ground acceleration time histories recorded on rock and on alluvium. The force-reduction factor spectrum represents the ratio of the elastic strength demand to the inelastic strength demand corresponding to a specific ductility demand for a range of periods of vibration. From Figure 1.7, it can be seen that the *R*-factor is period dependent. It demonstrates that soil conditions at the site can have a significant effect on the *R*-factor, particularly in very soft soil (Miranda and Bertero, 1994), and it also shows that the ductility demand is larger than the force-reduction factor for short-period structures, and the equal-displacement approximation is not appropriate. The method of developing the force-reduction factor spectrum is described in Appendix G.

2. *Constant member initial stiffness*: As shown in Figure 1.6, in the *R*-factor methodology, the ductility demand of a structural member is estimated by the equal-displacement assumption, which assumes a constant initial stiffness, K_e. Using this approach, it is assumed that the member's initial stiffness is independent of the member's strength, when, in reality, the opposite is the case. To demonstrate this, Figure 1.8 shows the moment–curvature relationship of a concrete column with cross section diameter of 48 in., subjected to different axial loads. INSTRUCT was used for the moment–curvature

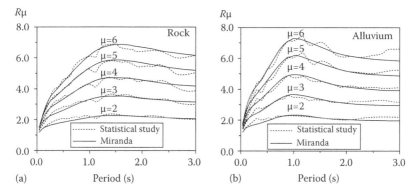

FIGURE 1.7 Mean force-reduction factors for (a) rock and (b) alluvium. (From Miranda, E. and Bertero, V., *Earthquake Spectra*, 10(2), 357, 1994.)

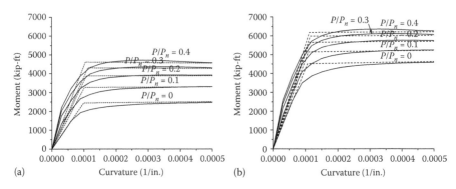

FIGURE 1.8 Moment–curvature curves of a 48″ circular column: (a) reinforcement ratio = 1.4% and (b) reinforcement ratio = 2.8%.

analysis. Two longitudinal reinforcement ratios of 1.4% and 2.8% are considered in the analysis The concrete compression strength, f'_c, is 4 ksi; steel yield stress, f_y, is 60 ksi, concrete cover is 2.6″ transverse reinforcement is No. 5 spirals with 3.25″ pitch; and the steel post-yield stress–strain slope is 1% of the elastic modulus. For each longitudinal steel ratio, the axial load ratios, defined as the ratio of column axial load, P, to the column axial compression nominal strength, $P_n = f'_c A_g$, of 0, 0.1, 0.2, 0.3, and 0.4 are considered in the analysis. The simplified bilinear moment–curvature $(M-\phi)$ curves are also plotted in the figure. The initial stiffness of the bilinear $M-\phi$ curve represents the cracked section flexural rigidity of the concrete member at which the first longitudinal steel reinforcement yield occurs. For bilinear $M-\phi$ curve, the point at which the line with initial stiffness intersects the line with post-yield stiffness defines the location of nominal moment M_n and nominal curvature ϕ_n. Figure 1.8 clearly indicates that the initial stiffness of the member is not a constant and is a function of the moment capacity. Figure 1.8 also shows that the nominal curvatures of the bilinear $M-\phi$ curves do not vary very much between the curves, where nominal curvature is about 0.0001 for this example. The moment capacity is strongly influenced by the axial load ratio and the amount of longitudinal reinforcement.

From the above discussion, Figure 1.9 compares the equal-displacement approximation with the more realistic condition of the reinforced concrete $M-\phi$ bilinear relationship (Priestley et al., 2007). It can be seen that the equal-displacement approximation correlates the strength poorly with the ductility demand (i.e., R-factor approach), due to the assumption that the nominal curvature will increase in proportion to the strength increase. In fact, the nominal curvature, ϕ_n, is independent of the strength (see Figure 1.9b) and is instead dependent on the column diameter and the yield strain, ε_y, of the longitudinal reinforcement. The column nominal curvature can be estimated by (Priestley et al., 1996)

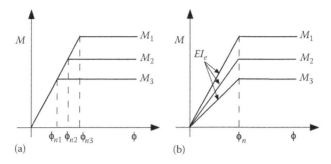

FIGURE 1.9 Moment–curvature relationship: (a) equal-displacement assumption and (b) realistic model.

$$\phi_n = \frac{2.45\varepsilon_y}{D} \quad \text{for circular concrete columns} \tag{1.16}$$

$$\phi_n = \frac{2.14\varepsilon_y}{h_c} \quad \text{for rectangular concrete columns} \tag{1.17}$$

where h_c=cross section depth. Figure 1.9b also shows that the initial bending stiffness, EI_e, increases as the strength increases.

3. *The use of elastic mode shapes to predict inelastic demand*: As mentioned previously, force-based design codes use the member stiffness at yield (i.e., cracked section stiffness for ductile members) in the elastic response spectrum analysis based on the code-provided design acceleration spectrum. However, this does not take into account the member inelastic stiffness distribution at the maximum inelastic response. For ductile structures, the inelastic mode shapes may be quite different from the elastic mode shapes used in the current design codes.

4. *Difficulty in predicting the bridge performance under strong ground motion:* As described above, the ductility demand of a ductile member cannot be accurately predicted, and, as such, the performance level of a bridge subjected to the design earthquake may not be achieved.

1.2.3.2 Capacity Design Concept

Normally, the strong beam–weak column design philosophy is used for bridge seismic design. In this strategy, plastic hinges are expected to occur in the columns but not in the beams or foundations. Whether or not a column can withstand a high ductility demand is dependent on the reinforcement details within and adjacent to the column plastic hinge zones. Columns with confined cores and sufficiently anchored reinforcement have been proven to have the necessary ductility capacity. Neither the AASHTO force-based standard specifications nor the LRFD design specifications provide detailed design criteria for estimating the ductility capacity of column subjected to the design earthquake. However, both specifications do require designers to use capacity design principles (i.e., strong beam–weak column design philosophy) to design cap beams and foundations. When designed according to capacity design principles, the nominal strength of the cap beam and foundation is greater than the column overstrength capacity, so that there is little or no damage to the cap beam and foundation. The column overstrength capacity is the result of the actual material strengths being greater than the minimum specified strength; confinement of concrete; and the strain hardening of steel reinforcement. The use of capacity design principles along with R-factors is intended to ensure that plastic hinges are developed at column ends.

In addition, the code also applies capacity design (or so-called capacity protection) principles to the column itself, with the intent of ensuring that column failure is governed by the flexural failure mode and not the brittle shear failure mode. However, the code shear design criteria only ensures that shear failure will not occur prior to the development of the plastic hinge, it does not provide shear capacity design criteria for columns subjected to large ductility demand. The concrete shear capacity within the plastic hinge region degrades as the ductility demand increases, and thus shear design criteria should be the function of column ductility demand. This issue was not addressed until the publication of the AASHTO Guide Specifications for LRFD Seismic Bridge Design.

1.2.4 *AASHTO Guide Specifications for **LRFD** Seismic Bridge Design* **(2009)**

After damaging earthquakes in the 1980s and 1990s (1989 Loma Prieta earthquake, CA; 1994 Northridge earthquake, CA; 1995 Kobe earthquake, Japan; 1999 Chi-Chi earthquake, Taiwan; 1999 Izmit earthquake, Turkey, etc.), further research efforts provided critical earthquake design recommendations, shifting design focus from the force-based R-factor design approach to the displacement-based design approach. In 2009, AASHTO published the Guide Specifications for LRFD Seismic Bridge Design (AASHTO LRFD, 2009), which mainly incorporates the research results published in ATC-32 (ATC, 1994), Caltrans Seismic Design Criteria (Caltrans, 1999), NCHRP 12 and 49 (ATC-MCEER, 2003), and the South Carolina Seismic Design Specifications for Highway Bridges (SCDOT, 2001).

This is the first AASHTO seismic design provision to incorporate displacement design principles for the design of ductile members. Compared with previous AASHTO standard specifications and LRFD specifications, several significant improvements are summarized as follows:

1. Discontinues use of R-factors for ductile column design.
2. While the equal-displacement approximation is still adopted for the estimation of inelastic displacement demand, the inelastic demand for short-period structures is increased by a modification factor, such that the more realistic equal energy approximation is applied to short-period structures.
3. As shown in Table 1.7, four SDCs (A, B, C, and D) are used instead of seismic zones 1, 2, 3, and 4 as in the previous LRFD specifications. For each SDC, the guide specifications describe the requirements for the

TABLE 1.7

Seismic Design Categories

Acceleration Coefficient, $S_{D1} = F_v S_1$	SDC
$S_{D1} \leq 0.15$	A
$0.15 < S_{D1} \leq 0.30$	B
$0.30 < S_{D1} \leq 0.50$	C
$0.50 < S_{D1}$	D

displacement demand analysis, displacement capacity check, minimum level of detailing, and consideration of liquefaction.

4. Uses capacity protection principles for the column shear capacity design. The degradation of shear capacity is recognized and evaluated for columns expected to have significant amounts of plastic deformation.

5. Uses capacity protection principles for the cap beam–column connection joint design.

6. Uses nonlinear pushover analysis to evaluate the displacement capacity of individual bents. The evaluation of displacement capacity involves determining the displacement at which the first column reaches its inelastic capacity, the point where concrete strain reaches the ultimate concrete compression strain, ε_{cu}, defined by the fracture of transverse reinforcement. The nonlinear pushover analysis procedure is briefly described in Section 1.2.4.1 and will be described in detail in Chapter 3.

7. Explicitly requires that the displacement capacity be greater than the displacement demand. The displacement capacity evaluation is required for individual bents, and the effect of foundation and cap beam flexibilities is considered in the displacement capacity evaluation.

8. The guide specifications also recognize that the inelastic displacement demand calculated by elastic response spectrum analysis with cracked section properties for concrete columns may not represent the realistic inelastic behavior of bridges under strong ground motion. With the bridge owner's concurrence, nonlinear time history analysis (see Appendix F) may be used to evaluate inelastic displacement demand, especially for bridges with distinct unequal column heights, different superstructure masses over bents, or bridges with sharp horizontal curves. Regardless of which analysis is used, the guide specifications require the pushover analysis be used to evaluate the displacement capacity of SDC D bridges. The following section provides a brief description of the nonlinear pushover analysis procedure.

1.2.4.1 Nonlinear Pushover Analysis Procedure

The nonlinear pushover analysis considers column nonlinear behavior, $P-\delta$ effects on the structure, and the flexibility of the foundation and soil system. In general, applicable permanent gravity loads are first applied to the structure, and then a horizontal lateral load or lateral displacement is incrementally applied to the mass center of the

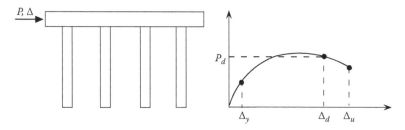

FIGURE 1.10 Pushover curve.

bridge (or individual bent) until the structural displacement capacity is reached. The AASHTO guide specifications define the structural displacement capacity as the displacement, Δ_d, at which the first column reaches its inelastic capacity (i.e., initiation of a failure mode) as shown in Figure 1.10. In the figure, Δ_y is the displacement corresponding to first yield of longitudinal reinforcement, and Δ_u represents the displacement at which a failure mechanism develops. During the analysis process, each member's inelastic deformation and corresponding forces are monitored. All possible member failure modes are checked at each incremental step. In INSTRUCT, either an incremental force or displacement could be applied to the structure.

For bridges designed with consideration of capacity design principles, the most likely column failure mode is confined to concrete compression failure due to fracture of the transverse reinforcement or tensile fracture of the longitudinal reinforcement. For existing bridges not designed using capacity design principles, the column failure mode could be one or a combination of the following:

1. Compression failure of unconfined concrete due to fracture of transverse reinforcement
2. Compression failure of confined concrete due to fracture of transverse reinforcement (Figure 1.11)
3. Compression failure due to buckling of the longitudinal reinforcement (Figure 1.12)
4. Longitudinal tensile fracture of reinforcing bar
5. Low-cycle fatigue of the longitudinal reinforcement
6. Failure in the lap-splice zone (Figure 1.13)
7. Shear failure of the member that limits ductile behavior (Figure 1.14)
8. Failure of the beam–column connection joint (Figure 1.15)

INSTRUCT is capable of checking all the possible concrete column failure modes described above. The analytical approach for checking individual failure modes is described in detail in Chapter 4.

1.3   DIRECT DISPLACEMENT-BASED DESIGN PROCEDURES

As mentioned in Section 1.2.4, since the Loma Prieta earthquake in 1989, extensive research has been conducted to develop improved seismic design criteria for

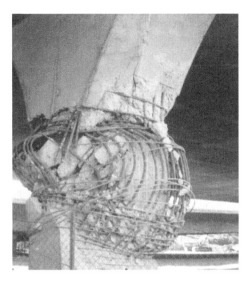

FIGURE 1.11   Compression failure of confined concrete. (With permission from Caltrans.)

FIGURE 1.12   Buckling of longitudinal reinforcement. (From Cheng, C.T., New paradigms for the seismic design and retrofit of bridges, PhD dissertation, Department of Civil Engineering, State University of New York, Buffalo, NY, 1997. With permission.)

concrete bridges. This research demonstrated that using displacement (or ductility) rather than force is a better measure of bridge performance. For example, the 2009 AASHTO guide specifications use displacement to quantify the demand and capacity of bridge bents. However, the guide specifications still use the acceleration (force) spectrum for the response spectrum analysis, and the displacement demand is still estimated based on the equal-displacement approximation with a modification for short-period structures. Using this approach, it is possible, in some cases (Suarez and Kowalsky, 2006), that the calculated displacement demands will not be

FIGURE 1.13   Lap-splice failure. (With permission from Caltrans.)

FIGURE 1.14   Column shear failure. (From Chile Earthquake, February 27, 2010.)

FIGURE 1.15   Cracking of beam–column connection. (With permission from Caltrans.)

in good agreement with results obtained from nonlinear time history analysis. This is due to the fact that the column cracked section stiffness distribution at yield in the response spectrum analysis is different from the stiffness distribution at the maximum demand response. In addition, the fundamental inelastic mode shape at the maximum demand response is different from the mode shape based on the cracked sectional stiffness distribution.

To overcome this problem, several researchers (Dwairi and Kowalsky, 2006; Suarez and Kowalsky, 2006; Priestley et al., 2007) have recommended using the direct displacement-based design (DDBD) method. Instead of using an acceleration spectrum and the equal-displacement approximation, DDBD uses the displacement spectrum (see Appendix G) at the design level of ground motion to obtain the inelastic structural period. DDBD uses an iterative approach to obtain the effective (secant) stiffnesses of individual bents and to calculate the target-displacement profile (i.e., displacement demand) of the inelastic structure. Depending on the importance of a bridge, the bridge can be designed for a certain level of performance in terms of target displacement, strain, or ductility. The DDBD approach will be briefly described in Chapter 2.

2 Pushover Analysis Applications

Several pushover analysis applications in bridge engineering are summarized here. From the previous overview of past American Association of State Highway and Transportation Officials (AASHTO) design code developments in the present, and for the future, it is clear that pushover analysis is a necessary tool for the evaluation of the displacement capacity of new bridges. This chapter also describes its application in the seismic retrofit of existing bridges and in the evaluation of bridge system redundancy.

2.1 DISPLACEMENT CAPACITY EVALUATION FOR THE SEISMIC DESIGN OF NEW BRIDGES

As described in Section 1.2.4, pushover analysis is required in the AASHTO load and resistance factors design (LRFD) guide specifications to check the displacement demand and evaluate the displacement capacity of seismic design category D (SDC D) bridges.

2.2 PERFORMANCE LEVEL VERIFICATION FOR NEW BRIDGES DESIGNED BY DDBD

Using the direct displacement-based design (DDBD) approach, a bridge is designed to meet an expected performance level (or so-called limit state), which is determined by the bridge owner. Normally, the performance level of a bridge could be the (1) serviceability limit state, (2) damage-control limit state, (3) life-safety limit state, or (4) survival limit state. In the serviceability limit state, the bridge should be in full operation with minor damage after a design ground motion. No major repair action is needed at this limit state. To avoid remedial actions after the earthquake, the column concrete cover should not be spalled. However, the first yield of longitudinal reinforcement of column is acceptable, and the maximum tension strain of the longitudinal reinforcement is limited to about 0.015. The concrete compression strain is limited to 0.02.

For the damage-control limit state, life safety is essentially protected and damage is moderate. In this state, spalling of column concrete cover is acceptable, but the damage is manageable and repair costs should be economically feasible. None of the possible concrete column failure modes mentioned in Section 1.2.4.1 should occur, as the column is designed conservatively for the ultimate concrete compression strain, ε_{cu}, at which the transverse reinforcement is close to fracture (see Equation 2.14).

Similarly, the maximum tension strain of longitudinal reinforcement, ε_{su}, should be less than but close to the actual ultimate tension strain limit of steel reinforcement. Conservatively, $\varepsilon_{su}=0.09$ is adopted in INSTRUCT. The conservatism of choosing ε_{cu} and ε_{su} is to control the damage so that the repair cost is acceptable.

In the life-safety limit state, significant structural damage occurs, but some margin against either partial or total structural collapse remains. Multiple column failures are expected. The overall risk of life threatening injury as a result of structural damage is expected to be low. It is possible to repair the structure, however, for economic reasons, this may not be practical.

In the survival limit state, avoiding structural collapse at the design level earthquake is the goal. Substantial damage such as multiple column failures is expected to occur, including significant degradation in the stiffness and strength reduction. Large permanent lateral deformation could occur, and live load carry capacity is reduced significantly. Bridge replacement will be required due to high repair costs.

Normally, the damage-control or the life-safety limit state is considered for the seismic design of bridges. However, the higher performance level such as serviceability limit state or the limit state between serviceability and damage-control limit states may be considered by the bridge owner. Depending on the importance of a bridge, the bridge can be designed for a certain level of performance in terms of target displacement, rotation, strain, or ductility.

The DDBD approach (Dwairi and Kowalsky, 2006; Suarez and Kowalsky, 2006; Priestley et al., 2007) is briefly described as follows:

Step 1: Obtain initial parameters such as column height (h) and diameter (D), superstructure mass, steel and concrete material properties (f_y, ε_y, f_c', etc.), and design elastic displacement spectrum.

Step 2: Per the bridge owner, define the desired performance level. For example, the performance level can be the limitation of column ductility ratio (say, $\mu \leq 4$), column plastic rotation capacity (say, $\theta_p \leq 0.035$ rad), or concrete strain level to ε_{cu}. Once the performance level is defined, the estimated critical target displacement of a critical bent i, Δ_i^c, can be calculated. For example, as bent No. 4 in Figure 2.1a has the shortest column height, it would be considered as the critical bent and its critical target displacement would be Δ_4^c.

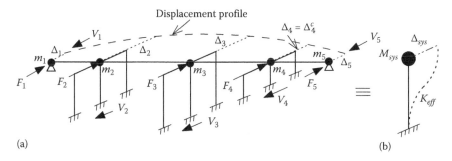

Displacement profile

(a)   (b)

FIGURE 2.1   (a) Target displacement profile; (b) substitute sdof structure.

Step 3: Estimate the column yield displacement, Δ_y, based on the column nominal curvature ϕ_n from analysis or Equations 1.16 or 1.17.

Step 4: Using eigen solution analysis, find the inelastic mode shape based on the secant stiffness, K_i, of each bent i, and estimate the target inelastic displacement profile (see Figure 2.1a). Since the secant stiffnesses of the bents will not be known for the first iteration, the initial displacement profile can be assumed to be the mode shape based on the column cracked section stiffnesses and calculated based on EI_e (see Figure 1.9b).

Step 5: Scale the displacement profile from Step 4 such that the displacement at the critical bent is equal to the estimated critical target displacement, Δ_i^c, from Step 2 (Δ_4^c in this example).

Step 6: Define a "substitute" single-degree-of-freedom (sdof) structure for the bridge (see Figure 2.1b) with equivalent system displacement and mass:

$$\Delta_{sys} = \frac{\sum_{i=1}^{n} m_i \Delta_i^2}{\sum_{i=1}^{n} m_i \Delta_i} \tag{2.1}$$

$$M_{sys} = \frac{1}{\Delta_{sys}} \sum_{i=1}^{n} m_i \Delta_i \tag{2.2}$$

where
$\quad m_i$ is the mass associated with bent i
$\quad \Delta_i$ is the target displacement of bent i obtained from Step 4
$\quad n$ is the total number of bents

Step 7: Estimate the equivalent viscous damping of the substitute sdof structure (Dwairi and Kowalsky, 2006):

$$\xi_{sys} = \alpha \sum_{i}^{m} \left(\frac{Q_i}{\sum_{k}^{m} Q_k} \xi_i \right) + \beta \xi_{abt} \tag{2.3}$$

where ξ_i is the equivalent viscous damping of individual column i

$$\xi_i = 0.05 + 0.5 \left(\frac{\mu_i - 1}{\mu_i \pi} \right) \quad \text{for concrete column } i \tag{2.4}$$

μ_i is the ductility demand of column i

$$Q_i = \frac{1}{h_i} \quad \text{for yield column } i, \text{ in which } \mu_i \geq 1 \tag{2.5}$$

$$Q_i = \frac{\mu_i}{h_i} \quad \text{for non-yield column } i, \text{ in which } \mu_i < 1 \tag{2.6}$$

where
> α is the fraction of the total base shear V_B (see Steps 9 and 11) taken by the inter-
> mediate bents
> β is the fraction of the total base shear V_B taken by the abutments

For example, α is given by $\alpha = (V_2 + V_3 + V_4)/V_B$ in Figure 2.1, and β is given by $\beta = (V_1 + V_5)/V_B$. The methodologies of calculating equivalent viscous damping are given in Appendix G.

Step 8: Determine the effective period, T_{eff}, of the substitute sdof structure from the design displacement spectrum (see Figure 2.2) based on Δ_{sys} and ξ_{sys} of the substitute structure. Appendix G provides the description of how to generate the displacement spectrum.

Step 9: Calculate the effective stiffness, K_{eff}, and base shear, V_B, of the substitute sdof system:

$$K_{eff} = 4\pi^2 \frac{M_{sys}}{T_{eff}^2} \tag{2.7}$$

$$V_B = K_{eff}\Delta_{sys} \tag{2.8}$$

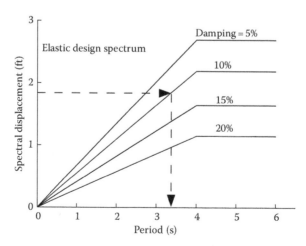

FIGURE 2.2 Obtain effective period from the Design Displacement Spectrum.

Step 10: Calculate the inertia forces by distributing the base shear V_B to the inertial mass location at each bent *i*:

$$F_i = V_B \frac{(m_i \Delta_i)}{\displaystyle\sum_k^n (m_k \Delta_k)} \tag{2.9}$$

Step 11: Calculate the base shear at each bent *i* (see Figure 2.1):

$$V_i = (V_B - F_{abt}) \frac{(\mu_i / h_i)}{\displaystyle\sum_{i=k}^n (\mu_k / h_k)} \tag{2.10}$$

in which

$$F_{abt} = F_1 + F_5 \tag{2.11}$$

$$V_B = \sum V_i = \sum F_i \tag{2.12}$$

Step 12: Find the secant stiffness of each bent based on Δ_i from Step 5 and V_i from Step 11:

$$K_i = \frac{V_i}{\Delta_i} \tag{2.13}$$

Step 13: Using the secant stiffness K_i for each bent *i*, perform a static structural analysis by applying lateral forces F_i at each bent to calculate the new estimated target displacement profile. Compare the calculated displacement at the critical bent (in this example, Δ_4) with the critical displacement Δ_4^c. If Δ_4 is not equal to Δ_4^c, scale the displacement profile such that the scaled $\Delta_4 = \Delta_4^c$.

Step 14: Iterate through Steps 4 through 14, until the target displacement profile converges in Step 13 (i.e., $\Delta_4 = \Delta_4^c$).

Step 15: Design the column longitudinal reinforcement based on the column axial load and moment from the static structural analysis in Step 13. Design the column transverse reinforcement to meet the performance level defined in Step 2 with the target displacement demand of Δ_i for each bent *i* in Step 13. For example, if the performance level is defined as when the column-confined concrete strain reaches ε_{cu}, the transverse reinforcement of confined column at bent No. 4 can be designed using Equation 3.24, reproduced here as Equation 2.14:

$$\varepsilon_{cu} = 0.004 + \frac{1.4 \rho_s f_{yh} \varepsilon_{su}}{f'_{cc}} \tag{2.14}$$

Note that Equation 2.14, for estimating the ultimate compression concrete strain, ε_{cu}, is based on a column subjected to axial compression without bending. For columns subjected to both axial compression and bending, the ultimate concrete compression strain estimated by Equation 2.14 is conservative (Mander et al., 1988). From Equation 2.14

$$\rho_s = 0.74(\varepsilon_{cu} - 0.004)\frac{f'_{cc}}{f_{yh}\varepsilon_{su}} \tag{2.15}$$

where
$\quad \varepsilon_{su}$ is the steel strain at maximum tensile stress (conservatively, $\varepsilon_{su}=0.09$ is adopted in INSTRUCT)
$\quad f_{yh}$ is the yield stress of the hoop or spiral bar
$\quad \rho_s$ is the volumetric ratio of the transverse reinforcement

ε_{cu} can be calculated as follows:

$$\varepsilon_{cu} = \phi_u c \tag{2.16}$$

where
$\quad \phi_u$ is the curvature corresponding to the moment at Δ_4^c
$\quad c$ is the neutral axis depth at ϕ_u and can be obtained from the moment–curvature analysis

It can also be estimated either by the formula in Appendix E or by the following approximated equation (Priestley et al., 2007):

$$\frac{c}{D} = 0.2 + 0.65\frac{P}{f'_{ce}A_g} \tag{2.17}$$

where
$\quad D$ is the column diameter
$\quad A_g$ is the gross cross-sectional area of column
$\quad f'_{ce}$ is the expected concrete compression strength

The value of $f'_{ce} = 1.3f'_c$ is usually adopted to take into account the material strength, which is generally greater than the specified strength of f'_c.

Once the columns are designed in Step 15, the pushover analysis can be used to verify that the expected performance level defined in Step 2 is achieved.

2.3   CAPACITY/DEMAND RATIOS FOR THE SEISMIC EVALUATION OF EXISTING BRIDGES

Another pushover analysis application is the seismic evaluation of existing bridges and the development of retrofitting strategies. In 2006, The U.S. Federal Highway

Administration (FHWA) published the *Seismic Retrofitting Manual for Highway Structures: Part I—Bridges* (FHWA, 2006). The manual specifies six evaluation methods. One of the evaluation methods is "Structure Capacity/Demand" Method (or so-called Method D2). In this method, the capacity assessment is based on the displacement capacity of individual bents as determined by pushover analysis with consideration of inelastic member behavior. The demand assessment is based on the multimode response spectrum analysis with consideration of cracked section properties. For each bent, the pushover analysis is performed independently in both the longitudinal and transverse directions. From the analysis, the displacement at which the first member reaches its inelastic capacity defines the displacement capacity of a bent. Since the pushover analysis is performed for each bent, the force distribution from bent-to-bent is neglected. The displacement capacity of a bent is then compared with the results from the elastic response spectrum analysis, which does consider the behavior of the whole bridge. The inelastic capacity of a column in the pushover analysis is determined by the maximum plastic hinge rotation corresponding to the governing column failure mode. The maximum plastic curvatures for possible governing failure modes can be estimated and are provided in Appendix E.

The capacity/demand ratio of a bent is determined as follows:

$$r = \frac{\Delta_C - \sum \Delta_{NS}}{\Delta_{EQ}} \tag{2.18}$$

where
   Δ_C is the displacement capacity of the bent from pushover analysis
   Δ_{NS} is the sum of any non-seismic displacement demands
   Δ_{EQ} is the seismic displacement demand from response spectrum analysis

If $r \geq 1.5$, no retrofit action is required.
If $1.0 \leq r \leq 1.5$, some remedial action may be required.
If $r < 1.0$, retrofit strategy that improves the ductility of bridge or reduces the seismic demand of bridge should be considered.

2.4 QUANTITATIVE BRIDGE SYSTEM REDUNDANCY EVALUATION

Both the AASHTO standard specifications and LRFD specifications require the consideration of redundancy for highway bridge design. However, both codes provide limited guidance on how to measure redundancy. This has led to a wide variation in the interpretation of the specifications and a need to develop a quantitative measure of bridge redundancy.

Bridge superstructure redundancy was investigated in NCHRP project 12-36 as reported in NCHRP Report 406 (Ghosn and Moses, 1998), while bridge substructure redundancy was investigated in NCHRP project 12-47 as reported in NCHRP Report 458 (Liu et al., 2001). In both studies, bridge redundancy is defined as the capability of a bridge to continue to carry loads after the failure of the first member. The failed member could be a superstructure girder or a substructure column. Member failure can be either in a ductile or brittle fashion and can be caused by lateral seismic load, overweight vehicular load, or collision by a vehicle or vessel.

Based on this definition, a system reserve ratio, R_u, can be used as a quantitative measure of substructure or superstructure redundancy, which is expressed as follows:

$$R_u = \frac{P_u}{P_f} \tag{2.19}$$

For superstructure redundancy, P_u is the maximum live load corresponding to the ultimate capacity (failure mechanism) of the superstructure, and P_f is the live load corresponding to the first main girder failure. For substructure redundancy, P_u is the lateral force corresponding to the collapse mechanism of a bent, and P_f is the lateral force corresponding to the first column failure. For example, when R_u of a substructure is equal to or less than one, the ultimate capacity of the substructure is equal to or less than the strength of the substructure at which the first column fails. In this case, the bridge with $R_u \leq 1$ is a nonredundant bridge. A two-column bent shown in Figure 2.3a is a nonredundant structure. A value of R_u greater than one means that the substructure has additional reserve strength such that the failure of one column does not result in the failure of the complete substructure system. The four-column bent shown in Figure 2.3b is a redundant structure. As the total number of columns increases, the level of redundancy increases. From Figure 2.3, it can be seen that pushover analysis is needed to generate the pushover curve from which the system substructure reserve ratio can be calculated. Similarly, pushover analysis is used to generate the live load–vertical superstructure displacement pushover curve, from which R_u of the superstructure can be calculated.

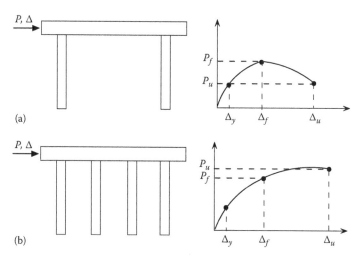

FIGURE 2.3 Pushover curves for bridge redundancy evaluation: (a) $R_u < 1$ and (b) $R_u > 1$.

2.5 MOMENT–CURVATURE CURVES AND AXIAL LOAD–MOMENT INTERACTION CURVES

Another pushover analysis application is to perform moment–curvature analysis for ductile members. Instead of using the conventional direct cross-sectional force-equilibrium analysis to generate moment–curvature curves, a simple structural model for the moment–curvature pushover analysis by INSTRUCT is used as shown in Figure 2.4.

In the figure, a simply supported member of length 2 has a constant axial load, P, applied at the ends of the member. The finite-segment finite-string material type is used for the member. This material type will be discussed in detail in Chapter 3. Incremental rotations are applied at both ends of the member with the same incremental magnitude. Since the member length is equal to 2, the end rotation represents the curvature of the cross section. Using this model, the moment–curvature curves generated by INSTRUCT are shown in Figure 2.5 for a reinforced concrete column. Once the family of moment–curvature curves is developed, the axial load–moment interaction curve can also be generated. Figure 2.5 shows the comparison of the moment–curvature curves generated by INSTRUCT and by the direct cross-sectional moment–curvature analysis (SEQMC, 1998). The column cross section and material details from the FHWA Seismic Design Example No. 4 (FHWA, 1996) were used herein, where column diameter = 48″, longitudinal reinforcement is 34 – #11, $f'_c = 4$ ksi, $f_y = 60$ ksi, spiral = #5 @3.5″, concrete cover = 2.63″ from the surface of longitudinal rebar to the surface of column, and the applied axial load = 660 kip. The post-yield modulus of the reinforcing steel stress–strain curve is assumed to be 1% of the elastic modulus. It can be seen that both curves are almost identical.

Similar to the moment–curvature analysis for reinforced concrete cross sections, INSTRUCT is also capable of performing moment–curvature analysis for various steel cross sections. Figure 2.6 shows a comparison of moment–curvature curves of a $W8 \times 31$ wide flange section, generated by INSTRUCT and from Chen and Lui (1991). Good agreement is observed.

2.6 OTHER APPLICATIONS

Another nonlinear pushover analysis application is to monotonically increase the invariant lateral load pattern to a building until a specific target displacement is

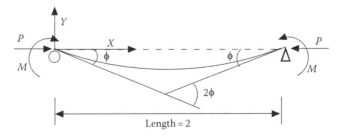

FIGURE 2.4   Structural model for moment–curvature analysis.

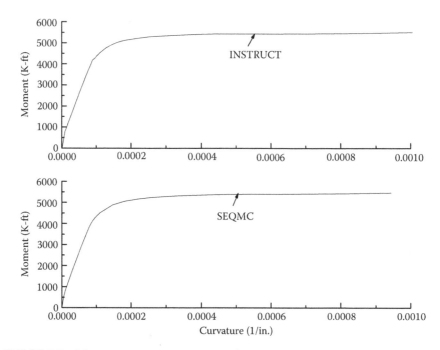

FIGURE 2.5   Moment–curvature curve comparison.

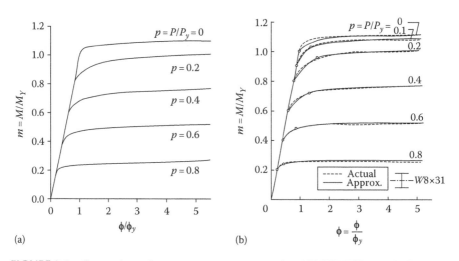

FIGURE 2.6   Comparison of moment–curvature curves by (a) INSTRUCT and (b) Chen and Lui (1991). (From Chen, W.F. and Lui, E.M., *Stability Design of Steel Frames*, CRC Press, Inc., Boca Raton, FL, 1991.)

exceeded (FEMA-273, 1997; FEMA-356, 2000). The specified target displacement is a function of structural fundamental period and the design earthquake level. Once a building is pushed to a target displacement, the design performance level (i.e., member strength, story drift limit, etc.) is checked for acceptance. The pushover analysis used in FEMA-273 and FEMA-356 does not include higher mode effects. To overcome the higher mode effect, several researchers (Gupta and Kunnath, 2000; Goel and Chopra, 2004; Chopra, 2005; Goel, 2005) used the earthquake design response acceleration spectrum as the basis for determining the incremental lateral forces to be applied to the building in the pushover analysis for each individual mode. The total structural response demand is the combination of responses from all modes using the SRSS rule (see Appendix H), and the response demand is checked with the performance level for acceptance. The above-mentioned approach is called modal pushover analysis. Strictly speaking, the lateral load pattern during the nonlinear pushover analysis is not invariant and is dependent on the instantaneous dynamic properties of the building at each increment step. Some adaptive pushover proce-dures, in which the applied load pattern continually changes during the pushover analysis, have been recommended (Bracci et al., 1997; Gupta and Kunnath, 2000).

The modal pushover analysis approach for estimating inelastic response demand is mainly used in the building seismic design and has not been adopted by AASHTO for bridge design. For a detailed description of the modal pushover analysis, see the above-mentioned references.

3   Nonlinear Pushover Analysis Procedure

3.1 INTRODUCTION

As mentioned in the preface of this book, INSTRUCT is capable of analyzing 3D structures. It was developed based on a microcomputer program INRESB-3D-SUPII (Cheng et al., 1996a,b) and mainframe program INRESB-3D-SUP (Cheng and Mertz, 1989a). INRESB-3D-SUPII was a modular computer program consisting of six primary blocks. The first block (STRUCT) defines the structural model. The remaining five blocks (SOL01, SOL02, SOL03, SOL04, and SOL05) are independent solutions for static loading, seismic loading, natural frequency or buckling load, static cyclic or pushover loading, and response spectrum analysis, respectively.

Since the purpose of this book is mainly performing nonlinear pushover analyses of reinforced concrete and steel bridge bents, it only includes SOL01 and SOL04. SOL04 is capable of performing not only unidirectional pushover but also cyclic pushover analysis. Depending on future needs, SOL02, SOL03, and SOL05 can be incorporated into the future versions of INSTRUCT. During the development of INSTRUCT, SOL04 was enhanced significantly, and it includes five different nonlinear element-bending stiffness formulation methods for pushover analysis. They are finite segment–finite string (FSFS), finite segment–moment curvature (FSMC), bilinear interaction axial load–moment (PM) interaction, plastic hinge length (PHL), and constant moment ratio (CMR) methods. These methods range from the most sophisticated to the simplest and will be discussed in Chapter 4.

To perform pushover analysis, the structural model must first be generated. The structural model consists of an assemblage of elements. The point where two or more elements connect is called a joint. A structure is modeled by first defining the location and orientation of each joint; then materials that describe the behavior of the elements, the elements that connect the joints, and the orientations of the elements are defined. All of these are defined in the STRUCT block in INSTRUCT program. The flowchart for STRUCT is shown in Figure 3.1.

Step 1: Define joints and determine the dofs. The coordinates of the joints and their orientation are defined by the user. The coordinates are defined in the global coordinate system (GCS). The GCS defines the location of a structure. The orientation of each structural joint defines its joint coordinate system (JCS). Each joint initially has six global degrees of freedom (Gdofs) in the JCS. The user also defines the joint's degrees of freedom (dofs) that are free, restrained, constrained, and condensed out. INSTRUCT generates the structural dof identification numbers for the user. The definitions of free, restrained, constrained, and condensed dofs are described in Chapter 5.

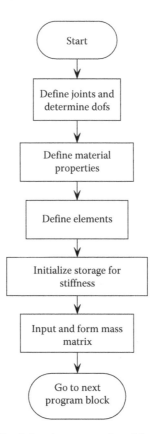

FIGURE 3.1   Block STRUCT—define the structural model.

Step 2: *Define material properties.* The material properties are input and initialized. There are different material behaviors available, which constitute the material library and are discussed later.

Step 3: Define elements. The element data is input. The element coordinate system (ECS), initial element structural stiffness, and the initial element geometric stiffness are calculated. There are different elements available in the program, which constitute the element library and are discussed later.

Step 4: Initialize storage for stiffness. The storage for the structural stiffness and geometric matrices is initialized.

Step 5: Input and store mass. The lumped mass at each joint is input. The structure mass matrix is stored. This is reserved for SOL02, SOL03, and SOL05. For SOL01 and SOL04, there is no need to generate the structure mass matrix.

To conduct pushover analysis, INSTRUCT first performs the elastic static analysis of bridge bents due to superstructure dead loads. The member forces and structural displacements from the static dead load analysis are then used as the initial condition for the pushover analysis. The elastic static analysis (SOL01) and the pushover analysis (SOL04) are described as follows.

3.2   SOL01—ELASTIC STATIC ANALYSIS

This block performs the elastic static analysis. The flowchart for SOL01 is shown in Figure 3.2.

Step 1: Input joint and element loadings. The joint loads and imposed displacements are input. Uniform or concentrated element loadings are input on the 3D beam or plate element.

Step 2: Form the structural stiffness and load matrices. The structural stiffness matrix is formed. Joint loadings are determined for the imposed displacements (support settlements) and combined with the input joint loadings and element loadings.

Step 3: Calculate displacements. The displacements are calculated by Gaussian elimination.

Step 4: Calculate reactions. The reactions at restrained dofs and the summation of reactions are calculated.

Step 5: Calculate element forces. The element forces are calculated.

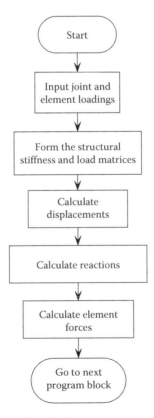

FIGURE 3.2   Block SOL01—static analysis.

3.3 SOL04—NONLINEAR STATIC PUSHOVER (CYCLIC OR MONOTONIC) ANALYSIS

3.3.1 FLOWCHART IN SOL04

Block SOL04 calculates the nonlinear static cyclic or monotonic structural response for a given loading pattern. A loading pattern consisting of joint loads, imposed displacements and element loads is defined and stored in the loading vector $\{Q\}$. $\{Q\}$ is multiplied by positive and negative load factors to generate loading cycles. Defining F_j as the loading factor for the current cycle, and F_i as the loading factor for the previous cycle, the total loads on the structure for cycles i and j are $F_i\{Q\}$ and $F_j\{Q\}$, respectively. The loading from $F_i\{Q\}$ to $F_j\{Q\}$ is carried out in a series of steps.

A fixed number of equal load steps is chosen to load from $F_i\{Q\}$ to $F_j\{Q\}$. The flowchart for SOL04 is shown in Figure 3.3.

Step 1: Input joint and element loadings. The joint loads and imposed displacements of the load pattern are input. Element loadings are also input.

Step 2: Input load factors. For each loading cycle, a load factor and the number of load steps are input.

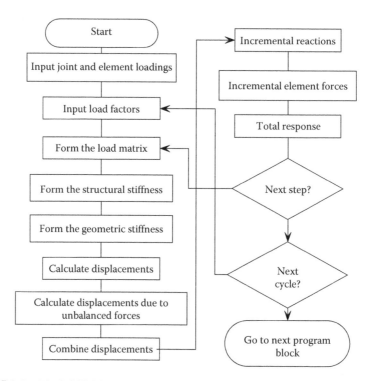

FIGURE 3.3 Block SOL04—nonlinear static cyclic response.

Step 3: Form the load matrix. The loading matrix is the incremental load. $(F_j - F_i)/N \times \{Q\}$, where N is the number of load steps.

Step 4: Form the structural stiffness. The structural stiffness is formed (1) for the first load step, (2) for every load step that an element's stiffness is modified, and (3) for every load step that the geometric stiffness is modified.

Step 5: Form the geometric stiffness. The geometric stiffness is formed (1) for the first load step and (2) for every load step if the actual element axial loads are used to calculate the geometric stiffness.

Step 6: Calculate displacements. The incremental displacements due to the applied loadings are calculated by Gaussian elimination.

Step 7: Calculate displacements due to unbalanced forces. The incremental displacements due to the unbalanced forces from the previous load step are calculated by Gaussian elimination.

Step 8: Combine displacements. The displacements due to the applied loading and the displacements due to the unbalanced loadings are added together.

Step 9: Incremental reactions. The incremental reactions are calculated.

Step 10: Incremental element forces. The hysteresis models in the material library are called to calculate the incremental element forces, given the incremental displacements and previous loading history. For nonlinear analysis, if the element's stiffness changes during the incremental displacement, (1) the element's unbalanced forces are calculated and (2) a flag to reform the structural stiffness in Step 4 is set for the next load step.

Step 11. Total response. The total displacements, reactions, and element forces are calculated. The unbalanced force vector for nonlinear analysis is also calculated. If desired, selected results may be written to output files.

Go to Step 3 for additional loading steps. Go to Step 2 for the next loading cycle.

3.3.2 NONLINEAR PUSHOVER PROCEDURE

The loadings described in Step 1 in the previous section may consist of joint loads (force control), imposed displacements (displacement control), or combination of joint loads and imposed displacements (Cheng and Mertz, 1989b; Cheng and Ger, 1992). The loading is divided into increments and applied to the structure in steps. At the beginning of each load step, the tangent stiffness of the structure is determined, and the structure is assumed to behave linearly for the duration of the load step. Unbalanced forces, when they exist, are calculated at the end of each load step and added to the incremental loads for the next load step (see Appendix D). The structural stiffness is updated at each load step, if necessary. Depending on the axial load, the geometric stiffness is updated for each load step. At the end of each load step, total forces and displacements are determined by summing the values for the previous load step and the incremental values. The analysis procedure is governed by the following equations:

$$[K]\{\Delta\delta\} = \{\Delta F\} + \{U\} \tag{3.1}$$

Partitioning the structural global stiffness $[K]$, displacement $\{\Delta\delta\}$, load $\{\Delta F\}$, and unbalanced force $\{U\}$ matrices between free (f) and restrained (r) dofs yields

$$\begin{bmatrix} [K_{ff}] & [K_{fr}] \\ [K_{rf}] & [K_{rr}] \end{bmatrix} \begin{bmatrix} \{\Delta\delta_f\} \\ \{\Delta\delta_r\} \end{bmatrix} = \begin{bmatrix} \{\Delta F_f\} + \{U_f\} \\ \{\Delta R\} + \{U_r\} \end{bmatrix} \tag{3.2}$$

where
   $\{\Delta\delta_r\}$ represents the imposed displacement vector (i.e., displacement control)
   $\{\Delta R\}$ represents the reaction vector
   $\{\Delta F_f\}$ is the incremental joint load vector (i.e., force control)

INSTRUCT can perform both force and displacement controls concurrently during the pushover analysis. Expanding Equation 3.2

$$[K_{ff}]\{\Delta\delta_f\} + [K_{fr}]\{\Delta\delta_r\} = \{\Delta F_f\} + \{U_f\} \tag{3.3}$$

$$[K_{rf}]\{\Delta\delta_f\} + [K_{rr}]\{\Delta\delta_r\} = \{\Delta R\} + \{U_r\} \tag{3.4}$$

Rewriting Equation 3.3 yields

$$[K_{ff}]\{\Delta\delta_f\} = \{\Delta F_f\} + \{U_f\} - [K_{fr}]\{\Delta\delta_r\} \tag{3.5}$$

which is solved for the free Gdofs $\{\Delta\delta_f\}$ by Gaussian elimination.
   Rewriting Equation 3.4 yields the reactions

$$\{\Delta R\} = [K_{rf}]\{\Delta\delta_f\} + [K_{rr}]\{\Delta\delta_r\} - \{U_r\} \tag{3.6}$$

The total structural global displacements, forces, and reactions at load step t are determined from

$$\{\delta^t\} = \{\delta^{t-1}\} + \{\Delta\delta^t\} \tag{3.7}$$

$$\{F^t\} = \{F^{t-1}\} + \{\Delta F^t\} \tag{3.8}$$

$$\{R^t\} = \{R^{t-1}\} + \{\Delta R^t\} \tag{3.9}$$

Once the total global displacement increment vector, $\{\Delta\delta^t\}$, is obtained, the individual member deformation increment vector, $\{\Delta\delta_e^t\}$, can be calculated by

$$\{\Delta\delta_e^t\} = [A]^T\{\Delta\delta^t\} \tag{3.10}$$

and

$$\{\Delta F_e^t\} = [k_e]\{\Delta\delta_e^t\} \tag{3.11}$$

where $[A]^T$ is the transformation matrix between the global displacements increment vector and the member deformation increment vector, which will be discussed in Section 5.2.3. $\{\Delta F_e^t\}$ is the element force increment vector at load step t. $[k_e]$ is the individual element stiffness matrix. The global unbalanced joint force vector, $\{U\}$, is assembled by the element unbalanced forces being transferred to the structural Gdofs. At the end of the load step t, the element unbalanced force vector is calculated, as is the difference between the calculated element force vector from Equation 3.11 and the element force vector calculated based on the element's hysteresis model or stress resultants from steel and concrete stress–strain relationships. These member unbalanced forces are transferred to the structural Gdofs to form the global unbalanced joint force vector $\{U\}$ for the next step. As mentioned above, the unbalanced forces are calculated at the end of each load step and added to the incremental loads for the next step to reduce drift-off errors. A detailed description of the nonlinear incremental solution scheme used in the program is provided in Appendix D. The unbalanced force vector $\{U\}$ will be discussed in Chapter 5.

3.4 MATERIAL LIBRARY

3.4.1 Elastic 3D Prismatic Beam Material (3D-BEAM)

This material consists of the elastic section properties of a 3D prismatic element, A_x, J, I_y, I_z, E, and G, representing the cross-sectional area of the element, torsional moment of inertia, moments of inertia in the element's Y_e and Z_e directions (see Section 3.5.1 for the definition of the ECS), elastic Young's modulus, and shear modulus, respectively.

3.4.2 Bilinear Hysteresis Model (BILINEAR)

A hysteretic material model that has a bilinear backbone curve and an elastic unloading and reloading curve is shown in Figure 3.4. This model is mainly used for spring elements. The model may represent the elastoplastic model by setting the post-yielding stiffness to zero.

3.4.3 Gap/Restrainer Model (GAP)

This hysteresis model simulates the restrainer's inelastic behavior, see Figure 3.5. This model is mainly used for spring elements. When a gap is opened and the

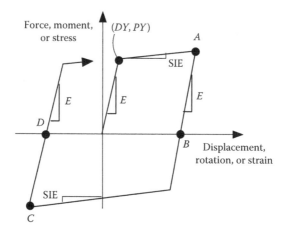

FIGURE 3.4 Bilinear hysteresis model.

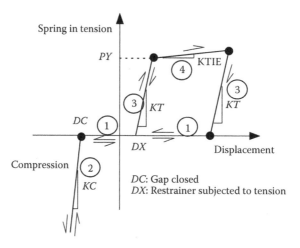

FIGURE 3.5 Gap/restrainer hysteresis model.

restrainer has not been engaged in tension, the program uses a very small stiffness (i.e., 0.001) to represent the gap opening. In the figure, DC represents the displacement at which the gap closes, and DX represents the displacement at which the restrainer is in tension.

3.4.4 TAKEDA HYSTERESIS MODEL (TAKEDA)

The Takeda model (Takeda et al., 1970), shown in Figure 3.6, is mainly used to model the bending deformation of reinforced concrete members subjected to cyclic loading. In the figure, three control points define the backbone curve, corresponding to the cracking moment (M_{cr}), nominal moment (M_n), and ultimate moment (M_u), respectively. The Takeda model consists of many hysteresis rules. These rules define

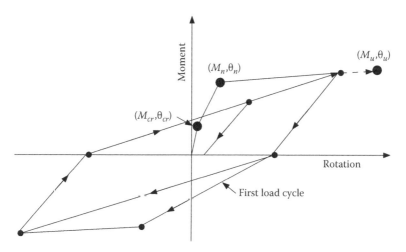

FIGURE 3.6   Takeda hysteresis model.

the loading, unloading, and load reversal paths. During the cyclic response analysis, the element's flexural stiffness is updated based on the hysteresis rules. These rules are described in the above-mentioned reference and are not explained in detail in this book.

3.4.5   BILINEAR MOMENT-ROTATION MODEL (HINGE)

This model is only used in the inelastic 3D beam (IE3DBEAM) element described in Section 3.5, when the PHL method is considered for the element-bending stiffness matrix formulation. The PHL method is described in Section 4.2. The axial load–nominal moment interaction in terms of a third-order polynomial equation can be considered, as is described in Section 4.8.

3.4.6   BILINEAR HYSTERESIS MODEL (IA_BILN)

This model is only used in the IE3DBEAM element. The model has a bilinear back-bone curve and an elastic unloading and reloading curve as shown in Figure 3.4. The axial load–nominal moment interaction can be considered in the pushover analysis. The IA_BILN material is used in the PM method for the element-bending stiffness matrix formulation, which is described in Section 4.1.

3.4.7   FINITE-SEGMENT STEEL STRESS–STRAIN HYSTERESIS MODEL (STABILITY1)

This model can be either a bilinear stress–strain relationship, as shown in Figure 3.4, or a Ramberg–Osgood stress–strain relationship (Ramberg and Osgood, 1943) as shown in Figure 3.7. The model is only used for the finite-segment element described in the element library later in this chapter. For the Ramberg–Osgood

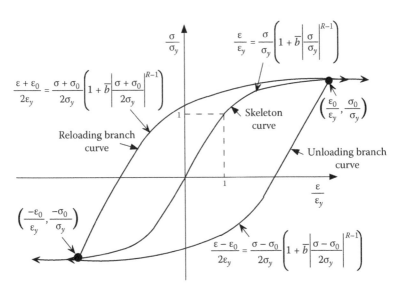

FIGURE 3.7 Ramberg–Osgood stress–strain relationship.

stress–strain relationship, the parameters R and $\bar{b}$ are positive constants chosen to fit the material nonlinear stress–strain curves. If R equals infinity, the stress–strain curve converges to the elastoplastic stress–strain relationship. When $R=1$ and $\bar{b}=0$, it represents the elastic condition. The STABILITY1 material is normally used for non-concrete members. Usually, the bilinear stress–strain relationship is sufficient to represent the steel stress–strain curve of most structural steel members. However, if the bilinear model cannot adequately represent the stress–strain relationship of a material, the Ramberg–Osgood stress–strain model may be considered. During the pushover analysis, the tangent modulus, TE, of the Ramberg–Osgood stress–strain model is calculated at each incremental step, which is given as follows:

$$TE = \frac{d\sigma}{d\varepsilon} = \frac{EM}{(1+R\bar{b}\,|\,\sigma/\sigma_y\,|^{R-1})} \quad \text{for the skeleton (or so-called backbone) curve}$$

(3.12)

$$TE = \frac{EM}{(1+R\bar{b}\,|\,(\sigma-\sigma_0)/2\sigma_y\,|^{R-1})} \quad \text{for the unloading branch curve} \quad (3.13)$$

$$TE = \frac{EM}{(1+R\bar{b}\,|\,(\sigma+\sigma_0)/2\sigma_y\,|^{R-1})} \quad \text{for the reloading branch curve} \quad (3.14)$$

in which $EM=\sigma_y/\varepsilon_y$.

3.4.8   FINITE-SEGMENT REINFORCED CONCRETE STRESS–STRAIN HYSTERESIS MODEL (R/CONCRETE1)

This model is based on Mander's concrete stress–strain relationship (Mander et al., 1988). However, the unloading curve is assumed to be linear in the program instead of using a nonlinear curve. The model is only used for the finite-segment element described later in the element library. The Mander's confined concrete stress–strain (f_c–ε) relationship is sketched in Figure 3.8 and given by Equation 3.15:

$$f_c = \frac{f_{cc}'\, xr}{r - 1 + x^r} \tag{3.15}$$

where

$$f_{cc}' = f_c'\left(2.254\sqrt{1 + \frac{7.94 f_l'}{f_c'}} - \frac{2 f_l'}{f_c'} - 1.254\right) \tag{3.16}$$

$$x = \frac{\varepsilon_c}{\varepsilon_{cc}} \tag{3.17}$$

$$\varepsilon_{cc} = 0.002\left[1 + 5\left(\frac{f_{cc}'}{f_c'} - 1\right)\right] \tag{3.18}$$

$$r = \frac{E_c}{E_c - E_{sec}} \tag{3.19}$$

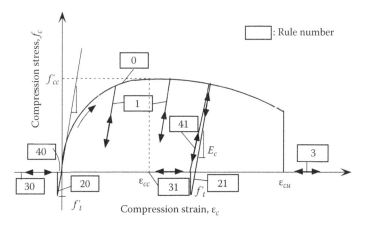

FIGURE 3.8   Confined concrete stress–strain hysteresis model.

$$E_c = 57,000\sqrt{f_c'} \text{ (psi)} \tag{3.20}$$

$$E_{sec} = \frac{f_{cc}'}{\varepsilon_{cc}} \tag{3.21}$$

$$f_l' = K_e f_l \tag{3.22}$$

$$f_l = \frac{2 f_{yh} A_{sp}}{D''s} \tag{3.23}$$

In the figure, f_{cc}' is the maximum confined concrete strength, and ε_{cc} is the concrete strain corresponding to f_{cc}'. f_l is the maximum effective stress around the hoop or spiral bar, and K_e is a confinement effectiveness coefficient. The typical values of K_e are 0.95 and 0.75 for circular and rectangular confined sections, respectively. K_e is zero for unconfined concrete such as concrete cover. f_l' is the effective lateral confined stress. D'' is the diameter of the hoop or spiral (measured to the centerline of the bar); A_{sp} is the cross-sectional area of the hoop or spiral bar; and s is the longitudinal spacing of the hoop or spiral. ε_{cu} is the ultimate confined concrete compression strain at which the hoop or spiral bar starts to fracture under axial compression load. ε_{cu} is conservatively expressed as follows:

$$\varepsilon_{cu} = 0.004 + \frac{1.4 \rho_s f_{yh} \varepsilon_{su}}{f_{cc}'} \tag{3.24}$$

where
   ε_{su} is the steel strain at maximum tensile stress (conservatively, $\varepsilon_{su} = 0.09$ is
       adopted in the program)
   f_{yh} is the yield stress of hoop or spiral bar
   ρ_s is the volumetric ratio of the hoop or spiral bar

ρ_s can be expressed as follows:

$$\rho_s = \frac{4 A_{sp}}{D''s} \quad \text{for circular sections} \tag{3.25}$$

$$\rho_s = \rho_X + \rho_Y \quad \text{for rectangular sections} \tag{3.26}$$

in which ρ_X and ρ_Y are the volumetric ratios of transverse hoops to core concrete in the X and Y directions (see Figure 3.9), respectively. ρ_X and ρ_Y can be expressed as follows:

$$\rho_X = \frac{N_X A_{sp}}{s h_y''} \tag{3.27}$$

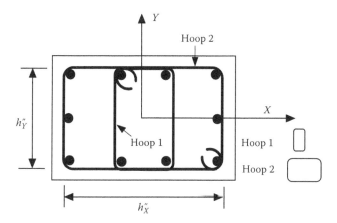

FIGURE 3.9 Rectangular cross section.

$$\rho_Y = \frac{N_Y A_{sp}}{sh''_x} \tag{3.28}$$

where
   h''_X and h''_Y are the confined core dimensions in the X and Y directions, respectively
   N_X and N_Y are the total number of transverse hoop legs in the X and Y directions, respectively

For example, the rectangular cross section in Figure 3.9 has $N_X = (2/3) \times 2 + (1/3) \times 4 = 2.67$, because the central one-third of the section has four hoop legs, and the other two-thirds of the section has only two hoop legs. Similarly, N_Y is equal to 4.

The maximum confined concrete strength, f'_{cc}, for rectangular cross section can be obtained from the ratio of f'_{cc}/f'_c, which can be found from Table 3.1. Table 3.1 was developed from Figure E.2 and incorporated into INSTRUCT for calculating f'_{cc}. In the table, f'_{lX} and f'_{lY} are the effective lateral confined stresses in X and Y directions, respectively. f'_{lX} and f'_{lY} can be calculated as follows:

$$f'_{lX} = K_e \rho_X f_{yh} \tag{3.29}$$

$$f'_{lY} = K_e \rho_Y f_{yh} \tag{3.30}$$

Equation 3.24 is derived based on a column subjected to axial compression without bending. For columns subjected to both axial compression and bending, the ultimate compression concrete strain estimated by Equation 3.24 is conservative (Mander et al., 1988). The actual ultimate confined concrete strain under combined axial force and moment is about 1.3–1.6 times ε_{cu} from Equation 3.24.

TABLE 3.1

Concrete Lateral Confined Stress for Rectangular Section

Confined Strength Ratio f'_{cc}/f'_c

f'_{lx}/f'_c \ f'_{ly}/f'_c	0	0.02	0.04	0.06	0.08	0.1	0.12	0.14	0.16	0.18	0.2	0.22	0.24	0.26	0.28	0.3
0	1															
0.02	1.04	1.13														
0.04	1.1	1.17	1.26													
0.06	1.13	1.22	1.3	1.38												
0.08	1.15	1.25	1.34	1.43	1.48											
0.1	1.19	1.28	1.37	1.45	1.52	1.57										
0.12	1.21	1.31	1.4	1.49	1.55	1.62	1.66									
0.14	1.23	1.33	1.43	1.52	1.58	1.66	1.7	1.76								
0.16	1.24	1.35	1.45	1.53	1.62	1.68	1.74	1.79	1.83							
0.18	1.25	1.37	1.48	1.56	1.64	1.72	1.77	1.83	1.88	1.9						
0.2	1.26	1.39	1.49	1.58	1.66	1.74	1.79	1.85	1.9	1.94	1.97					
0.22	1.27	1.4	1.5	1.59	1.68	1.75	1.82	1.88	1.93	1.97	2	2.05				
0.24	1.28	1.41	1.52	1.61	1.7	1.77	1.83	1.9	1.96	2	2.04	2.07	2.12			
0.26	1.29	1.42	1.53	1.62	1.72	1.78	1.85	1.92	1.97	2.02	2.06	2.1	2.14	2.18		
0.28	1.295	1.43	1.54	1.64	1.73	1.8	1.87	1.94	2	2.05	2.08	2.13	2.17	2.2	2.25	
0.3	1.3	1.43	1.55	1.65	1.74	1.82	1.9	1.95	2.02	2.06	2.1	2.15	2.2	2.23	2.27	2.3

The slope (i.e., tangent stiffness) of the backbone ($f_c-\varepsilon$) curve in Figure 3.8 can be expressed as follows:

$$\frac{df_c}{d\varepsilon} = AA[B - C \times D] \tag{3.31}$$

where

$$AA = \frac{f'_{cc}r}{\varepsilon_{cc}} \tag{3.32}$$

$$B = \frac{1}{a + (\varepsilon/\varepsilon_{cc})^r} \tag{3.33}$$

$$C = \frac{\varepsilon}{[a + (\varepsilon/\varepsilon_{cc})^r]^2} \tag{3.34}$$

$$D = \frac{r}{\varepsilon_{cc}} \left(\frac{\varepsilon}{\varepsilon_{cc}} \right)^{r-1} \tag{3.35}$$

and

$$a = r - 1 \tag{3.36}$$

The cracking stress, $f'_t = 9\sqrt{f'_c}$ (psi), is used in the program.

Example 3.1

Find the ρ_X, ρ_Y, and the maximum confined concrete strength, f'_{cc}, of a rectangular section shown in Figure 3.10. The cross-sectional area of the transverse reinforcement, A_{sp}, is 0.122 in.^{2} The yield stress of the transverse reinforcement is 67,570 psi, and the spacing of the hoops is 4.61 in. Concrete $f'_c = 4104$ psi.

Solution

From Figure 3.10, $N_X = (2/3) \times 4 + (1/3) \times 6 = 4.67$; $N_Y = 4.67$.
ρ_X and ρ_Y can be calculated from Equations 3.27 and 3.28 as follows:

$$\rho_X = \frac{N_X A_{sp}}{sh''_Y} = \frac{4.67 \times 0.122}{4.61 \times 14.33} = 0.0086$$

$$\rho_Y = \frac{N_Y A_{sp}}{sh''_X} = \frac{4.67 \times 0.122}{4.61 \times 19.33} = 0.0064$$

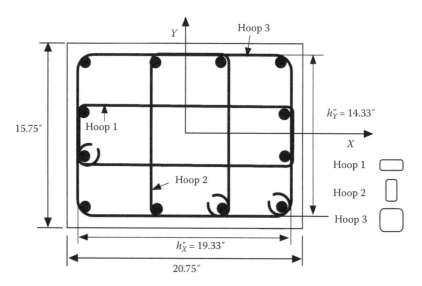

FIGURE 3.10 Rectangular cross section for Example 3.1.

$$f'_{lX} = K_e \rho_X f_{yh} = 0.75 \times 0.0086 \times 67,570 = 435.8 \text{ psi}$$

$$f'_{lY} = K_e \rho_Y f_{yh} = 0.75 \times 0.0064 \times 67,570 = 324.3 \text{ psi}$$

$$\frac{f'_{lX}}{f'_c} = \frac{435.8}{4104} = 0.106; \quad \frac{f'_{lY}}{f'_c} = \frac{324.3}{4104} = 0.079$$

From Table 3.1,

$$\frac{f'_{cc}}{f'_c}(0.1,0.06) = 1.45 \quad \text{at} \quad \frac{f'_{lX}}{f'_c} = 0.1 \quad \text{and} \quad \frac{f'_{lY}}{f'_c} = 0.06 \qquad (3.37)$$

$$\frac{f'_{cc}}{f'_c}(0.1,0.08) = 1.52 \quad \text{at} \quad \frac{f'_{lX}}{f'_c} = 0.1 \quad \text{and} \quad \frac{f'_{lY}}{f'_c} = 0.08 \qquad (3.38)$$

$$\frac{f'_{cc}}{f'_c}(0.12,0.06) = 1.49 \quad \text{at} \quad \frac{f'_{lX}}{f'_c} = 0.12 \quad \text{and} \quad \frac{f'_{lY}}{f'_c} = 0.06 \qquad (3.39)$$

The interpolation ratios in the X and Y directions are

$$Ratio\ X = \frac{0.106 - 0.1}{0.12 - 0.1} = 0.3 \qquad (3.40)$$

$$Ratio\ Y = \frac{0.079 - 0.06}{0.08 - 0.06} = 0.95 \qquad (3.41)$$

From Equations 3.37 through 3.41

$$\frac{f'_{cc}}{f'_c}(0.106, 0.079) \cong \left[\frac{f'_{cc}}{f'_c}(0.1, 0.08) - \frac{f'_{cc}}{f'_c}(0.1, 0.06)\right] \times Ratio\ Y$$

$$+ \left[\frac{f'_{cc}}{f'_c}(0.12, 0.06) - \frac{f'_{cc}}{f'_c}(0.1, 0.06)\right] \times Ratio\ X + \frac{f'_{cc}}{f'_c}(0.1, 0.06)$$

$$= 0.0665 + 0.012 + 1.45 = 1.53 \qquad (3.42)$$

Therefore, $f'_{cc} = 1.53 f'_c = 1.53 \times 4104 = 6279$ psi, which also can be obtained from Figure E.2.

The unconfined concrete stress–strain model is shown in Figure 3.11. In the figure, f'_c is the maximum concrete strength and ε_{c0} is the concrete strain corresponding to f'_c. $\varepsilon_{c0} = 0.002$ is used in the program. ε_{sp} is the concrete spalling strain of unconfined cover concrete. A straight line is assumed between $2\varepsilon_{c0}$ and ε_{sp}.

The hysteresis rules shown in Figures 3.8 and 3.11 are

- *Rule 0*: Compression backbone curve
- *Rule 1*: Unloading compression curve from compression backbone curve

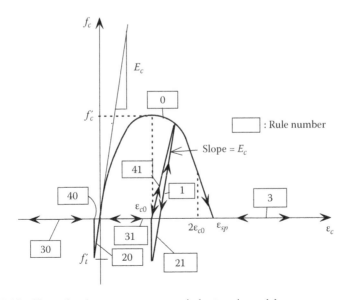

FIGURE 3.11   Unconfined concrete stress–strain hysteresis model.

- *Rule 20*: Tension backbone curve
- *Rule 21*: Unloading tension curve from compression backbone curve
- *Rule 31*: Tensile crack after unloading from compression
- *Rule 41*: Concrete compression after reloading from tensile crack
- *Rule 30*: Tensile crack after tensile strain greater than crack strain, ε_t, corresponding to f_t'
- *Rule 40*: Tensile strain between ε_t and zero
- *Rule 3*: Compression strain greater than ε_{cu} or ε_{sp}

For reinforcing steel, the bilinear hysteresis model shown in Figure 3.4 is adopted.

3.4.9   FINITE SEGMENT–MOMENT CURVATURE MODEL (MOMCURVA1)

This model is based on user-defined backbone moment–curvature curves. The model is only used for the finite-segment element described in Section 3.5. INSTRUCT uses a bilinear curve for the unloading and reloading conditions; therefore, this model is mainly for monotonic pushover. The moment–curvature model is sketched in Figure 3.12a using the multiple segment backbone curve. The minimum number of control points is two, representing a bilinear moment–curvature model as shown in Figure 3.12b. During pushover analysis, if a segment's curvature exceeds $D(n)$, the program uses the slope of the last two control points (i.e., $M(n-1)$ and $M(n)$) to calculate tangent bending rigidity EI.

To consider the axial load–moment interaction, the user can input multiple moment–curvature curves as illustrated in Figure 3.12c. Each curve corresponds to a certain magnitude of axial load. At each incremental load step during the pushover analysis, the program will calculate the member's axial load first and then the

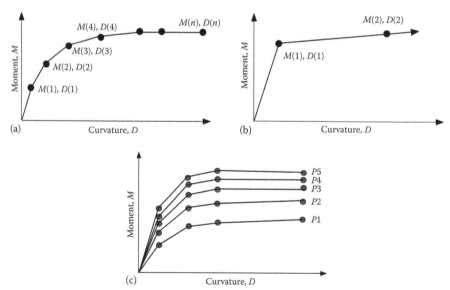

FIGURE 3.12   (a) Finite segment–moment curvature model; (b) moment–curvature model (two control points); (c) multiple moment–curvature curves.

member's tangent bending rigidity, EI, by interpolating EIs from the two adjacent moment–curvature curves.

3.4.10   PLATE MATERIAL (PLATE)

This material defines the elastic material properties for the rectangular plate element. The plate element is described later in the element library.

3.4.11   POINT MATERIAL (POINT)

This material defines the elastic material properties of the point element. The point element is described later in the element library.

3.4.12   BRACE MATERIAL (BRACE)

This material defines the hysteresis rule of Jain–Goel–Hanson's model (Jain et al., 1980). The model is mainly for struts with angle or rectangular tube sections. For I-shape sections, several control points in the model are modified in order to fit the experimental results achieved by Black et al. (1980).

As shown in Figure 3.13, the normalized axial load–axial deformation coordinate, P/P_y, Δ/Δ_y, is used, and tension and compression are treated as positive and negative, respectively. When a member is subjected to compression loading, the axial force–axial displacement relationship will follow path OA elastically. Point A represents the initial buckling load, which can be expressed as follows:

$$P_{max} = \left[1 - \frac{(KL/r)^2}{2C_c^2}\right]\sigma_y A \quad \text{for} \quad \frac{KL}{r} < C_c \tag{3.43}$$

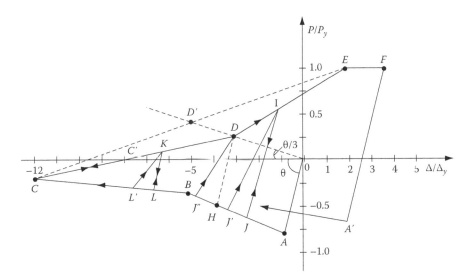

FIGURE 3.13   Hysteresis model for brace members.

$$P_{max} = \frac{\pi^2 AE}{(KL/r)^2} \quad \text{for } C_c < \frac{KL}{r} \leq 200 \tag{3.44}$$

where $C_c = \sqrt{(2\pi^2 E)/\sigma_y}$ and σ_y represents yield stress. Continued negative axial displacement results in the path ABC. The coordinates of control point B are $[-18/(KL/r), -5]$ for angle or rectangular tube sections and $[-11.3/(KL/r), -5]$ for I-shape sections. The coordinates of control point C are $[-12/(KL/r), -12]$ for angle or rectangular tube sections and $[-8.5/(KL/r), -12]$ for I-shape sections. If the reverse point occurs at C, the member will follow path CDE. If the negative axial displacement is beyond point C, the axial stiffness is assumed to be zero. The coordinates of control point E are $(1,1 + \varepsilon L/\Delta_y)$ in which ε is the member residual strain, given by

$$\varepsilon = 1.75 \left[\frac{0.55\Delta}{KL/r} + 0.0002\Delta^2 \right] \quad \text{for angle or rectangular tube sections} \tag{3.45}$$

$$\varepsilon = 1.4 \left[\frac{0.55\Delta}{KL/r} + 0.0002\Delta^2 \right] \quad \text{for I-shape section} \tag{3.46}$$

in which Δ is the current maximum compressive displacement. As shown in Equations 3.45 and 3.46, the residual strain of a member is influenced by the effective slenderness ratio, KL/r, of a member and the maximum compressive displacement of a member, Δ. The coordinates of control point D can be obtained by first drawing a line OD' with a slope of one-third times the initial elastic slope, θ. The point located at $60/(KL/r)$ times the distance OD' is taken as control point D for angle or rectangular tube sections and $20/(KL/r)$ times the distance OD' for I-shape sections. When the member starts unloading from the tension side on the branch EF, the maximum compression in the second cycle is given by $-(30P_y)/(KL/r)$ and the third and subsequent cycles by $-(25P_y)/(KL/r)$.

Using this model, Figure 3.14 shows the hysteresis loops based on experimental results (Black et al., 1980; Popov and Black, 1981) and numerical calculation from INSTRUCT, for a $W6 \times 20$ member with $KL/r = 80$. It can be seen that the analytical result is in favorable agreement with the experimental curves.

3.5 ELEMENT LIBRARY

3.5.1 ELASTIC 3D PRISMATIC ELEMENT (3D-BEAM)

The elastic 3D prismatic beam element is shown in Figure 3.15.

This element connects a start and an end joint. At the start end of the element, a rigid body (or so-called rigid zone) of length XS is used to model the structural joint. A similar rigid body of length XE is used at the end joint. The ECS X_e axis goes from end "A" toward end "B." The orientation of the ECS Y_e axis is defined by a vector V_{XY}, which lies on the ECS XY-plane. The ECS Z_e axis is perpendicular to the X_e and

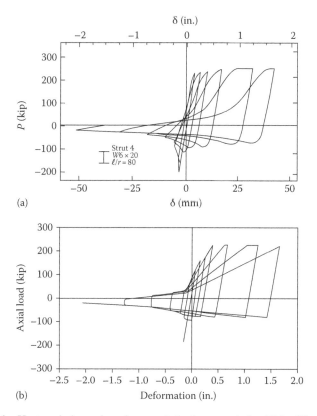

FIGURE 3.14 Hysteresis loops based on analytical approach for $KL/r=80$: (a) test result and (b) analytical result. (From Cheng, F.Y. et al., Observations on behavior of tall steel building under earthquake excitations, *Proceedings of Structural Stability Research Council*, Pittsburgh, PA, pp. 15–26, 1992.)

Y_e axes, oriented according to the right-hand rule. There are six internal forces F_X, F_Y, F_Z, M_X, M_Y, M_Z at end "A" in the ECS. Similarly, six internal forces also exist at end "B." All of the internal forces are positive in the direction of the ECS. Therefore, the 3D-BEAM element has 12 dofs. The formulation of the 12×12 element stiffness matrix is described in Chapter 5.

The element considers axial deformation, torsional deformation, and bending deformations about the Y_e and Z_e axes. Warping torsion and shear deformation are not considered. The geometric stiffness for $P-\delta$ effects is also available. The 3D-BEAM material is used with this element (see Table 3.2).

3.5.2 Spring Element (SPRING)

The spring element consists of an isolated spring that connects the start and end joints. At the start end of the spring, a rigid body of length XS is used to model the joint depth. A similar rigid body, of length XE is used at the end joint. The spring

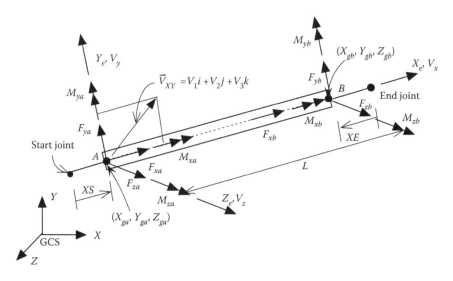

FIGURE 3.15   3D prismatic element.

TABLE 3.2
Material-Element Cross Reference

Element		Material Availability	
Type	**Subroutine Name in INSTRUCT**	**Type**	**Subroutine Name in INSTRUCT**
3D-BEAM	ELE01	3D-BEAM	MAT01
SPRING	ELE02	3D-BEAM	MAT01 (for axial spring only)
		BILINEAR	MAT07
		GAP	MAT15B
		TAKEDA	MAT06
BRACE	ELE08	BRACE	MAT08
IE3DBEAM	ELE09	IA_BILN	MAT10 (PM method)[a]
		HINGE	MAT19 (PHL method)[a]
		TAKEDA	MAT06 (CMR method)[a]
STABILITY	ELE12	STABILITY1	MAT12 (bilinear and Ramberg–Osgood models)
		R/CONCRETE1	MAT17
		MOMCURVA1	MAT18
PLATE	ELE16	PLATE	MAT16
POINT	ELE20	POINT	MAT20

[a] For PM-, PHL-, and CMR-bending stiffness formulations, see Chapter 4.

ECS X_e axis goes from end "A" toward end "B." The orientation of the ECS Y_e axis is defined by the user. The ECS Z_e axis is perpendicular to the X_e and Y_e axes, oriented according to the right-hand rule. When the distance between the start and end joints is zero, the orientation of the ECS is identical to the start joint's JCS. Subtracting the length of the rigid bodies from the distance between the start and end joints gives the length of the spring element. Optionally, the user may define the length of the spring element. The spring element may behave elastically or nonlinearly depending on the material properties used and the magnitude of forces acting on the spring. Second-order $P-\delta$ forces are not calculated for the spring element.

The spring may be orientated in one of six positions as shown in Figure 3.16.

The axial spring is parallel to the element's X_e axis. The rigid bodies at the ends of the spring reduce the length of the axial spring. The spring's axial force, F_X, at end "A" is positive in the X_e direction.

The Y-axis shear spring and Z-axis shear spring are orientated parallel to the element's Y_e and Z_e axes, respectively. The rigid bodies at the ends of the spring reduce the length of the shear spring. The spring's internal shears, F_Y and F_Z, at end "A" are positive in the Y_e and Z_e directions.

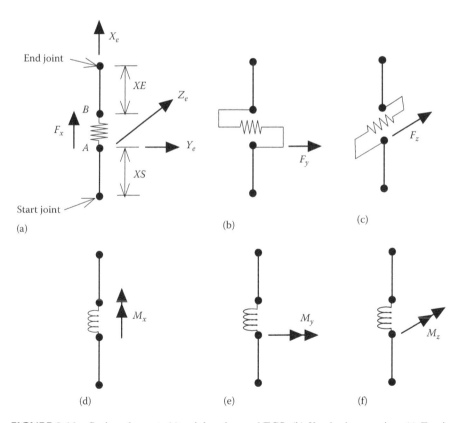

FIGURE 3.16 Spring element: (a) axial spring and ECS, (b) Y-axis shear spring, (c) Z-axis shear spring, (d) torsional spring, (e) Y-axis rotational spring, and (f) Z-axis rotational spring.

The torsional spring is parallel to the element's X_e axis. The rigid bodies reduce the length of the torsional spring. The spring's internal torsion, M_X at end "A" is positive in the X_e direction.

The Y-axis rotational spring and the Z-axis rotational spring are rotational springs about the Y_e and Z_e axes, respectively. The rigid bodies at the ends of the spring reduce the length of the rotational spring. The spring's internal moments, M_Y and M_Z, at end "A" are positive in the Y_e and Z_e directions. A nonlinear rotational spring, with material type of BILINEAR or TAKEDA, could be applied to the end of a structural member (e.g., attached to an elastic 3D-BEAM element) to evaluate the inelastic behavior of the member.

3.5.3 INELASTIC 3D BEAM ELEMENT (IE3DBEAM)

The inelastic 3D beam-column element is shown in Figure 3.17 and has the same ECS as elastic 3D-BEAM element.

Similar to the elastic 3D-BEAM element, this element connects a start and an end joint. At the start end of the element, a rigid body of length XS is used to model the structural joint. A similar rigid body of length XE is used at the end joint. The ECS X_e axis goes from end "A" toward end "B." The orientation of the ECS Y_e axis is defined by a vector V_{XY}, which lies on the ECS XY-plane. The ECS Z_e axis is perpendicular to the X_e and Y_e axes, oriented according to the right-hand rule. There are six internal forces F_X, F_Y, F_Z, M_X, M_Y, M_Z at end "A" in the ECS. Similarly, six internal forces also exist at end "B." All of the internal forces are positive in the direction of the ECS. Therefore, the IE3DBEAM element has 12 dofs.

The element considers axial, torsional, and bending deformations. Warping torsion and shear deformation are not considered. The geometric stiffness is also available. The HINGE, IA_BILN, or TAKEDA material can be used for bending

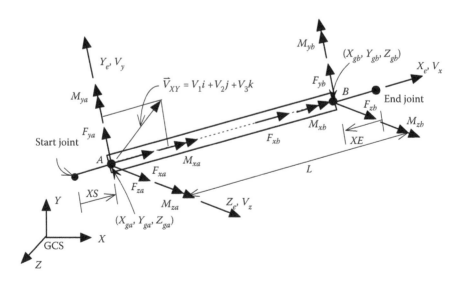

FIGURE 3.17   Inelastic 3D beam element.

deformation. Only IA_BILN material can be used for torsion and axial deformations. The formulation of element stiffness matrix is described in Chapter 5.

3.5.4 FINITE-SEGMENT ELEMENT (STABILITY)

The finite-segment element (Ger and Cheng, 1993; Ger et al., 1993) is shown in Figure 3.18, in which the member is divided into several segments. Each segment has 12 dofs, and the cross section is divided into many small elements (or so-called strings) as shown in Figure 3.19.

The finite-segment element connects a start and an end joint. At the start end of the element, a rigid body of length XS is used to model the structural joint. A similar rigid body of length XE is used at the end joint. The ECS X_e axis goes from end "A" toward end "B." The orientation of the ECS Y_e axis is defined by a vector V_{XY}, which lies on the ECS XY plane. The ECS Z_e axis is perpendicular to the X_e and Y_e axes, oriented according to the right-hand rule. There are six internal forces F_X, F_Y, F_Z, M_X, M_Y, M_Z at end "A" in the ECS. Similarly, six internal forces also exist at end "B."

All of the internal forces are positive in the direction of the ECS. An initial imperfection of sinusoidal shape can be considered for the finite-segment element (see Figure B.3).

The element considers nonlinear axial and bending deformations. Warping torsion and shear deformation are not considered. The member is divided into several segments. The cross section of each segment is further divided into many small elements, and U_0, V_0, and W_0 represent an individual segment's sectional reference coordinates as shown in Figure 3.19. If a member is perfectly straight without initial imperfection, the relationship between the ECS (X_e, Y_e, Z_e) and the segment's reference coordinate system (U_0, V_0, W_0) is that X_e, Y_e, and Z_e axes are parallel to W_0, U_0, and V_0 axes, respectively (see Figure 3.18).

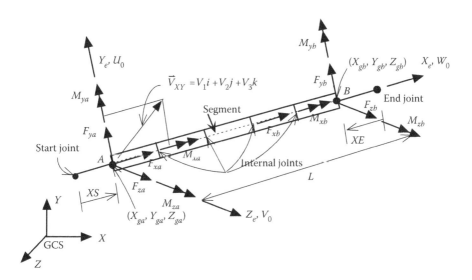

FIGURE 3.18 Finite-segment element.

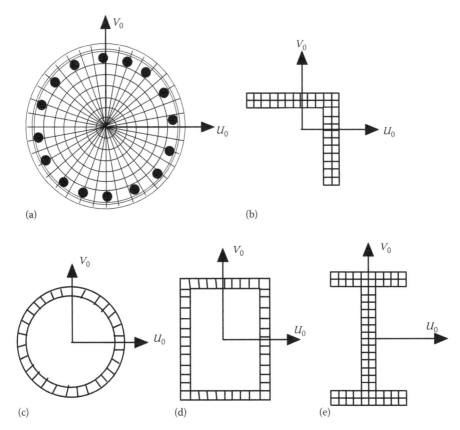

FIGURE 3.19 Segment reference coordinates (U_0, V_0, W_0). (a) Reinforced concrete circular or rectangular section, (b) steel angle section, (c) steel tube section, (d) steel box section, and (e) steel wide-flange section.

In the analysis of a structural system, the dofs of an individual member should be reduced so that a computational efficiency can be achieved. A substructural technique is applied to the finite-segment element for which the internal dofs is condensed out by Gaussian elimination, and only the dofs at both ends of the member are maintained. The STABILITY1, R/CONCRETE1, or MOMCURVA1 hysteresis material model described in the previous section can be used for the finite-segment element. The second-order $P-\delta$ forces are considered in the element stiffness matrix formulation. The formulation of the element stiffness matrix is described in Chapter 5.

3.5.5 PLATE ELEMENT (PLATE)

The plate element consists of a plate linking four joints as shown in Figure 3.20. The ECS X_e axis goes from joint "3" toward joint "4." The orientation of the ECS Y_e axis goes from joint "3" toward joint "2." The ECS Z_e axis is perpendicular to X_e and Y_e axes, oriented according to the right-hand rule. The joint rigid body zone is not

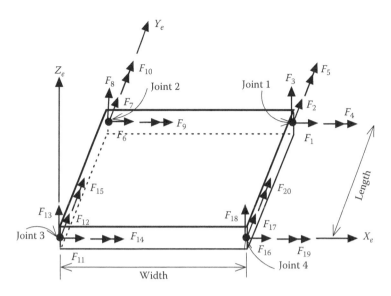

FIGURE 3.20 Plate element.

available for the plate element. The plate element has 20 dofs at the corner joints as shown in Figure 3.20. At each joint, there are five internal forces F_X, F_Y, F_Z, M_X, and M_Y in the ECS. The PLATE material is used with this element.

3.5.6 POINT ELEMENT (POINT)

The point element (see Figure 3.21) is a point consisting of a 6×6 stiffness matrix. For example, bridge foundation stiffnesses can be modeled by point elements. The ECS's X_e and Y_e axes of a point element are determined by the users. The 6×6

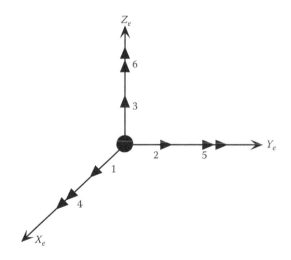

FIGURE 3.21 Point element.

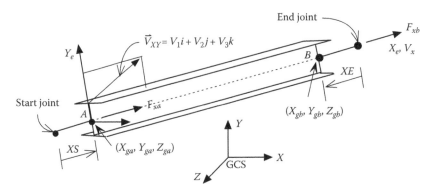

FIGURE 3.22   Brace element.

stiffness matrix is also determined by the users, and the coupling effect of any two dofs can be input. The POINT material is used with this element.

3.5.7   BRACE ELEMENT (BRACE)

As shown in Figure 3.22, the brace element (Ger and Cheng, 1992) is similar to the axial spring element shown in Figure 3.16a. At the start end of the brace element, a rigid body of length XS is used to model the joint depth. A similar rigid body of length XE is used at the end joint. The brace ECS X_e axis goes from end "A" toward end "B." The orientation of the ECS Y_e axis is defined by the user. The ECS Z_e axis is perpendicular to the X_e and Y_e axes, oriented according to the right-hand rule. The cross section of the brace element could be an angle, rectangular tube, or I-shape. The BRACE material is used with this element (see Table 3.2). The formulation of the element stiffness matrix is described in Chapter 5.

3.6   MATERIAL-ELEMENT CROSS REFERENCE

Table 3.2 shows which materials can be used for each element type.

4 Nonlinear Bending Stiffness Matrix Formulations

As described in Chapter 2, pushover analysis is a verification tool for new bridge design. It is also a useful tool to estimate the capacities of existing bridge bents. However, due to material and geometrical nonlinearities of bridge bents, it is imperative to have a computer program to perform the pushover analysis. It is equally important to understand the methodologies used in the pushover analysis. This chapter briefly describes how INSTRUCT formulates the element bending stiffness matrix using five different methods: (1) bilinear interaction axial load–moment (PM), (2) plastic hinge length (PHL), (3) constant moment ratio (CMR), (4) finite segment–finite string (FSFS), and (5) finite segment–moment curvature (FSMC) methods. The PM, PHL, and CMR methods are simpler than FSFS and FSMC, with FSFS method being the most sophisticated.

4.1 BILINEAR INTERACTION AXIAL LOAD–MOMENT METHOD

The bilinear moment–curvature curve shown in Figure 4.1a is used to generate the nonlinear member bending stiffness matrix. The moment–curvature curve is composed of two imaginary components shown in Figure 4.1b. In these figures, the slopes of the linear and elastoplastic components are $a_1 = p \times EI$, $a_2 = q \times EI$, and $p + q = 1$, where p is the fraction of flexural rigidity apportioned to the linear component and q is the fraction of flexural rigidity apportioned to the elastoplastic component. The post-yield slope of the elastoplastic component is equal to zero.

The nonlinear member shown in Figure 4.2 is used to formulate the nonlinear bending member stiffness matrix (Cheng, 2000). As described previously, the nonlinear member has two components, linear and elastoplastic. θ_i and θ_j are member-end total rotations; α_i and α_j are plastic rotations at each end of the elastoplastic component. The member stiffness matrix at any incremental step can be formulated according to the state of yield. Appendix A provides the derivation of the member stiffness matrix at different yield states. The state of yield may be one of the following four conditions: (a) both ends linear, (b) i end nonlinear and j end linear, (c) i end linear and j end nonlinear, and (d) both ends nonlinear.

The flexural stiffness matrix for condition (a) is

$$\begin{Bmatrix} \Delta M_i \\ \Delta M_j \end{Bmatrix} = \begin{bmatrix} a & b \\ b & a \end{bmatrix} \begin{Bmatrix} \Delta\theta_i \\ \Delta\theta_j \end{Bmatrix} \tag{4.1}$$

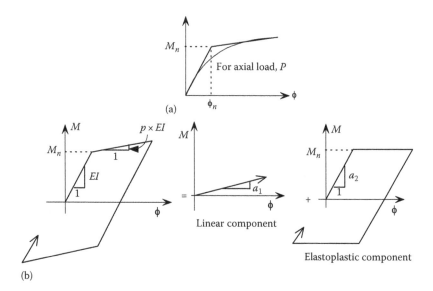

FIGURE 4.1 (a) Bilinear moment–curvature model; (b) linear and elastoplastic components.

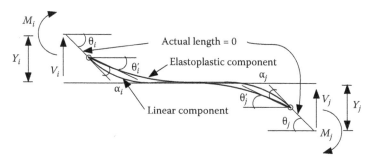

FIGURE 4.2 Nonlinear member.

The stiffness matrix for condition (b) is

$$\begin{Bmatrix} \Delta M_i \\ \Delta M_j \end{Bmatrix} = \begin{bmatrix} pa & pb \\ pb & pa + qe \end{bmatrix} \begin{Bmatrix} \Delta \theta_i \\ \Delta \theta_j \end{Bmatrix} \tag{4.2}$$

The stiffness matrix for condition (c) is

$$\begin{Bmatrix} \Delta M_i \\ \Delta M_j \end{Bmatrix} = \begin{bmatrix} pa + qe & pb \\ pb & pa \end{bmatrix} \begin{Bmatrix} \Delta \theta_i \\ \Delta \theta_j \end{Bmatrix} \tag{4.3}$$

The stiffness matrix for condition (d) is

$$\begin{Bmatrix} \Delta M_i \\ \Delta M_j \end{Bmatrix} = p \begin{bmatrix} a & b \\ b & a \end{bmatrix} \begin{Bmatrix} \Delta\theta_i \\ \Delta\theta_j \end{Bmatrix} \tag{4.4}$$

in which $a=4EI/L$, $b=2EI/L$, and $e=3EI/L$. As shown in Figure 4.1, the nominal moment capacity, M_n, is used for judging the member end's yield condition. During monotonic pushover analysis, the nominal moment capacity is influenced by the magnitude of axial load. If the variation of axial load is large, an interaction axial load–nominal moment (P–M) curve can be input into INSTRUCT for the analysis. Note that the program does not consider material isotropic hardening or kinematic hardening (i.e., the interaction P–M yield surface will not move outward). When the column axial load is small, the nominal moment capacity, M_n, will not change much due to axial load effects. In this case, consideration of axial load–moment interaction may not be necessary.

The dimension of the member flexural stiffness matrix in Equations 4.1 through 4.3 or 4.4 is 2×2. The actual member stiffness matrix incorporated into the INSTRUCT program is 12×12, which includes bending, axial, and torsional loads for the inelastic 3D-beam (IE3DBEAM) element shown in Figure 3.17. The derivation of the 12×12 element stiffness matrix is described in Chapter 5.

4.2 PLASTIC HINGE LENGTH METHOD

One of the popular methods used for the nonlinear pushover analysis of bridges with concrete columns is the PHL method. In this method, the stiffness matrix of a column is formulated by the combination of an elastic column element and a nonlinear rotational spring connected at each end of the element. As shown in Figure 4.3, the

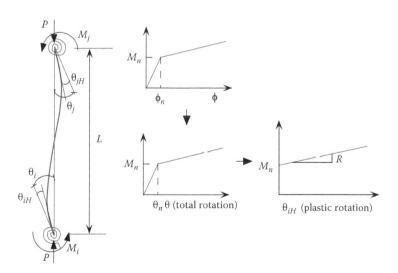

FIGURE 4.3 PHL method.

rotational spring and the elastic bending stiffness of the column element behave as two springs in series. The stiffness of the rotational spring is governed by the moment–rotational curve of a hinge with length L_p, which is called the PHL. L_p can be calculated as follows:

$$L_p = 0.08L + 0.15 f_y d_{bl} \geq 0.3 f_y d_{bl} \tag{4.5}$$

where
   L (in in.) is the distance from the critical section of the column plastic hinge to the point of contraflexure of the column
   d_{bl} (in in.) is the diameter of the longitudinal reinforcement
   f_y (in ksi) is the yield stress of the longitudinal reinforcement

Equation 4.5 has been calibrated using experimental data from large-scale test columns, which limit the PHL method to bridge bents with reinforced concrete columns. The PM method mentioned previously could be used for either pile cap bents with steel piles or bents with reinforced concrete columns.

The member stiffness matrix can be derived using the modified slope-deflection theory as follows:

$$\Delta M_i = \frac{EI}{L}[4(\Delta\theta_i - \Delta\theta_{iH}) + 2(\Delta\theta_j - \Delta\theta_{jH})] \tag{4.6}$$

$$\Delta M_j = \frac{EI}{L}[2(\Delta\theta_i - \Delta\theta_{iH}) + 4(\Delta\theta_j - \Delta\theta_{jH})] \tag{4.7}$$

where

$$\Delta\theta_{iH} = \frac{\Delta M_i}{R_i} \quad \text{and} \quad \Delta\theta_{jH} = \frac{\Delta M_j}{R_j} \tag{4.8}$$

$\Delta\theta_{iH}$ and $\Delta\theta_{jH}$ are the incremental plastic rotations at ends "a" and "b," respectively. As shown in Figure 4.3, the moment–plastic rotation ($M - \theta_H$) curve can be generated from the column moment–curvature ($M - \phi$) relationship. R_i and R_j are the inelastic stiffnesses of the plastic hinges at ends "a" and "b," respectively, and can be obtained from the slopes of the $M - \theta_H$ curves corresponding to ends "a" and "b," respectively. EI is the bending rigidity of the elastic column element. Solving Equations 4.6 and 4.7 for M_i and M_j gives

$$\Delta M_i = \frac{EI}{L}[S_{ii}\Delta\theta_i + S_{ij}\Delta\theta_j] \tag{4.9}$$

$$\Delta M_j = \frac{EI}{L}[S_{ij}\Delta\theta_i + S_{jj}\Delta\theta_j] \tag{4.10}$$

or

$$\begin{Bmatrix} \Delta M_i \\ \Delta M_j \end{Bmatrix} = \frac{EI}{L} \begin{bmatrix} S_{ii} & S_{ij} \\ S_{ij} & S_{jj} \end{bmatrix} \begin{Bmatrix} \Delta\theta_i \\ \Delta\theta_j \end{Bmatrix} \tag{4.11}$$

where

$$S_{ii} = \left[4 + \frac{12EI}{LR_j} \right] \Big/ R^* \tag{4.12}$$

$$S_{ii} = \left[4 + \frac{12EI}{LR_i} \right] \Big/ R^* \tag{4.13}$$

$$S_{ij} = \frac{2}{R^*} \tag{4.14}$$

$$R^* = \left(1 + \frac{4EI}{LR_i} \right)\left(1 + \frac{4EI}{LR_j} \right) - \left(\frac{EI}{L} \right)^2 \left(\frac{4}{R_i R_j} \right) \tag{4.15}$$

To calculate the plastic rotation of a reinforced concrete column for a pushover analysis using the PHL method, the bilinear moment–curvature ($M-\phi$) relationship and PHL (L_p) of the column need to be defined first. For illustration, a bilinear $M-\phi$ relationship as shown in Figure 4.4 is used here. In the figure, nominal moment, M_n, is defined as the moment where either the extreme compression concrete strain reaches 0.004 or the first longitudinal reinforcement's tensile strain reaches 0.015. M_u is the ultimate moment capacity of the column cross section. M_y is the initial yield moment defined as the moment where the extreme tensile rebar reaches initial yield. EI_e is the effective bending rigidity, which is the slope of elastic segment OM_y.

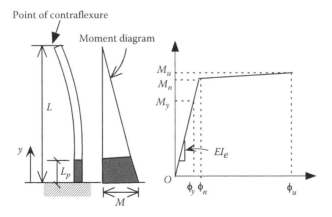

FIGURE 4.4   Calculation of $M-\theta_H$ curve based on $M-\phi$ curve and L_p.

The total rotation at the top of the column plastic hinge can be expressed as $\theta = \theta_n + \theta_H$ in which θ_n is the yield rotation at the hinge top and θ_H is the plastic rotation at the hinge top. Based on the $M-\phi$ relationship, θ_n and θ_H can be calculated as follows:

$$\theta_n = \int_0^{L_p} \phi\, dy = \int_0^{L_p} \frac{M}{EI_e}\, dy = \int_0^{L_p} \frac{M_n}{EI_e}\left(\frac{L-y}{L}\right) dy = \frac{M_n}{EI_e}\left[\int_0^{L_p} dy - \frac{1}{L}\int_0^{L_p} y\, dy\right]$$

$$= \frac{M_n}{EI_e}\left[L_p - \frac{1}{L}*\frac{L_p^2}{2}\right] = \phi_n\left[L_p - \frac{1}{L}*\frac{L_p^2}{2}\right] \tag{4.16}$$

$$\theta_H = (\phi - \phi_n)L_p \tag{4.17}$$

where ϕ is the total curvature corresponding to the moment, M, at the bottom of the column (see Figure 4.4). Equation 4.17 assumes that the plastic curvature, $(\phi - \phi_n)$, is uniformly distributed along the plastic length, L_p.

Example 4.1

Table 4.1 shows the moment–curvature relationship of a column with $L=270$ in. and $L_p=34.3$ in. The corresponding moment–curvature curve is shown in Figure 4.5. A bilinear model is also shown in the figure. Points (limit states) 1, 2, and 3 in the bilinear model correspond to moments at M_y, M_n, and M_u, respectively. $M_y=2510$ k-ft, $\phi_y=7.71E-5$, $M_n=3296$ k-ft, $\phi_n=0.0001$, $M_u=3370.6$ k-ft, and $\phi_u=0.00108$. The curvature ϕ_n at point 2 can be calculated as $\phi_n=M_n/EI_e$ and $EI_e=M_y/\phi_y$. Calculate the moment–total rotation $(M-\theta)$ and moment–plastic rotation $(M-\theta_H)$ curve based on (1) the moment–curvature curve and (2) the bilinear model.

TABLE 4.1
Moment–Curvature Relationship

Curvature (1/in.)	Moment (k-ft)
0	0
8.5E−6	686
1.4E−5	906
5.86E−5	2091.2
7.83E−5	2540.1
1E−4	2807.4
1.16E−4	2920.4
1.67E−4	3126
2.54E−4	3265
3.39E−4	3311.1
0.00108	3370.6

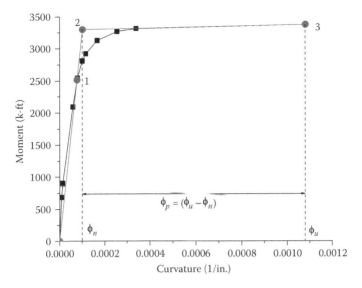

FIGURE 4.5 Moment–curvature curves.

Solution

(1) Based on the moment–curvature curve:

$$\phi_n = 0.0001\,(\text{rad})$$

$$\theta_n = \phi_n\left[L_p - \frac{1}{L}*\frac{L_p^2}{2}\right] = 0.0001\left[34.3 - \frac{1}{270}*\frac{(34.3)^2}{2}\right] = 3.24E-3\,(\text{rad})$$

At $M=686\,\text{k-ft}$

$$\theta = \theta_{elas} = \phi\left[L_p - \frac{1}{L}*\frac{L_p^{\,2}}{2}\right] = (8.5E-6)(32.12) = 2.74E-4\,(\text{rad}) < \theta_n$$

Therefore, no plastic rotation developed, $\theta_H = 0$ (rad)

At $M=2091.2\,\text{k-ft}$

$$\theta = \theta_{elas} = \phi\left[L_p - \frac{1}{L}*\frac{L_p^{\,2}}{2}\right] = (5.86E-5)(32.12) = 1.88E-3\,(\text{rad}) < \theta_n$$

Therefore, no plastic rotation developed, $\theta_H = 0$ (rad)

At $M=2807.4\,\text{k-ft}$

$$\theta = \theta_{elas} = \phi\left[L_p - \frac{1}{L}*\frac{L_p^{\,2}}{2}\right] = (1.0E-4)(32.12) = 3.24E-3\,(\text{rad}) = \theta_n$$

Therefore, no plastic rotation developed, $\theta_H=0$ (rad)

At $M=2920.4$ k-ft

$\theta_H = (\phi - \phi_n) * L_p = ((1.16E - 4) - (1.0E - 4)) * 34.3 = 5.14E - 4 \, (\text{rad})$

Total rotation $\theta = \theta_n + \theta_H = 3.24E - 3 + 5.14E - 4 = 3.75E - 3$ (rad)

At $M=3126$ k-ft

$\theta_H = (\phi - \phi_n) * L_p = ((1.67E - 4) - (1.0E - 4)) * 34.3 = 2.30E - 3 \, (\text{rad})$

Total rotation $\theta = \theta_n + \theta_H = 3.24E - 3 + 2.30E - 3 = 5.54E - 3$ (rad)

At $M=3265$ k-ft

$\theta_H = (\phi - \phi_n) * L_p = ((2.54E - 4) - (1.0E - 4)) * 34.3 = 5.28E - 3 \, (\text{rad})$

Total rotation $\theta = \theta_n + \theta_H = 3.24E - 3 + 5.28E - 3 = 8.52E - 3$ (rad)

At $M=3311.1$ k-ft

$\theta_H = (\phi - \phi_n) * L_p = ((3.39E - 4) - (1.0E - 4)) * 34.3 = 8.16E - 3 \, (\text{rad})$

Total rotation $\theta = \theta_n + \theta_H = 3.24E - 3 + 8.16E - 3 = 1.14E - 2$ (rad)

At $M=3370.6$ k-ft: Since $M=M_u$ at point 3, the plastic-curvature capacity is $\phi_p = \phi_u - \phi_n$.

$\theta_H = \phi_p L_p = (\phi_u - \phi_n) * L_p = ((1.08E - 3) - (1.0E - 4)) * 34.3 = 3.36E - 2 \, (\text{rad})$

Total rotation $\theta = \theta_n + \theta_H = 3.24E - 3 + 3.36E - 2 = 3.68E - 2$ (rad)

(2) Based on the bilinear model:

$\phi_n = 0.0001 \, (\text{rad})$

$$\theta_n = \phi_n \left[L_p - \frac{1}{L} * \frac{L_p^2}{2} \right] = 0.0001 \left[34.3 - \frac{1}{270} * \frac{(34.3)^2}{2} \right] = 3.24E - 3 \, (\text{rad})$$

For $M < M_n = 3296$ k-ft, $\theta_H=0$ (rad)
At $M=M_n=3296$ k-ft, $\theta_H=0$ (rad)
Total rotation $\theta = \theta_n + \theta_H = 3.24E - 3 + 0 = 3.24E - 3$ (rad)
At $M=3370.6$ k-ft

$\theta_H = (\phi_u - \phi_n) * L_p = ((1.08E - 3) - (1.0E - 4)) * 34.3 = 3.36E - 2 \, (\text{rad})$

Total rotation $\theta = \theta_n + \theta_H = 3.24E - 3 + 3.36E - 2 = 3.68E - 2$ (rad)

From the above calculation, the moment–total rotation and the moment–plastic rotation curves are shown in Figures 4.6 and 4.7, respectively.

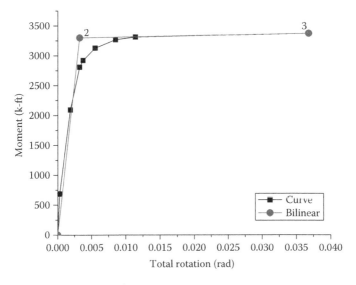

FIGURE 4.6 Moment–total rotation curves.

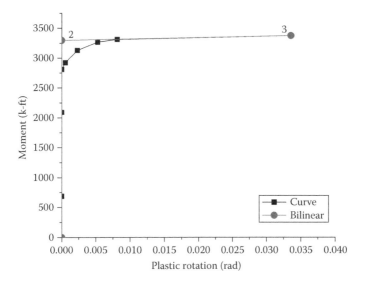

FIGURE 4.7 Moment–plastic rotation curves.

Note that the PHL, L_p, in Equation 4.5 is a function of the distance from the column plastic hinge to the point of contraflexure of the column. Once the user defines the location of the contraflexure point along column, it is assumed that the contraflexure point is fixed during the pushover analysis. In reality, the contraflexure point could shift if the ratio of two column end moments, M_i/M_j, changes.

4.3 CONSTANT MOMENT RATIO METHOD

In the CMR method (Cheng and Ger, 1992), the nonlinear bending stiffness matrix is derived based on a simply supported structural model as shown in Figure 4.8. Given a member of length L, the end moments M_i, M_j, and the moment–curvature relationship, the end moment–rotation relationship at each end can be obtained by the conjugate beam theory. However, using the PHL, L_p, to generate the moment–rotation curve, is recommended for concrete columns because L_p in Equation 4.5 is based on the experiment results of large-scale test columns. The Takeda moment–rotation material described in Chapter 3 can also be used to formulate the nonlinear stiffness matrix of the concrete member by CMR method. Normally, the conjugate beam theory is used to generate the moment–rotation curve for steel members.

As mentioned previously, if the moment ratio, M_i/M_j is close to a constant, the location of the contraflexure point is assumed to be fixed. The flexibility of the inelastic rotation can be lumped at the member ends. Therefore, the total rotation, elastic rotation, and the plastic rotation expressed in terms of incremental forms are

$$\begin{bmatrix} \Delta\theta_i \\ \Delta\theta_j \end{bmatrix} = \begin{bmatrix} \Delta\theta_{iE} \\ \Delta\theta_{jE} \end{bmatrix} + \begin{bmatrix} \Delta\theta_{iIE} \\ \Delta\theta_{jIE} \end{bmatrix} \tag{4.18}$$

$$\begin{bmatrix} \Delta\theta_{iE} \\ \Delta\theta_{jE} \end{bmatrix} = \frac{L}{EI} \begin{bmatrix} 1/3 & -1/6 \\ -1/6 & 1/3 \end{bmatrix} \begin{bmatrix} \Delta M_i \\ \Delta M_j \end{bmatrix} \tag{4.19}$$

$$\begin{bmatrix} \Delta\theta_{iIE} \\ \Delta\theta_{jIE} \end{bmatrix} = \begin{bmatrix} f_i & 0 \\ 0 & f_j \end{bmatrix} \begin{bmatrix} \Delta M_i \\ \Delta M_j \end{bmatrix} \tag{4.20}$$

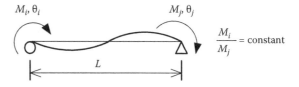

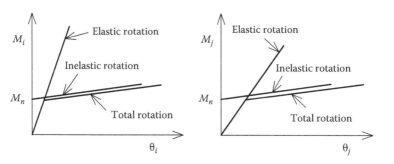

FIGURE 4.8   Moment–rotation relationship.

where f_i and f_j are the flexibilities of the plastic rotations at member ends i and j, respectively, obtained from the moment–inelastic rotation curves in Figure 4.8. Note that the inelastic rotations calculated from Equation 4.20 are approximate because it is assumed that the inelastic rotation increment at one end is not influenced by the moment increment at the other end. From Equations 4.18 through 4.20, the member stiffness matrix can be derived as follows:

$$\begin{Bmatrix} \Delta M_i \\ \Delta M_j \end{Bmatrix} = \frac{1}{D} \begin{bmatrix} \dfrac{L}{3EI} + f_j & \dfrac{L}{6EI} \\ \dfrac{L}{6EI} & \dfrac{L}{3EI} + f_i \end{bmatrix} \begin{Bmatrix} \Delta\theta_i \\ \Delta\theta_j \end{Bmatrix} \tag{4.21}$$

in which

$$D = \frac{L^2}{12(EI)^2} + \frac{L}{3EI}(f_i + f_j) + f_i f_j \tag{4.22}$$

For a multiple-column bent subjected to earthquake, if a column deforms in a double-curvature shape, it may be assumed that the point of contraflexure is at the middle point of the column. In this case, $M_i/M_j \cong 1$ and $f_i = f_j$ can be used for the stiffness matrix formulation.

The following example demonstrates how to generate the moment–rotation curve from the moment–curvature curve by conjugate beam theory.

Example 4.2

A bilinear moment–curvature curve of a W8×31 steel cross section is shown in Figure 4.9. The length of the steel member is 10 ft. It deforms in a double-curvature shape when subjected to lateral load. The contraflexure point is located near the mid height of the column (i.e., $M_a/M_b = 1$). Find the moment–total rotation and moment–plastic rotation curves of the member. For demonstration purposes, assume the ultimate curvature capacity, ϕ_u, is equal to 0.001.

Solution

a. At point 2 (i.e., $M = M_n = 700$ k-ft):
   Since the curvature distribution is symmetric, the rotation θ_a can be calculated by taking the moment at the midpoint of the member (i.e., point "c") with consideration of half of the member length (Figure 4.10).

$$\sum M_c = 0:$$

$$\theta_a(L) - \frac{\phi_n L}{2}\left(\frac{2L}{3}\right) = 0 \quad \therefore \theta_a = \frac{\phi_n L}{3} = \frac{(0.00025)(60)}{3} = 0.005 \text{ (rad)}$$

b. At point 3 (i.e., $M = M_u = 745$ k-ft):

From Figure 4.10 and Table 4.2, the rotation θ_a can be calculated as

$$\sum M_c = 0:$$

$$\theta_a(L) - A\bar{X} = 0 \quad \therefore \theta_a = \frac{A\bar{X}}{L} = \frac{0.4}{60} = 6.7E - 3 \text{ (rad)}$$

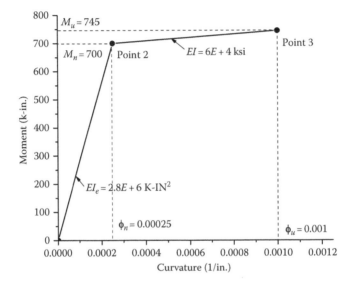

FIGURE 4.9   Bilinear moment–curvature curve of $W8 \times 31$.

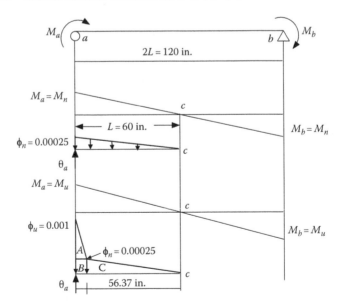

FIGURE 4.10   Curvature distribution at ϕ_n and ϕ_u.

TABLE 4.2

Conjugate Load

Section	Area (A)	Arm from Point $c(\bar{X})$	$A\bar{X}$
A	$1.36E-3$	58.79	0.08
B	$9.08E-4$	58.19	0.053
C	$7.04E-3$	37.58	0.265
Total			0.4

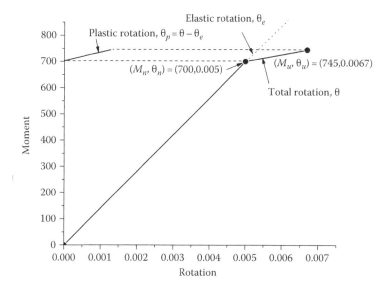

FIGURE 4.11 Moment–rotation curves.

Figure 4.11 shows the moment–total rotation, moment–elastic rotation, and moment–plastic rotation curves.

Although the flexural stiffness matrix formulation for the PHL method shown in Equation 4.11 is different from that for the CMR method shown in Equation 4.21, the numerical values of the stiffness matrices calculated from both methods are identical, if the same moment–rotation curve is used for both methods. This is demonstrated in the following example.

Example 4.3

Use the moment–rotation curve generated in Example 4.1 to calculate the flexural stiffness matrix based on the PHL and CMR methods. Assuming the contraflexure point is at the mid height of a column with length $L=45$ ft and bending rigidity of $EI=2,729,000$ (k-ft²).

Solution

From Example 4.1, the moment–total rotation and moment–plastic rotation curves are shown in Figure 4.12. In the figure, slopes 1 and 2 are the effective and post-yield slopes of the moment–total rotation curve. Slope 3 is the slope corresponding to the moment–plastic rotation curve. The values of slopes 1–3 are 1,098,667, 2,223, and 2,194 (k-ft²/rad), respectively.

1. Using the PHL method:
   a. Elastic case

$$R_i = R_j = \infty$$

From Equation 4.15,

$$R^* = \left(1 + \frac{4EI}{LR_i}\right)\left(1 + \frac{4EI}{LR_j}\right) - \left(\frac{EI}{L}\right)^2 \left(\frac{4}{R_i R_j}\right) = 1 - 0 = 1$$

From Equations 4.12 through 4.14,

$$S_{ii} = \left[4 + \frac{12EI}{LR_j}\right] \bigg/ R^* = \frac{(4+0)}{1} = 4$$

$$S_{ii} = S_{ii} = 4$$

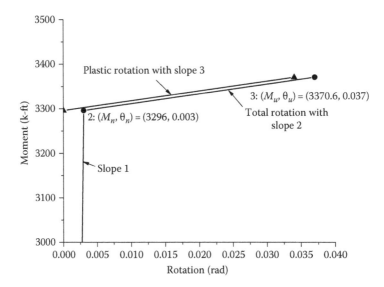

FIGURE 4.12   Moment–rotation curves from Example 4.1.

$$S_{ij} = \frac{2}{R^*} = 2$$

Therefore,

$$\begin{Bmatrix} \Delta M_i \\ \Delta M_j \end{Bmatrix} = \frac{EI}{L} \begin{bmatrix} S_{ii} & S_{ij} \\ S_{ij} & S_{jj} \end{bmatrix} \begin{Bmatrix} \Delta\theta_i \\ \Delta\theta_j \end{Bmatrix} = (60,644) \begin{bmatrix} 4 & 2 \\ 2 & 4 \end{bmatrix} \begin{Bmatrix} \Delta\theta_i \\ \Delta\theta_j \end{Bmatrix}$$

(a)

$$= \begin{bmatrix} 242,576 & 121,288 \\ 121,288 & 242,576 \end{bmatrix} \begin{Bmatrix} \Delta\theta_i \\ \Delta\theta_j \end{Bmatrix}$$

b. Inelastic case

Since the contraflexure point is at the mid height of the column, $R_i = R_j = slope\ 3 = 2194$ (k-ft²/rad).

$$R^* = \left(1 + \frac{4EI}{LR_i}\right)\left(1 + \frac{4EI}{LR_j}\right) - \left(\frac{EI}{L}\right)^2 \left(\frac{4}{R_i R_j}\right)$$

$$= (111.49)(111.49) - (60,604)^2 (8.309E - 7) = 9,374.2$$

$$S_{ii} = \left[4 + \frac{12EI}{LR_j}\right] / R^* = \frac{(4 + 727,728 / 2,194)}{9,374.2} = 0.0358$$

$$S_{jj} = S_{ii} = 0.0358$$

$$S_{ij} = \frac{2}{R^*} = 0.00021$$

$$\begin{Bmatrix} \Delta M_i \\ \Delta M_j \end{Bmatrix} = \frac{EI}{L} \begin{bmatrix} S_{ii} & S_{ij} \\ S_{ij} & S_{jj} \end{bmatrix} \begin{Bmatrix} \Delta\theta_i \\ \Delta\theta_j \end{Bmatrix} = \begin{bmatrix} 2171 & 12.73 \\ 12.73 & 2171 \end{bmatrix} \begin{Bmatrix} \Delta\theta_i \\ \Delta\theta_j \end{Bmatrix}$$

(b)

2. Using the CMR method:
   a. Elastic case

   The flexibilities of the inelastic rotation at member ends i and j are

   $$f_i = f_j = 0$$

   From Equation 4.22,

   $$D = \frac{L^2}{12(EI)^2} + \frac{L}{3EI}(f_i + f_j) + f_i f_j = (2.26E - 11) + 0 + 0 = 2.26E - 11$$

From Equation 4.21,

$$\begin{Bmatrix} \Delta M_i \\ \Delta M_j \end{Bmatrix} = \frac{1}{D}\begin{bmatrix} k_{ii} & k_{ij} \\ k_{ij} & k_{jj} \end{bmatrix}\begin{Bmatrix} \Delta\theta_i \\ \Delta\theta_j \end{Bmatrix}$$

in which

$$k_{ii} = \frac{1}{D}\left(\frac{L}{3EI} + f_j\right) = \frac{1}{D}(5.4965E-6+0) = 242,576$$

$$k_{jj} = k_{ii} = 242,576$$

$$k_{ij} = \frac{1}{D}\left(\frac{L}{6EI}\right) = 121,288$$

$$\begin{Bmatrix} \Delta M_i \\ \Delta M_j \end{Bmatrix} = \begin{bmatrix} 242,576 & 121,288 \\ 121,288 & 242,576 \end{bmatrix}\begin{Bmatrix} \Delta\theta_i \\ \Delta\theta_j \end{Bmatrix} \qquad (c)$$

Comparing Equations (a) and (c), the elastic flexural stiffness matrices by the PHL and CMR methods are identical.

b. Inelastic case

$$f_i = f_j = \frac{1}{R_i} = \frac{1}{2194} = 4.558E-4$$

$$D = \frac{L^2}{12(EI)^2} + \frac{L}{3EI}(f_i + f_j) + f_i f_j = 2.127E-7$$

$$k_{ii} = \frac{1}{D}\left(\frac{L}{3EI} + f_j\right) = 2169$$

$$k_{jj} = k_{ii} = 2169$$

$$k_{ij} = \frac{1}{D}\left(\frac{L}{6EI}\right) = 12.9$$

From Equation 4.21,

$$\begin{Bmatrix} \Delta M_i \\ \Delta M_j \end{Bmatrix} = \frac{1}{D}\begin{bmatrix} k_{ii} & k_{ij} \\ k_{ij} & k_{jj} \end{bmatrix}\begin{Bmatrix} \Delta\theta_i \\ \Delta\theta_j \end{Bmatrix} = \begin{bmatrix} 2169 & 12.9 \\ 12.9 & 2169 \end{bmatrix}\begin{Bmatrix} \Delta\theta_i \\ \Delta\theta_j \end{Bmatrix} \qquad (d)$$

Comparing Equations (b) and (d), the inelastic flexural stiffness matrices by the PHL and CMR methods are essentially the same with minor difference between (b) and (d) being due to numerical truncation errors.

4.4 FINITE SEGMENT–FINITE STRING METHOD

Another common method of nonlinear pushover analysis is the use of the distributed plasticity model. Using this method, a structural member (e.g., a bridge column) is divided into several segments (Chen and Atsuta, 1977). Each segment has 12 degrees of freedom, and its cross section is divided into many finite elements (or so-called finite strings) along the segment's longitudinal direction as shown in Figure 4.13. When a load or displacement increment is applied to a member in the pushover analysis, each segment is deformed and may become partially plastic as sketched in Figure 4.13. The plastification of the cross section can be detected by the steel and concrete stress–strain relationships. For simplicity, the segment's cross-sectional plastification and strains are calculated based on the average curvature along the segment length.

For each small element (string) on the segment's cross section, the strain increment can be expressed as follows:

$$\Delta\varepsilon_c^{ij} = \Delta\varepsilon_c^j + V_i\Delta\varphi_u^j - U_i\Delta\varphi_v^j \tag{4.23}$$

in which

$$\Delta\varepsilon_c^j = \frac{(\Delta W_b^j - \Delta W_a^j)}{L} \tag{4.24}$$

$$\Delta\varphi_u^j = \frac{(\Delta\theta_{ub}^j - \Delta\theta_{ua}^j)}{L} \tag{4.25}$$

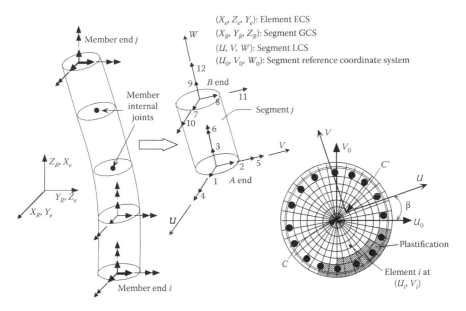

(X_e, Z_e, Y_e): Element ECS
(X_R, Y_R, Z_R): Segment GCS
(U, V, W): Segment LCS
(U_0, V_0, W_0): Segment reference coordinate system

FIGURE 4.13 FSFS method based on the distributed plasticity model.

$$\Delta \varphi_v^j = \frac{(\Delta \theta_{vb}^j - \Delta \theta_{va}^j)}{L} \tag{4.26}$$

where

U_i and V_i equal the location of the ith cross-sectional element in the segment local coordinates U and V, respectively

W equals the segment local coordinate along the longitudinal direction of the segment

The subscripts a and b represent the two ends of the segment

i equals the ith cross-sectional element

j equals the jth segment

L equals the segment length

$\Delta \varepsilon_c^{ij}$ equals the strain increment of cross-sectional element i in the jth segment

$\Delta \varepsilon_c^j$ equals the normal strain at the centroid of the jth segment

ΔW_a^j equals the longitudinal deformation increment at a end of the segment j

$\Delta \theta_{ua}^j$ and $\Delta \theta_{va}^j$ equal the rotational increments at a end of the segment j in the segment local U and V directions, respectively

$\Delta \varphi_u$ and $\Delta \varphi_v$ equal the average bending curvature increments about the U and V axes, respectively

The current total strain for element i is

$$\varepsilon^{ij} = \varepsilon_p^{ij} + \Delta \varepsilon^{ij} \tag{4.27}$$

where ε_p^{ij} is the ith element total strain in the previous deformation state. As shown in the flowchart in Figure 4.14, once the plastification of the cross section is known, the current principal axes, sectional properties, and the stiffness matrix of individual segments can be calculated. The procedures for calculating principal axes and sectional properties are described in Appendix B.

The direction of the segment local coordinate system (U, V, W) is updated in each load step, in the direction of the segment's instantaneous principal U, V, and W axes. Each segment's material and geometric stiffness matrices are transferred from the segment local coordinate system (U, V, W) to the segment global coordinate system (X_R, Y_R, Z_R). The member's stiffness matrix is established by stacking up the segmental stiffness matrices, for which a rotation matrix $[\bar{R}]_{12 \times 12}$ is required for each segment, by transferring the segment stiffness matrix from the segment local coordinate system (U, V, W) to the segment global coordinate system (X_R, Y_R, Z_R). The calculation of the rotation matrix $[\bar{R}]_{12 \times 12}$ is described in Appendix B. The more segments assigned, the more accurate the element stiffness matrix will be.

In order to provide computational efficiency, the member internal degrees of freedom at the member's internal joints (see Figure 4.13) are condensed out by Gaussian elimination, and only the degrees of freedom at both ends of the member are maintained. Thus, the condensed member stiffness matrix has a dimension of 12×12. This condensed member stiffness matrix will be transformed from the segment global coordinate system (X_R, Y_R, Z_R) to the member coordinate system (X_e, Y_e, Z_e), as shown in Figure 4.13.

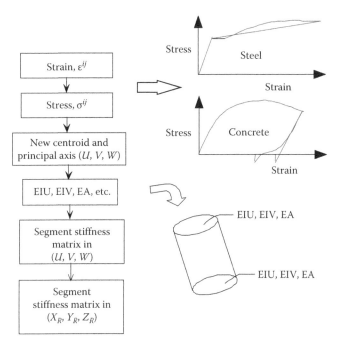

FIGURE 4.14 Segment stiffness matrix formulation based on average curvature.

The disadvantage of using this method is that the plastification at each end of the segment is not actually calculated, and a uniform plastification distribution along the segment based on the average curvature increment, as shown in Equations 4.24 through 4.26, is assumed. Because of this assumption, the unbalanced force calculation at each end of a segment (see Appendix C) is also approximated. The program only calculates the unbalanced forces for segments in single curvature. For a segment deformed in a double-curvature shape, the unbalanced forces are ignored by the program. In order to improve numerical accuracy, it is recommended that (1) more segments are used for each column, so the curvature distribution along each segment is close to a uniform distribution and (2) smaller incremental steps be used. If a numerical instability exists, try to use the simple Euler incremental approach (see Appendix D) with small incremental steps and without consideration of unbalanced forces (i.e., choose UNBAL = .FALSE. option in the SOL04 input data block as described in Chapter 6). Future program enhancement will include the plastification at each end of the segment according to the curvature increments at each end, so more accurate unbalanced force calculation can be incorporated in the program.

4.5 FINITE SEGMENT–MOMENT CURVATURE METHOD

This method is similar to the FSFS method except that the cross section of each segment is not divided into many elements. The segment stiffness matrix at each incremental step is calculated based on the cross-sectional axial load–moment–curvature family of curves from which the flexural property, EI (i.e., the slope of moment–curvature curve) can be obtained (see Section 3.4.9). The total curvature at each step is the accumulation of the incremental curvatures from the previous steps based on Equations 4.24 through 4.27. Similar to the FSFS method, the member stiffness matrix is established by stacking up the segment stiffness matrix with consideration of the segmental rotation matrix $[\bar{R}]_{12\times12}$ and $P-\delta$ effects.

4.6   CONCRETE COLUMN FAILURE MODES

For a reinforced concrete column, the plastic-curvature capacity, ϕ_p, is controlled by the governing failure mode. The possible failure modes include

1. Compression failure of unconfined concrete
2. Compression failure of confined concrete
3. Compression failure due to buckling of the longitudinal reinforcement
4. Longitudinal tensile fracture of the reinforcing bar
5. Low-cycle fatigue of the longitudinal reinforcement
6. Failure in the lap-splice zone
7. Shear failure of the member that limits ductile behavior
8. Failure of the connection joint

INSTRUCT can perform moment–curvature analysis by the FSFS method to calculate the ultimate curvature capacity, ϕ_u, and plastic-curvature capacity, ϕ_p, corresponding to the governing failure mode (see Figure 4.21). The ϕ_p obtained from the FSFS method can be used to calculate the plastic rotation capacity, θ_p, with which the column plastic rotation is checked at each incremental step during pushover analysis, when either the PM or PHL bending stiffness formulation method is considered. In addition, a column shear failure and the joint shear failure modes are also checked during the pushover analysis if the column is modeled by the IE3DBEAM element. Using the FSFS method to calculate plastic-curvature capacities due to failure modes 1–6 is briefly described as follows:

1. *Compression failure of unconfined concrete*
   The concrete stress–strain relationship for unconfined concrete is shown in Figure 3.11. Conservatively, it is assumed that the compression failure of the unconfined concrete column occurs when the ultimate concrete compression strain, ε_{cu}, at the column extreme fiber is equal to

$$\varepsilon_{cu} = 2 \times \varepsilon_{c0} = 0.004 \qquad (4.28)$$

   in which ε_{c0} is the concrete strain corresponding to f_c'.
2. *Compression failure of confined concrete*
   As shown in Equation 3.24, the ultimate confined concrete compression strain is equal to

$$\varepsilon_{cu} = 0.004 + \frac{1.4\rho_s f_{yh}\varepsilon_{su}}{f_{cc}'} \qquad (4.29)$$

3. *Compression failure due to buckling of the longitudinal reinforcement*
   The buckling of longitudinal rebar is based on the following equation
   (Dutta and Mander, 1998):

$$\frac{L_b}{d_b} = (N_h + 1)\left(\frac{s}{d_b}\right)$$

$$= 10\sqrt{\frac{f_{su}}{f_{cr}}\frac{\left(1 - \left|f_{cr}/f_{su}\right|^2\right)}{\left(1 - 0.4\left|f_{cr}/f_{su}\right|^2\right)}\left[1 + \frac{0.3}{1 - \left|f_{cr}/f_{su}\right|^2}\frac{N_h}{K}\left(\frac{f_{yh}}{f_{su}}\right)\left(\frac{d_{bh}}{d_b}\right)^2\frac{L_b}{d_b}\right]}$$

$$(4.30)$$

where
   N_h is the number of hoop(s) within the buckling region (excluding bound-
      ary hoops as shown in Figure 4.15)
   d_b is the longitudinal rebar diameter
   d_{bh} is the transverse rebar (hoop) diameter
   K is the restrained coefficient
   $K = 1$ for rectangular section
   $K = N/2\pi$ for circular section
   N is the number of longitudinal reinforcing steel bars
   f_{su} is the ultimate stress of longitudinal reinforcing steel bars
   f_{cr} is the buckling stress of longitudinal reinforcing steel bars
   f_y is the yield stress of longitudinal reinforcing steel bars
   f_{yh} is the yield stress of hoop
   s is the spacing of hoops
   L_b is the buckling region length $= (N_h + 1)(s)$

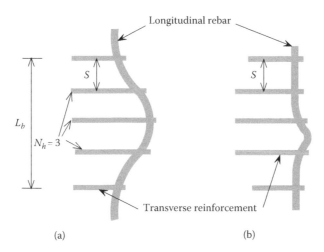

(a)   (b)

FIGURE 4.15 Buckling of longitudinal reinforcing steel: (a) global buckling and (b) local
buckling.

INSTRUCT finds each f_{cr} corresponding to each individual N_h (from $N_h=0$ through $N_h=15$). The least f_{cr} represents the buckling stress of the longitudinal bars. Equation 4.30 considers both local and global buckling of the longitudinal bars (see Figure 4.15). The program only considers the buckling at the post-yield stress level of the bar (i.e., $f_{cr} \geq f_y$) at the plastic hinge location. In the pushover analysis, when the stress of a longitudinal bar reaches f_{cr} at a certain incremental step, the program output will report the reinforcing steel buckling failure mode occurrence.

4. *Longitudinal tensile fracture of reinforcing bar*
   As described in Section 3.4.8, the maximum tensile strain of $\varepsilon_{su}=0.09$ is used for reinforcing steel in INSTRUCT.

5. *Low-cycle fatigue of the longitudinal reinforcement*
   Low-cycle fatigue of the longitudinal reinforcement is dependent on the fundamental natural period, T_n, of the bridge. Once T_n is input by the user, the program will calculate the plastic strain amplitude, ε_{ap}, of the steel from Equation 4.31:

$$\varepsilon_{ap} = 0.08(2N_f)^{-0.5} \tag{4.31}$$

in which N_f is the effective number of equal-amplitude cycles of loading that lead to fracture, which can be approximated by

$$N_f = 3.5(T_n)^{-1/3} \tag{4.32}$$

provided that $2 \leq N_f \leq 10$. The corresponding plastic-curvature capacity can be obtained from the following equation (Dutta and Mander, 1998):

$$2\varepsilon_{ap} = \phi_p D\left(1 - \frac{2d'}{D}\right) \tag{4.33a}$$

or

$$\phi_p = \frac{0.113(N_f)^{-0.5}}{D(1-2d'/D)} \tag{4.33b}$$

where
   D is the overall depth of the section
   d' is the depth from the extreme concrete compression fiber to the center of the compression reinforcement

INSTRUCT only considers low-cycle fatigue of the longitudinal reinforcement for confined concrete columns.

6. *Failure of the lap-splice in the plastic hinge zone (Priestley et al., 1996)*
   INSTRUCT uses the following equation to calculate the stress of longitudinal steel bar at which the splice failure occurs and flexural strength starts degrading:

$$f_{slap} = \frac{f_t p l_s}{A_b} \qquad (4.34)$$

where

   f_{slap} represents the tensile stress at which lap-splice failure occurs

   $f_t = 4\sqrt{f_c'}$ psi
   l_s represents the splice length, input by user
   A_b represents the cross-sectional area of longitudinal bar
   p represents the perimeter of crack for each spliced pairs of longitudinal bars (see Figure 4.16), which is defined as

$$p = \frac{\pi D'}{2n} + 2(d_b + c) \leq 2\sqrt{2}(d_b + c) \quad \text{for circular columns} \qquad (4.35)$$

$$p = \frac{s'}{2} + 2(d_b + c) \leq 2\sqrt{2}(d_b + c) \quad \text{for rectangular columns} \qquad (4.36a)$$

where

   n is the total number of longitudinal bars
   D' is the concrete core diameter
   c is the concrete cover
   s' is the average spacing between spliced pairs of longitudinal bars

If f_{slap} is less than f_y, the flexural strength degradation occurs at the curvature corresponding to f_{slap}. If f_{slap} is greater than f_y, the flexural strength degradation starts when the concrete extreme fiber compression strain reaches 0.002. For a confined concrete column, it is possible that the flexural

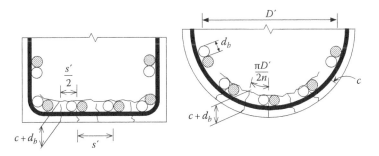

FIGURE 4.16   Splice failure of longitudinal reinforcement.

strength of a splice will degrade under cyclic loading if the volumetric ratio of transverse reinforcement ρ_s is less than the following:

$$\rho_s = \frac{2.42 A_b f_y}{p l_s f_{yh}} \quad \text{for circular sections} \tag{4.36b}$$

$$\rho_X = \frac{1.21 A_b f_y}{p l_s f_{yh}} \quad \text{and} \quad \rho_Y = \frac{1.21 A_b f_y}{p l_s f_{yh}} \quad \text{for rectangular sections} \tag{4.37}$$

Conservatively, the program uses Equation 4.34 to check the flexural strength degradation of both poor-confined and confined concrete columns. The maximum plastic-curvature capacity due to lap-splice failure can be estimated as follows:

$$\phi_p = \phi_{lap} + 7\phi_n \tag{4.38}$$

where ϕ_n is the nominal curvature corresponding to M_n; $\phi_{lap} = 0$ when f_{slap} is less than f_y. If f_{slap} is greater than f_y, ϕ_{lap} is the plastic curvature at which the concrete extreme fiber compression strain reaches 0.002. If the volumetric ratio of transverse reinforcement is greater than that shown in Equations 4.36b and 4.37, the transverse reinforcement can provide enough clamping stress across the concrete fracture surface, and the flexural strength degradation will not occur. Although the program uses the above-mentioned f_{slap} criteria to check the flexural strength degradation for both poor-confined and confined concrete columns, the user should check whether or not there is a sufficient volumetric ratio of transverse reinforcement to prevent the flexural strength degradation per Equations 4.36b and 4.37, which provide enough transverse reinforcement to ensure that the concrete dilation strain is less than 0.0015, and the coefficient of friction of $\mu = 1.4$ is appropriately achieved.

The FHWA publication entitled *Seismic Retrofitting Manual for Highway Structures* (FHWA, 2006) provides a closed form formula (see Appendix E) to estimate the plastic-curvature capacities for the six above-mentioned failure modes. Note that these formulas are approximate. To obtain the more accurate plastic-curvature capacities, Equations 4.28 through 4.38 should be used.

As mentioned previously, INSTRUCT also checks shear and joint shear failure modes during the pushover analysis when the PM or PHL method is used. The analytical approaches of checking shear and joint shear failure modes are described as follows:

7. *Column shear failure (Priestley et al., 1996)*
The shear strength capacity of the member is equal to

$$V_{cap}(\mu) = V_c + V_s + V_p \tag{4.39}$$

where

V_c is the concrete shear strength
V_s is the shear strength due to transverse reinforcement
V_p is the shear strength due to compressive axial load

The value of V_c depends on the rotational ductility of the member end (μ):

$$V_c = V_{ci} = k\sqrt{f_c'}\,A_e; \quad k = 3.5; \quad A_e = 0.8A_{gross} \tag{4.40}$$

$$V_c = V_{cf} = k\sqrt{f_c'}\,A_e; \quad k = 0.6; \quad A_e = 0.8A_{gross} \tag{4.41}$$

where

f_c' is in psi
A_{gross} is the gross cross-sectional area
V_{ci} is the initial concrete shear strength
V_{cf} is the concrete shear strength when μ is greater than or equal to 15

As shown in Figure 4.17, the coefficient k is a function of member-end rotational ductility.

The shear strength due to transverse reinforcement, V_s, is

$$V_s = \frac{\pi A_{sp} f_{yh} D' \cot(\theta)}{2s} \quad \text{for circular sections} \tag{4.42}$$

$$V_s = \frac{A_v f_{yh} D' \cot(\theta)}{s} \quad \text{for rectangular sections} \tag{4.43}$$

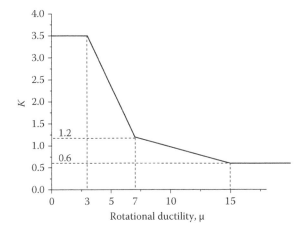

FIGURE 4.17 Concrete shear strength capacity in terms of member-end rotational ductility.

where

$\theta = 30°$

D' is the core dimension measured to the centerline of hoop or width between the centerline of the rectangular transverse reinforcement

s is the spacing of transverse reinforcement

A_{sp} is the cross-sectional area of transverse reinforcement

A_v is the effective area of transverse reinforcement, calculated as $A_v = N_x A_{sp}$ or $N_y A_{sp}$ based on Equations 3.27 and 3.28

The shear strength, V_p, due to compressive axial load can be calculated as follows:

$$V_p = P \tan(\alpha) \tag{4.44}$$

where

P is the axial load (compression is positive)

α is the angle between the column axis and the line joining the centers of flexural compression of concrete at the top and bottom of the column

As shown in Figure 4.18, $\tan(\alpha)$ can be calculated as follows:

$$\tan(\alpha) = \frac{D-c}{L} \quad \text{for multiple column bents} \tag{4.45}$$

or

$$\tan(\alpha) = \frac{D-c}{2L} \quad \text{for single column bents} \tag{4.46}$$

As described previously, INSTRUCT checks shear and connection joint shear failure modes if the IE3DBEAM element is used with either the PM or PHL method. To help the user in preparing input data for the pushover analysis using the PM or PHL method, the parameters V_c, V_s, and c can be calculated from the moment–curvature analysis by the FSFS method. Alternatively, an approximate value of c can be obtained using Equation E.13 for rectangular sections or Equation E.14 for circular sections as described in Appendix E. At each incremental step, the member shear force, V, will be compared with the corresponding shear strength capacity, $V_{cap}(\mu)$, calculated from Equation 4.39. If $V \geq V_{cap}(\mu)$, shear failure occurs, and a message will be shown in the INSTRUCT output file.

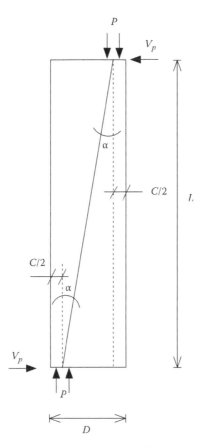

FIGURE 4.18 Column shear strength due to axial load.

8. *Connection joint shear failure (Priestley et al., 1996)*

The joint shear stress capacity of a joint is equal to

$$v_j(\mu) = \sqrt{p_t^2 - p_t(f_v + f_h) + f_v f_h} \qquad (4.47)$$

where
f_v is the average column axial stress (compression is negative)
f_h is the average horizontal joint stress (compression is negative)
p_t is the principal tensile stress of the joint (tension is positive)

$$f_v = \frac{P}{(h_c + h_b)b_{je}} \qquad (4.48)$$

$$f_h = \frac{P_b}{b_b h_b} \qquad (4.49)$$

where

h_c is the column width

b_c is the column depth

h_b is the beam depth

b_b is the beam width

P_b is the prestress force from cap beam

P is the column axial force

b_{je} is the effective beam width of the joint, which is calculated as

$$b_{je} = \sqrt{2}D \leq b_b \quad \text{for circular columns} \tag{4.50}$$

$$b_{je} = h_c + b_c \leq b_b \quad \text{for rectangular sections} \tag{4.51}$$

The principal tensile stress, p_t, is a function of the member-end rotational ductility, μ, and can be calculated as follows:

$$p_t = 5\sqrt{f_c'}\,\text{psi} \quad \text{for } 0 < \mu \leq 3 \tag{4.52}$$

$$p_t = 3.5\sqrt{f_c'}\,\text{psi} \quad \text{for } 7 < \mu \tag{4.53}$$

p_t can be obtained from Figure 4.19. During the pushover analysis, if the principal tensile stress is less than $3.5\sqrt{f_c'}$ psi, the initial joint diagonal crack is not expected. However, when the principal tensile stress reaches $3.5\sqrt{f_c'}$ psi, the joint diagonal crack is initiated. As long as the principal tensile stress is under the principal tensile stress capacity envelope as shown in Figure 4.19,

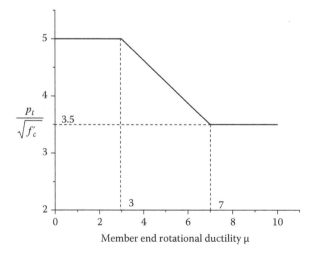

FIGURE 4.19 Principal tensile stress in terms of member-end rotational ductility.

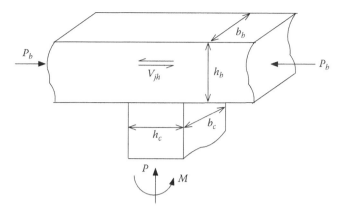

FIGURE 4.20 Stresses acting on joint.

the joint strength does not degrade although the initial diagonal crack may have been developed. Once the principal tensile stress exceeds the envelope, the joint strength is degraded, which defines the connection shear failure.

The joint shear stress demand, v_{jh} (Figure 4.20), can be calculated as follows:

$$v_{jh} = \frac{(M/h_b)}{(h_c b_{je})} \tag{4.54}$$

where M is the column moment adjacent to the joint. At each incremental step, the joint shear stress, v_{jh}, will be compared with the corresponding joint shear stress capacity, $v_j(\mu)$. If $v_{jh} \ge v_j(\mu)$, joint shear failure occurs, and a message will be shown in the INSTRUCT output file. When $p_t \ge 3.5\sqrt{f_c'}$ psi, joint shear reinforcement needs to be provided according to seismic design codes such as AASHTO *Guide Specifications for LRFD Seismic Bridge Design* (AASHTO, 2009).

4.7 BILINEAR MOMENT–CURVATURE CURVES

As described in Section 2.5, moment–curvature curves can be generated by the FSFS method using a single simply supported FSFS element with a length of 2 and one segment. Section 4.6 has discussed that the ultimate curvature capacity, ϕ_u, and plastic-curvature capacity, ϕ_p, are determined by one of the six failure modes as shown in Table 4.3.

Figure 4.21 shows that the bilinear moment–curvature curve can be defined using the idealized nominal curvature, ϕ_n (i.e., point 2 in the figure). ϕ_n can be calculated as follows:

$$\phi_n = \frac{M_n}{EI_e} \tag{4.55}$$

TABLE 4.3
Control Points for Bilinear Moment–Curvature Curves

Point 1 (M_y, ϕ_y)   First tensile rebar reaches yield

Point 2 (M_n, ϕ_n)   Concrete extreme fiber compression strain reaches 0.004 or tensile reinforcing steel strain reaches 0.015

Point 3 (M_u, ϕ_u)   Compression failure of unconfined concrete
Compression failure of confined concrete
Compression failure due to buckling of the longitudinal reinforcement
Longitudinal tensile fracture of reinforcing bar
Low-cycle fatigue of the longitudinal reinforcement
Failure in the lap-splice zone

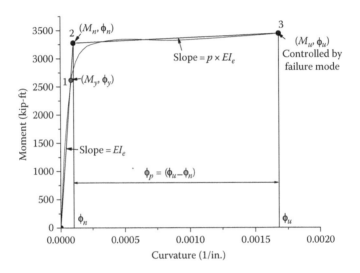

FIGURE 4.21   Bilinear moment–curvature expression.

in which

$$EI_e = \frac{M_y}{\phi_y} \tag{4.56}$$

where

M_n is defined as the nominal moment where either the extreme compression concrete strain reaches 0.004 or the first longitudinal reinforcement tensile strain reaches 0.015

M_y is the initial yield moment defined as the moment where the first longitudinal tensile reinforcement reaches initial yield

The ultimate curvature capacity, ϕ_u, in Figure 4.21 is governed by the failure modes mentioned in Section 4.6.

4.8   COLUMN AXIAL LOAD–MOMENT INTERACTION

As shown in Figures 4.1 and 4.3, the bilinear moment–curvature curves are used in the PM and PHL methods to formulate the element flexural stiffness matrix. For single column bents, normally, the column axial load does not vary much when subjected to lateral load, and there is no need to consider axial load–nominal moment interaction effects. For multicolumn bents subjected to lateral load, column axial load may vary a lot, and the effect of axial load on the nominal moment may be significant (see Figure 1.8). In this case, the nominal moment, M_n, in the bilinear moment–curvature curve needs to be adjusted in accordance with the axial load–moment interaction curve.

The axial load–moment interaction curve can be generated by performing several moment–curvature analyses with different magnitudes of axial load. As shown in Figure 4.22, a set of bilinear moment–curvature curves can be generated by the FSFS method. Plotting M_n s (i.e., point 2 of each bilinear curve) and corresponding axial loads in Figure 4.22, the axial load–moment interaction curve is obtained as shown in Figure 4.23.

In the figure, a third-order polynomial fitting curve representing the axial load–moment interaction is used in INSTRUCT. The polynomial is given by

$$M_n = a_0 + a_1 P + a_2 P^2 + a_3 P^3 \tag{4.57}$$

in which a_0, a_1, a_2, and a_3 are the coefficients for zero-order, first-order, second-order, and third-order terms, respectively. The user can input either these coefficients into INSTRUCT or data points (i.e., points 2 shown in Figure 4.23) directly into

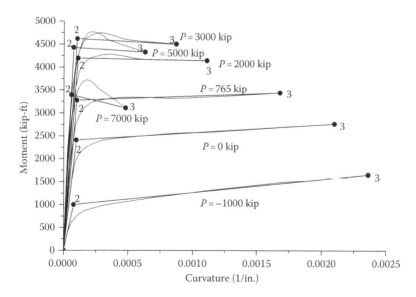

FIGURE 4.22   Moment–curvature curves.

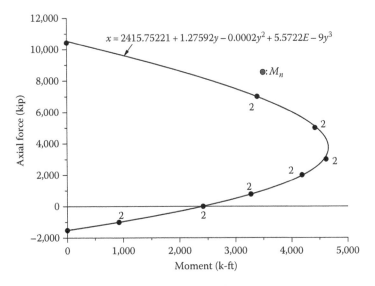

FIGURE 4.23 Axial load–nominal moment interaction.

INSTRUCT. If data points are input by the user, INSTRUCT will perform polynomial curve fitting to obtain a_0–a_3. The numerical analysis of polynomial curve fitting
is described in Appendix I. For the PM and PHL methods, at each incremental load
step, INSTRUCT will adjust M_n based on the polynomial equation in Equation 4.57.
Note that the axial load–nominal moment interaction is not considered in the CMR
method, because the CMR method is mainly used with the Takeda hysteresis model
to predict the cyclic behavior of concrete members. It is difficult to modify the Takeda
hysteretic rules to account for the variation of member axial load due to cyclic loading.

4.9 COLUMN AXIAL LOAD–PLASTIC CURVATURE CAPACITY CURVE

As shown in Figure 4.21, the plastic-curvature capacity is

$$\phi_p = (\phi_u - \phi_n) \tag{4.58}$$

Since the ultimate curvature, ϕ_u (point 3 in Figures 4.21 and 4.22), is dependent
on the column axial load, P, the column plastic curvature also depends on P. From
Figure 4.22 and Equation 4.58, the column axial load–plastic curvature capacity
(i.e., P–ϕ_p) curve is plotted in Figure 4.24. Multiplying Equation 4.58 by the PHL,
L_p, gives the column axial load–plastic rotation capacity (i.e., P–θ_p) curve shown in
Figure 4.25. Similar to Equation 4.57, INSTRUCT uses a polynomial to represent the
P–θ_p interaction, given as

$$\theta_p = b_0 + b_1 P + b_2 P^2 + b_3 P^3 \tag{4.59}$$

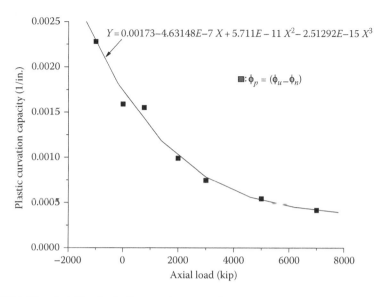

FIGURE 4.24   Axial load–plastic curvature capacity curve.

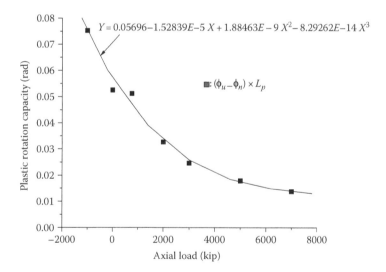

FIGURE 4.25   Axial load–plastic rotation capacity curve.

The user can either input coefficients, b_0, b_1, b_2, and b_3 into INSTRUCT, or enter data points (i.e., points as shown in Figure 4.25) into INSTRUCT so that INSTRUCT will perform polynomial curve fitting to obtain b_0–b_3. Equation 4.59 is used in the PM and PHL methods to calculate plastic rotation capacity due to column axial load effects.

5 Analytical Formulation for Structures

This chapter describes how to combine bending, shear, axial, and torsional stiffnesses to form the element stiffness matrices for bridge columns and cap beams. The bending stiffness matrices of column and cap beams are presented in Chapter 4 and based on the hysteresis models described in Chapter 3. The stiffness matrix formulation for other elements such as brace and plate elements are introduced in this chapter. Once all the element stiffness matrices are formulated, a 3D structural system subjected to both static and nonlinear pushover loadings can be analyzed. To perform static and nonlinear pushover analyses, the structural joints and degrees of freedom (dofs) need to be defined first; a process is described below.

5.1 JOINT DEFINITION AND DEGREES OF FREEDOM

A joint is defined as the point where two or more elements are connected. The assemblage of all elements becomes a structural model. The structural model is built by first defining the location and orientation of each joint. Then, the elements that connect the joints and their orientations are defined.

5.1.1 GLOBAL COORDINATE SYSTEM

The global coordinate system (GCS) defines the location of a structure. The GCS is a Cartesian coordinate system with three perpendicular axes X_g, Y_g, and Z_g. Z_g is defined as X_g cross Y_g (right-hand rule), as shown in Figure 5.1. The location of the GCS's origin is arbitrary and usually taken at the base of the structure.

5.1.2 JOINT COORDINATE SYSTEM

The location of a joint is defined by its X_g, Y_g, and Z_g coordinates in the GCS. Each joint is assumed to have six dofs. The first three dofs are translational and correspond to the joint's X_j, Y_j, and Z_j axes. The remaining three dofs are rotational about the joint's X_j, Y_j, and Z_j axes as shown in Figure 5.1. The X_j, Y_j, and Z_j axes define a joint coordinate system (JCS) for a given joint. The JCS need not be parallel to the GCS, and the JCS may vary for different joints. Thus, the orientation of the JCS for a given joint is defined by two vectors $\vec{V}_{xj}$ and $\vec{V}_{yj}$. The origin of the JCS is at the joint. A third vector is then $\vec{V}_{zj} = \vec{V}_{xj} \times \vec{V}_{yj}$. The three vectors are written in matrix form as follows:

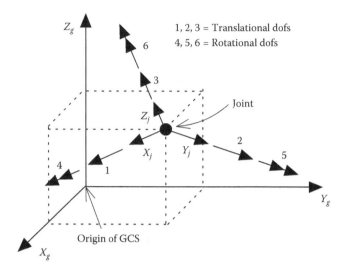

FIGURE 5.1 Global and joint coordinate systems.

$$
\{V_j\} = \begin{bmatrix} \vec{V}_{xj} \\ \vec{V}_{yj} \\ \vec{V}_{zj} \end{bmatrix} = \begin{bmatrix} C_{j11} & C_{j12} & C_{j13} \\ C_{j21} & C_{j22} & C_{j23} \\ C_{j31} & C_{j32} & C_{j33} \end{bmatrix} \begin{bmatrix} \vec{i} \\ \vec{j} \\ \vec{k} \end{bmatrix} = [C_j] \begin{bmatrix} \vec{i} \\ \vec{j} \\ \vec{k} \end{bmatrix} \tag{5.1}
$$

where $\vec{i}$, $\vec{j}$, and $\vec{k}$ are unit vectors parallel to the X_g, Y_g, and Z_g axes. Note that the orientation of the JCS determines the orientation of the global degrees of freedom (Gdofs).

5.1.3 RIGID BODY CONSTRAINTS

In general, the deformation of one structural component (e.g., a beam cap with integral concrete diaphragm) may be very small relative to the deformations of other components (e.g., columns). The component with very small deformation may be idealized as a rigid body. Two joints on the rigid body are constrained, such that the deformation of one joint (the "slave" joint) can be represented by the deformation of the other joint (the "master" joint). Thus, the dofs for the slave joint are transferred to the master joint, and the number of dofs in a structural model is reduced. The reduced set of dofs is referred to as the Gdofs. Transformations for a 3D constraint and a planar constraint are described below.

Let joint m be the master joint and joint s be the slave joint. Also, let the orientation of both joints be identical, $\{V_j\}_m = \{V_j\}_s$. Assume that the two joints are connected by a rigid body. Thus, the forces at the slave joint are transferred to the master joint, and the displacement of the slave joint is expressed in terms of the master joint. Examining Figure 5.2, for the typical notation, F_{jmx} represents the force

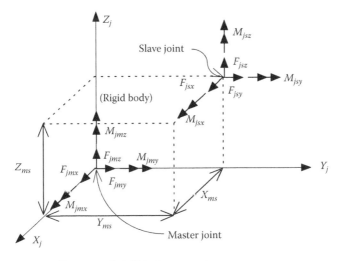

FIGURE 5.2 Three-dimensional rigid body constraint.

at the master joint m in the JCS X-direction, and M_{jmz} represents the moment at the master joint m about the JCS Z-axis. Likewise, F_{jsx} represents the force at the slave joint about the JCS X-direction, and M_{jsz} represents the moment at the slave joint about the JCS Z-axis.

Summing the forces acting on the slave joint about the master joint, in three dimensions, yields the force transformation for a 3D rigid body as follows:

$$
\begin{Bmatrix} F_{jmx} \\ F_{jmy} \\ F_{jmz} \\ M_{jmx} \\ M_{jmy} \\ M_{jmz} \end{Bmatrix} = \begin{bmatrix} 1 & 0 & 0 & 0 & 0 & 0 \\ 0 & 1 & 0 & 0 & 0 & 0 \\ 0 & 0 & 1 & 0 & 0 & 0 \\ 0 & -Z_{ms} & Y_{ms} & 1 & 0 & 0 \\ Z_{ms} & 0 & -X_{ms} & 0 & 1 & 0 \\ -Y_{ms} & X_{ms} & 0 & 0 & 0 & 1 \end{bmatrix} \begin{Bmatrix} F_{jsx} \\ F_{jsy} \\ F_{jsz} \\ M_{jsx} \\ M_{jsy} \\ M_{jsz} \end{Bmatrix}
\tag{5.2}
$$

or

$$
\{F_{jm}\} = [T_{ms}]\{F_{js}\}
\tag{5.3}
$$

where
   $\{F_{jm}\}$ represents the forces acting on the master joint
   $\{F_{js}\}$ represents the forces acting on the slave joint

A similar transformation for displacements can be derived

$$
\{\delta_{js}\} = [T_{ms}]^T \{\delta_{jm}\}
\tag{5.4}
$$

where

$\{\delta_{jm}\}$ represents the displacements of the master joint
$\{\delta_{js}\}$ represents the displacements of the slave joint

The distances X_{ms}, Y_{ms}, and Z_{ms} are in the master joint's JCS.

Recall the joint's coordinates are defined in the GCS. Transferring the coordinates of both joints from the GCS into the JCS yields

$$\begin{Bmatrix} X_{ms} \\ Y_{ms} \\ Z_{ms} \end{Bmatrix} = [C_j] \begin{Bmatrix} X_{gs} - X_{gm} \\ Y_{gs} - Y_{gm} \\ Z_{gs} - Z_{gm} \end{Bmatrix} \quad (5.5)$$

where

the typical notation X_{gm} represents the global X coordinate of the master joint m
Z_{gs} represents the global Z coordinate of the slave joint s

A bridge deck is relatively stiff in its in-plane direction, yet it is flexible out of plane. Thus, a planar constraint could be used to treat the deck's in-plane stiffness as a rigid body.

Let the plane of the deck be in the joint's $X_j - Y_j$ plane as shown in Figure 5.3. The moments about the X_j and Y_j axes and the force in the Z_j axis cannot be transferred from the slave to the master joint because the floor is flexible in the out of plane direction. Thus, the force transformation is

$$\begin{Bmatrix} F_{jmx} \\ F_{jmy} \\ F_{jmz} \\ M_{jmx} \\ M_{jmy} \\ M_{jmz} \end{Bmatrix} = \begin{bmatrix} 1 & 0 & 0 & 0 & 0 & 0 \\ 0 & 1 & 0 & 0 & 0 & 0 \\ 0 & 0 & 1 & 0 & 0 & 0 \\ 0 & 0 & 0 & 1 & 0 & 0 \\ 0 & 0 & 0 & 0 & 1 & 0 \\ -Y_{ms} & X_{ms} & 0 & 0 & 0 & 1 \end{bmatrix} \begin{Bmatrix} F_{jsx} \\ F_{jsy} \\ F_{jsz} \\ M_{jsx} \\ M_{jsy} \\ M_{jsz} \end{Bmatrix} \quad (5.6)$$

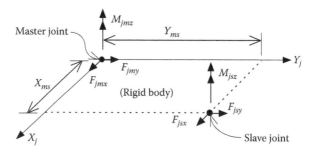

FIGURE 5.3 X–Y planar constraint.

The slave joint's translations in the X_j and Y_j axes and rotation about the Z_j axis are transferred to the master joint. The slave joint's translation in the Z_j axis and rotations about the X_j and Y_j axes remain at the slave joint.

5.1.4 CONDENSED DEGREES OF FREEDOM

In Section 3.3.2, the structural global stiffness matrix, $[K]$, is partitioned between free and restrained dofs for the pushover analysis. For a typical bridge bent, the dimension of the stiffness matrix corresponding to free dofs, $[K_{ff}]$, shown in Equation 3.2, is usually not a very large number, and the computation time for the matrix inversion of $[K_{ff}]$ in Equation 3.5 is not significant. However, if a pushover analysis is performed for an entire bridge with many intermediate bents or a high rise building, the total number of free dofs may be very large. In this case, significant computation time for the matrix inversion of $[K_{ff}]$ may be needed in order to solve the unknown displacements corresponding to the free dofs, $\{\Delta\delta_f\}$. In this case, a common Gaussian elimination procedure can be used to partition the original free dofs into condensed dofs and remaining free dofs. Hence, the dimension of $[K_{ff}]$ is reduced to the total number of remaining free dofs, which decreases the computation time for the matrix inversion at each load step during the pushover analysis.

5.1.5 GLOBAL DEGREES OF FREEDOM

The Gdofs are in the JCS, and these dofs describe the structural response. Once the joints have been defined and the constraints have been identified, the Gdofs are numbered by INSTRUCT. Any Gdof not condensed out, restrained, or eliminated by a constraint is a free degree of freedom. It is advantageous to partition the stiffness matrix along the following boundaries: dofs to be condensed out, free dofs, and restrained degree of freedom. Thus, in INSTRUCT, the dofs are assigned in the following order:

1. dofs to be condensed out are assigned first
2. Free dofs are assigned second
3. Restrained dofs are assigned third

The Gdof numbers for each joint are stored in the array $\{Lm_j\}$, in which subscript j represents the jth joint.

5.2 INELASTIC IE3DBEAM ELEMENT

The elastic 3D-BEAM and inelastic IE3DBEAM elements are shown in Figures 3.15 and 3.17, respectively. For IE3DBEAM, the element bending stiffness is determined based on the PM, PHL, or CMR methods as described in Chapter 4. The axial and torsional stiffnesses are based on the bilinear hysteresis model (IA_BILN) described in Chapter 3. Since the elastic 3D-BEAM and the inelastic IE3DBEAM have the same element coordinate system (ECS) and dofs, the formulation of their structural and geometric stiffness matrices is similar, and only the IE3DBEAM element stiffness formulation is described here.

5.2.1 ELEMENT COORDINATE SYSTEM AND DEGREES OF FREEDOM

Let X_{ga}, Y_{ga}, and Z_{ga} be the coordinates of the start joint A in the GCS, and X_{gb}, Y_{gb}, and Z_{gb} be the coordinates of the end joint B in the GCS. The distance between the start joint and end joint is given by

$$L = \sqrt{(X_{ga} - X_{gb})^2 + (Y_{ga} - Y_{gb})^2 + (Z_{ga} - Z_{gb})^2} \tag{5.7}$$

Define $\vec{V}_x$ as a unit vector from the start to the end joints,

$$\vec{V}_x = \frac{(X_{gb} - X_{ga})\vec{i} + (Y_{gb} - Y_{ga})\vec{j} + (Z_{gb} - Z_{ga})\vec{k}}{L} \tag{5.8}$$

The vector $\vec{V}_x$ defines the orientation of the element's local X_e axis. Choose a vector, $\vec{V}_{xy}$, such that both $\vec{V}_x$ and $\vec{V}_{xy}$ lie on the element local XY plane.

$$\vec{V}_z = \frac{\vec{V}_x \times \vec{V}_{xy}}{|\vec{V}_{xy}|} \tag{5.9}$$

and

$$\vec{V}_y = \vec{V}_z \times \vec{V}_x \tag{5.10}$$

where $\vec{V}_y$ and $\vec{V}_z$ are unit vectors, which define the orientation of the element's local Y_e and Z_e axes, respectively.

The three unit vectors $\vec{V}_x$, $\vec{V}_y$, and $\vec{V}_z$ define the ECS, denoted X_e, Y_e, Z_e, with the origin at the joint A. The three unit vectors that define the orientation of the ECS are written in matrix form as follows:

$$\{V_e\} = \begin{bmatrix} \vec{V}_x \\ \vec{V}_y \\ \vec{V}_z \end{bmatrix} = \begin{bmatrix} C_{11} & C_{12} & C_{13} \\ C_{21} & C_{22} & C_{23} \\ C_{31} & C_{32} & C_{33} \end{bmatrix} \begin{bmatrix} \vec{i} \\ \vec{j} \\ \vec{k} \end{bmatrix} = [C_e] \begin{bmatrix} \vec{i} \\ \vec{j} \\ \vec{k} \end{bmatrix} \tag{5.11}$$

where $[C_e]$ is the direction cosine matrix for the ECS. The element has 12 dofs as shown in Figure 3.17 and is reproduced here as Figure 5.4. In matrix form, the local forces and displacements in the ECS are given by

$$\{F_e\} = \{F_{xa}, F_{ya}, F_{za}, M_{xa}, M_{ya}, M_{za}, F_{xb}, F_{yb}, F_{zb}, M_{xb}, M_{yb}, M_{zb}\}^T \tag{5.12a}$$

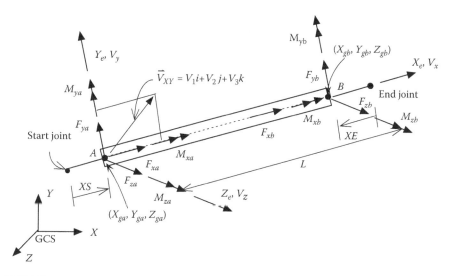

FIGURE 5.4 Elastic 3D beam and inelastic IE3DBEAM elements.

$$\{\delta_e\} = \{\delta_{xa}, \delta_{ya}, \delta_{za}, \theta_{xa}, \theta_{ya}, \theta_{za}, \delta_{xb}, \delta_{yb}, \delta_{zb}, \theta_{xb}, \theta_{yb}, \theta_{zb}\}^T \qquad (5.12b)$$

5.2.2 Element Stiffness Matrix in ECS

The bending stiffness matrices of an IE3DBEAM element corresponding to the $\vec{V}_y$ and $\vec{V}_z$ directions can be expressed as follows:

$$\begin{bmatrix} M_{ya} \\ M_{yb} \end{bmatrix} = \begin{bmatrix} B & D \\ D & B' \end{bmatrix} \begin{bmatrix} \theta_{ya} \\ \theta_{yb} \end{bmatrix} \qquad (5.13)$$

$$\begin{bmatrix} M_{za} \\ M_{zb} \end{bmatrix} = \begin{bmatrix} A & C \\ C & A' \end{bmatrix} \begin{bmatrix} \theta_{za} \\ \theta_{zb} \end{bmatrix} \qquad (5.14)$$

where

θ_{ya} and θ_{yb} represent the rotations in the $\vec{V}_y$ direction

θ_{za} and θ_{zb} represent the rotations in the $\vec{V}_z$ direction

The nonlinear bending stiffness coefficients, A, A', B, B', C, and D, are described in Chapter 4 and are dependent on the hysteresis model used in the analysis. The torsional stiffness is given by

$$\begin{bmatrix} M_{xa} \\ M_{xb} \end{bmatrix} = \frac{G_t J}{L} \begin{bmatrix} 1 & -1 \\ -1 & 1 \end{bmatrix} \begin{bmatrix} \theta_{xa} \\ \theta_{xb} \end{bmatrix} = \begin{bmatrix} Q & -Q \\ -Q & Q \end{bmatrix} \begin{bmatrix} \theta_{xa} \\ \theta_{xb} \end{bmatrix} \qquad (5.15a)$$

and

$$G_t = \begin{bmatrix} G & \text{for elastic condition} \\ SIG & \text{for inelastic condition} \end{bmatrix} \tag{5.15b}$$

where
    J is the polar moment of inertia of the cross section
    G_t is the tangent torsional rigidity
    G is the elastic torsional rigidity
    SIG is the inelastic torsional rigidity

Similarly, the axial stiffness is given by

$$\begin{bmatrix} F_{xa} \\ F_{xb} \end{bmatrix} = \frac{E_t A}{L} \begin{bmatrix} 1 & -1 \\ -1 & 1 \end{bmatrix} \begin{bmatrix} \delta_{xa} \\ \delta_{xb} \end{bmatrix} = \begin{bmatrix} H & -H \\ -H & H \end{bmatrix} \begin{bmatrix} \delta_{xa} \\ \delta_{xb} \end{bmatrix} \tag{5.16}$$

and

$$E_t = \begin{bmatrix} E & \text{for elastic condition} \\ SIE & \text{for inelastic condition} \end{bmatrix} \tag{5.17}$$

where
    E_t is the tangent modulus
    E is the elastic modulus
    SIE is the inelastic modulus
    A is the cross-sectional area

The bilinear hysteresis model, IA_BILN, described in Chapter 3 is employed in Equations 5.15 and 5.16. Combining stiffness terms from Equations 5.13 through 5.16 yields the local element stiffness matrix as follows:

$$\begin{Bmatrix} F_{xa} \\ M_{xa} \\ M_{ya} \\ M_{za} \\ F_{xb} \\ M_{xb} \\ M_{yb} \\ M_{zb} \end{Bmatrix} = \begin{bmatrix} H & 0 & 0 & 0 & -H & 0 & 0 & 0 \\ 0 & Q & 0 & 0 & 0 & -Q & 0 & 0 \\ 0 & 0 & B & 0 & 0 & 0 & D & 0 \\ 0 & 0 & 0 & A & 0 & 0 & 0 & C \\ -H & 0 & 0 & 0 & H & 0 & 0 & 0 \\ 0 & -Q & 0 & 0 & 0 & Q & 0 & 0 \\ 0 & 0 & D & 0 & 0 & 0 & B' & 0 \\ 0 & 0 & 0 & C & 0 & 0 & 0 & A' \end{bmatrix} \begin{Bmatrix} \delta_{xa} \\ \theta_{xa} \\ \theta_{ya} \\ \theta_{za} \\ \delta_{xb} \\ \theta_{xb} \\ \theta_{yb} \\ \theta_{zb} \end{Bmatrix} \tag{5.18}$$

or

$$\{\bar{F}\} = [S_I]\{\bar{\delta}\} \tag{5.19}$$

Examining Figure 5.4, the following relationships are derived by summing moments about ends A and B:

$$\sum (M_y)_A = 0 = M_{ya} + M_{yb} - F_{zb}L \Rightarrow F_{zb} = \frac{M_{ya} + M_{yb}}{L} \tag{5.20}$$

$$\sum (M_y)_B = 0 = M_{ya} + M_{yb} + F_{za}L \Rightarrow F_{za} = \frac{-M_{ya} - M_{yb}}{L} \tag{5.21}$$

$$\sum (M_z)_A = 0 = M_{za} + M_{zb} + F_{yb}L \Rightarrow F_{yb} = \frac{-M_{za} - M_{zb}}{L} \tag{5.22}$$

$$\sum (M_z)_B = 0 = M_{za} + M_{zb} - F_{ya}L \Rightarrow F_{ya} = \frac{M_{za} + M_{zb}}{L} \tag{5.23}$$

Rewriting Equations 5.20 through 5.23 in matrix form yields

$$\{F_e\} = \begin{Bmatrix} F_{xa} \\ F_{ya} \\ F_{za} \\ M_{xa} \\ M_{ya} \\ M_{za} \\ F_{xb} \\ F_{yb} \\ F_{zb} \\ M_{xb} \\ M_{yb} \\ M_{zb} \end{Bmatrix} = \frac{1}{L} \begin{bmatrix} L & 0 & 0 & 0 & 0 & 0 & 0 & 0 \\ 0 & 0 & 0 & 1 & 0 & 0 & 0 & 1 \\ 0 & 0 & -1 & 0 & 0 & 0 & -1 & 0 \\ 0 & L & 0 & 0 & 0 & 0 & 0 & 0 \\ 0 & 0 & L & 0 & 0 & 0 & 0 & 0 \\ 0 & 0 & 0 & L & 0 & 0 & 0 & 0 \\ 0 & 0 & 0 & 0 & L & 0 & 0 & 0 \\ 0 & 0 & 0 & -1 & 0 & 0 & 0 & -1 \\ 0 & 0 & 1 & 0 & 0 & 0 & 1 & 0 \\ 0 & 0 & 0 & 0 & 0 & L & 0 & 0 \\ 0 & 0 & 0 & 0 & 0 & 0 & L & 0 \\ 0 & 0 & 0 & 0 & 0 & 0 & 0 & L \end{bmatrix} \begin{Bmatrix} F_{xa} \\ M_{xa} \\ M_{ya} \\ M_{za} \\ F_{xb} \\ M_{xb} \\ M_{yb} \\ M_{zb} \end{Bmatrix} = [A_e] \begin{Bmatrix} F_{xa} \\ M_{xa} \\ M_{ya} \\ M_{za} \\ F_{xb} \\ M_{xb} \\ M_{yb} \\ M_{zb} \end{Bmatrix} \tag{5.24}$$

Substituting Equation 5.19 into Equation 5.24 leads to

$$\{F_e\} = [A_e][S_I][A_e]^T \{\delta_e\} = [k_e]\{\delta_e\} \tag{5.25}$$

Therefore, the IE3DBEAM element stiffness matrix in the ECS is given by

$$[k_e] = \begin{bmatrix}
H & 0 & 0 & 0 & 0 & 0 & -H & 0 & 0 & 0 & 0 & 0 \\
0 & S_{22} & 0 & 0 & 0 & S_{26} & 0 & -S_{22} & 0 & 0 & 0 & S_{212} \\
0 & 0 & S_{33} & 0 & -S_{35} & 0 & 0 & 0 & -S_{33} & 0 & -S_{311} & 0 \\
0 & 0 & 0 & Q & 0 & 0 & 0 & 0 & 0 & -Q & 0 & 0 \\
0 & 0 & -S_{35} & 0 & B & 0 & 0 & 0 & S_{35} & 0 & D & 0 \\
0 & S_{26} & 0 & 0 & 0 & A & 0 & -S_{26} & 0 & 0 & 0 & C \\
-H & 0 & 0 & 0 & 0 & 0 & H & 0 & 0 & 0 & 0 & 0 \\
0 & -S_{22} & 0 & 0 & 0 & -S_{26} & 0 & S_{22} & 0 & 0 & 0 & -S_{212} \\
0 & 0 & -S_{33} & 0 & S_{35} & 0 & 0 & 0 & S_{33} & 0 & S_{311} & 0 \\
0 & 0 & 0 & -Q & 0 & 0 & 0 & 0 & 0 & Q & 0 & 0 \\
0 & 0 & -S_{311} & 0 & D & 0 & 0 & 0 & S_{311} & 0 & B' & 0 \\
0 & S_{212} & 0 & 0 & 0 & C & 0 & -S_{212} & 0 & 0 & 0 & A'
\end{bmatrix}$$

$$(5.26)$$

where

$$S_{22} = \frac{A + 2C + A'}{L^2} \tag{5.27}$$

$$S_{26} = \frac{A + C}{L} \tag{5.28}$$

$$S_{212} = \frac{A' + C}{L} \tag{5.29}$$

$$S_{33} = \frac{B + 2D + B'}{L^2} \tag{5.30}$$

$$S_{35} = \frac{B + D}{L} \tag{5.31}$$

$$S_{311} = \frac{B' + D}{L} \tag{5.32}$$

5.2.3 ELEMENT STIFFNESS MATRIX IN TERMS OF GLOBAL DEGREES OF FREEDOM

The transformation of dofs from the ECS to Gdofs consists of two steps. First, the dofs at each of the two joints are rotated from the ECS to JCS at joints A and B. Second, the constraint transformation moves the dofs from each of the slave joints to the master joints, if constrained dofs are considered.

Recall the transformation between the global forces and forces in an ECS is given by

$$\{F_e\} = \begin{bmatrix} [C_e] & 0 & 0 & 0 \\ 0 & [C_e] & 0 & 0 \\ 0 & 0 & [C_e] & 0 \\ 0 & 0 & 0 & [C_e] \end{bmatrix} \{F_{GCS}\} = [\bar{C}_e]\{F_{GCS}\} \tag{5.33}$$

where
  $\{F_e\}$ is the force vector in the ECS
  $\{F_{GCS}\}$ is the force vector in the GCS
  $[\bar{C}_e]$ is the direction cosine matrix of the ECS

Thus, rotating the element forces, $\{F_e\}$, to global forces, $\{F_{GCS}\}$, is achieved by

$$\{F_{GCS}\} = [\bar{C}_e]^T\{F_e\} \tag{5.34}$$

and rotating the global forces to joint forces, $\{F_j\}$, is achieved by

$$\{F_j\} = \begin{bmatrix} [C_j]_A & 0 & 0 & 0 \\ 0 & [C_j]_A & 0 & 0 \\ 0 & 0 & [C_j]_B & 0 \\ 0 & 0 & 0 & [C_j]_B \end{bmatrix} \{F_{GCS}\} = [\bar{C}_j]\{F_{GCS}\} \tag{5.35}$$

Substituting Equation 5.34 into Equation 5.35 leads to

$$\{F_j\} = [\bar{C}_j][\bar{C}_e]^T\{F_e\} \tag{5.36}$$

The third transformation is the constraint transformation for each joint, which transfers forces from the slave joint to the master joint. Let $[T_{ms}]_A$ and $[T_{ms}]_B$ be the constraint transformation matrices for joint A and joint B, respectively. The forces acting on the master joints, $\{F_{jm}\}$, can be expressed as follows:

$$\{F_{jm}\} = \begin{bmatrix} [T_{ms}]_A & 0 \\ 0 & [T_{ms}]_B \end{bmatrix} \{F_j\} = [\bar{T}_m]\{F_j\} \tag{5.37}$$

If constraints are not present, the transformation matrix $[\bar{T}_m]$ reduces to an identity matrix. Combining Equations 5.36 and 5.37 yields the transformation from internal element forces, $\{F_e\}$, to forces acting on the master joints, $\{F_{jm}\}$, at the Gdof:

$$\{F_{jm}\} = [\bar{T}_m][\bar{C}_j][\bar{C}_e]^T\{F_e\} = [A]\{F_e\} \tag{5.38}$$

Similarly, the transformation for the deformation is given by

$$\{\delta_e\} = [A]^T \{\delta_{jm}\} \tag{5.39}$$

Recall $\{Lm_j\}$ described in Section 5.1.5 is a vector containing the Gdof numbers at joint j. For an element e with two end joints A and B, the vector $\{Lm\}_e$ that contains the Gdof numbers at ends A and B is

$$\{Lm\}_e = \begin{bmatrix} \{Lm_A\} \\ \{Lm_B\} \end{bmatrix} \tag{5.40}$$

The vector of Gdof numbers is used in the assembly of the global stiffness matrix. The stiffness matrix is transformed from the member stiffness, Equation 5.25, to the Gdof by

$$[k_e]_G = [A][k_e][A]^T \tag{5.41}$$

In structural analysis, we assume that members are directly connected to a joint. However, in reality, only center lines of structural members are intersected at the connection joint, and the end of a member is connected to the rigid zone, as shown in Figure 5.5. If the rigid zone effect is considered in the structural analysis, the column or beam element stiffness matrix, $[k_e]$, should be transferred from the member ends to the joints in the rigid zones by rigid body transformation.

The member force vector, $\{F_e\}$, shown in Equation 5.25, is transferred to the start and end joints (see Figure 5.4) in the rigid zones by the transformation matrix, $[T]$:

$$\{F_e''\} = [T]\{F_e\} \tag{5.42}$$

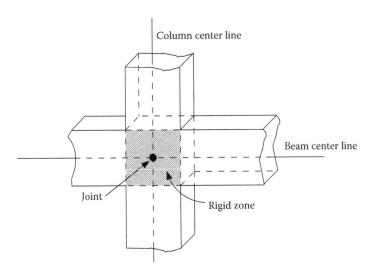

FIGURE 5.5   Rigid zone of structural connection.

and

$$\{\delta_e\} = [T]^T \{\delta_e''\} \tag{5.43}$$

in which

$$[T] = \begin{bmatrix} [T_{SA}] & 0 \\ 0 & [T_{EB}] \end{bmatrix} \tag{5.44}$$

$\{F_e''\}$ and $\{\delta_e''\}$ represent the member force and displacement vectors at the rigid zone joints. The subscripts S and E in Equation 5.44 represent the start and end joints, respectively. $[T_{SA}]$ and $[T_{EB}]$ are the transformation matrices corresponding to the start and end joints, respectively. From Equation 5.2, $[T_{SA}]$ and $[T_{ED}]$ are

$$[T_{SA}] = \begin{bmatrix} 1 & 0 & 0 & 0 & 0 & 0 \\ 0 & 1 & 0 & 0 & 0 & 0 \\ 0 & 0 & 1 & 0 & 0 & 0 \\ 0 & 0 & 0 & 1 & 0 & 0 \\ 0 & 0 & -XS & 0 & 1 & 0 \\ 0 & XS & 0 & 0 & 0 & 1 \end{bmatrix} \tag{5.45}$$

and

$$[T_{EB}] = \begin{bmatrix} 1 & 0 & 0 & 0 & 0 & 0 \\ 0 & 1 & 0 & 0 & 0 & 0 \\ 0 & 0 & 1 & 0 & 0 & 0 \\ 0 & 0 & 0 & 1 & 0 & 0 \\ 0 & 0 & XE & 0 & 1 & 0 \\ 0 & -XE & 0 & 0 & 0 & 1 \end{bmatrix} \tag{5.46}$$

Substituting Equation 5.25 into Equation 5.42 leads to

$$\{F_e''\} = [T][k_e][T]^T \{\delta_e''\} = [k_e'']\{\delta_e''\} \tag{5.47}$$

Once $[k_e'']$ is formed, it can be transformed to the Gdof by the same procedure described by Equations 5.38 through 5.41, except that $\{F_e\}$, $[k_e]$, and $\{\delta_e\}$ are replaced by $\{F_e''\}$, $[k_e'']$, and $\{\delta_e''\}$, respectively.

5.2.4 ELEMENT GEOMETRIC STIFFNESS MATRIX IN GDOF

The "lumped mass" geometric stiffness matrix is formulated with consideration of the effect of axial load on the member's lateral deflections. The element is idealized as a rigid bar with an axial load P as shown in Figure 5.6.

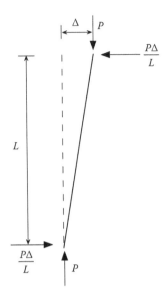

FIGURE 5.6 $P - \delta$ force for the IN3DBEAM or 3D-BEAM element.

The axial load P is positive when the member is in compression. The shear at each end of the member is equal to $P\Delta/L$. Thus, the element geometric stiffness matrix is

$$
[G_e] = \frac{P}{L}
\begin{bmatrix}
0 & 0 & 0 & 0 & 0 & 0 & 0 & 0 & 0 & 0 & 0 & 0 \\
0 & 1 & 0 & 0 & 0 & 0 & 0 & -1 & 0 & 0 & 0 & 0 \\
0 & 0 & 1 & 0 & 0 & 0 & 0 & 0 & -1 & 0 & 0 & 0 \\
0 & 0 & 0 & 0 & 0 & 0 & 0 & 0 & 0 & 0 & 0 & 0 \\
0 & 0 & 0 & 0 & 0 & 0 & 0 & 0 & 0 & 0 & 0 & 0 \\
0 & 0 & 0 & 0 & 0 & 0 & 0 & 0 & 0 & 0 & 0 & 0 \\
0 & 0 & 0 & 0 & 0 & 0 & 0 & 0 & 0 & 0 & 0 & 0 \\
0 & -1 & 0 & 0 & 0 & 0 & 0 & 1 & 0 & 0 & 0 & 0 \\
0 & 0 & -1 & 0 & 0 & 0 & 0 & 0 & 1 & 0 & 0 & 0 \\
0 & 0 & 0 & 0 & 0 & 0 & 0 & 0 & 0 & 0 & 0 & 0 \\
0 & 0 & 0 & 0 & 0 & 0 & 0 & 0 & 0 & 0 & 0 & 0 \\
0 & 0 & 0 & 0 & 0 & 0 & 0 & 0 & 0 & 0 & 0 & 0
\end{bmatrix}
\tag{5.48a}
$$

The "consistent mass" geometric stiffness matrix is formulated with consideration of the effect of axial load on the member's rotations and lateral deflections, which can be expressed as follows:

$$[G_e] = P \begin{bmatrix} 0 & 0 & 0 & 0 & 0 & 0 & 0 & 0 & 0 & 0 & 0 & 0 \\ 0 & F & 0 & 0 & 0 & D & 0 & -F & 0 & 0 & 0 & E \\ 0 & 0 & F & 0 & -D & 0 & 0 & 0 & -F & 0 & -E & 0 \\ 0 & 0 & 0 & 0 & 0 & 0 & 0 & 0 & 0 & 0 & 0 & 0 \\ 0 & 0 & -D & 0 & A & 0 & 0 & 0 & D & 0 & -C & 0 \\ 0 & D & 0 & 0 & 0 & A & 0 & -D & 0 & 0 & 0 & C \\ 0 & 0 & 0 & 0 & 0 & 0 & 0 & 0 & 0 & 0 & 0 & 0 \\ 0 & -F & 0 & 0 & 0 & -D & 0 & F & 0 & 0 & 0 & -E \\ 0 & 0 & -F & 0 & D & 0 & 0 & 0 & F & 0 & E & 0 \\ 0 & 0 & 0 & 0 & 0 & 0 & 0 & 0 & 0 & 0 & 0 & 0 \\ 0 & 0 & -E & 0 & -C & 0 & 0 & 0 & E & 0 & B & 0 \\ 0 & E & 0 & 0 & 0 & -C & 0 & -E & 0 & 0 & 0 & B \end{bmatrix}$$

$$(5.48b)$$

in which $A = 2L/15$, $B = A$, $C = L/30$, $D = E = 0.1$, $F = 1.2/L$, and L is the element length. The geometric stiffness is transferred from element to Gdof by the transformation

$$[G_e]_G = [A][G_e][A]^T \qquad (5.49)$$

5.3   FINITE-SEGMENT ELEMENT

As described in Section 3.5.4, the finite-segment element consists of two joints, A and B, as shown in Figures 3.18 and 4.13. Figure 3.18 is reproduced here as Figure 5.7a. The member is divided into several segments and each segment has 12 dofs. The element's cross section is divided into many sectional elements as shown in Figure 3.19.

5.3.1   ELEMENT COORDINATE SYSTEM AND DEGREES OF FREEDOM

The equations used to define the ECS for a finite-segment element are identical to Equations 5.7 through 5.12.

5.3.2   ELEMENT STIFFNESS MATRIX IN ECS

As described in Appendix B, the element stiffness matrix, $[\bar{k}]$, corresponding to a segment's Gdof direction (X_R, Y_R, Z_R) is formulated first, in which

$$\{\bar{F}_e\} = [\bar{k}]\{\bar{\delta}_e\} \qquad (5.50)$$

and

$$\{\bar{F}_e\} = \{F_1, F_2, F_3, F_4, F_5, F_6, F_7, F_8, F_9, F_{10}, F_{11}, F_{12}\}^T \qquad (5.51)$$

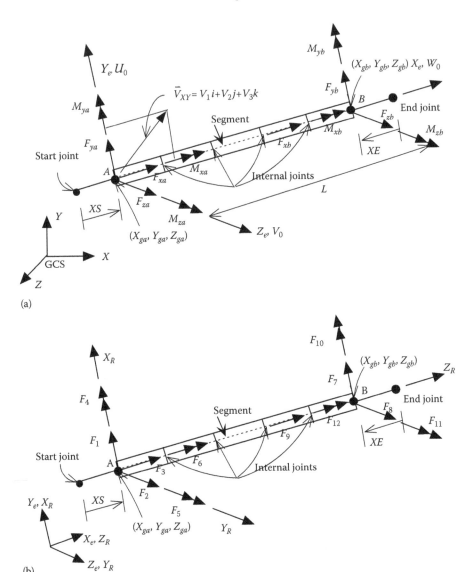

FIGURE 5.7 Finite-segment element forces corresponding to (a) (X_e, Y_e, Z_e) and (b) (X_R, Y_R, Z_R).

$$\{\bar{\delta}_e\} = \{\delta_1, \delta_2, \delta_3, \delta_4, \delta_5, \delta_6, \delta_7, \delta_8, \delta_9, \delta_{10}, \delta_{11}, \delta_{12}\}^T \qquad (5.52)$$

The element force directions, $\{\bar{F}_e\}$, are shown in Figure 5.7b. Since (X_R, Y_R, Z_R) and (X_e, Y_e, Z_e) are not identical, the transformation between the element forces, $\{\bar{F}_e\}$, and the forces in ECS, $\{F_e\}$, is given by

$$\{F_e\} = \begin{bmatrix} 0 & 0 & 1 & 0 & 0 & 0 & 0 & 0 & 0 & 0 & 0 & 0 \\ 1 & 0 & 0 & 0 & 0 & 0 & 0 & 0 & 0 & 0 & 0 & 0 \\ 0 & 1 & 0 & 0 & 0 & 0 & 0 & 0 & 0 & 0 & 0 & 0 \\ 0 & 0 & 0 & 0 & 0 & 1 & 0 & 0 & 0 & 0 & 0 & 0 \\ 0 & 0 & 0 & 1 & 0 & 0 & 0 & 0 & 0 & 0 & 0 & 0 \\ 0 & 0 & 0 & 0 & 1 & 0 & 0 & 0 & 0 & 0 & 0 & 0 \\ 0 & 0 & 0 & 0 & 0 & 0 & 0 & 0 & 1 & 0 & 0 & 0 \\ 0 & 0 & 0 & 0 & 0 & 0 & 1 & 0 & 0 & 0 & 0 & 0 \\ 0 & 0 & 0 & 0 & 0 & 0 & 0 & 1 & 0 & 0 & 0 & 0 \\ 0 & 0 & 0 & 0 & 0 & 0 & 0 & 0 & 0 & 0 & 0 & 1 \\ 0 & 0 & 0 & 0 & 0 & 0 & 0 & 0 & 0 & 1 & 0 & 0 \\ 0 & 0 & 0 & 0 & 0 & 0 & 0 & 0 & 0 & 0 & 1 & 0 \end{bmatrix} \{\bar{F}_e\} \qquad (5.53)$$

or

$$\{F_e\} = [\bar{A}]\{\bar{F}_e\} \qquad (5.54)$$

Substituting Equation 5.50 into Equation 5.54

$$\{F_e\} = [\bar{A}][\bar{k}][\bar{A}]^T\{\delta_e\} = [k_e]\{\delta_e\} \qquad (5.55)$$

where $[k_e]$ represents the element stiffness matrix in the ECS. The transformation of dofs from the ECS to Gdofs for the finite-segment element is same as that for the IE3DBEAM element. Since the geometric stiffness matrix for each segment has been included in the segmental stiffness matrix formation as described in Appendix B, the geometric stiffness effect has been included in the member stiffness matrix $[k_e]$.

5.4 BRACE ELEMENT

The brace element consists of two joints, A and B, as shown in Figure 3.22 and reproduced here as Figure 5.8. The orientation of the brace element is defined by the ECS. The element stiffness is governed by the bracing member's hysteresis model as described in Section 3.4.12.

5.4.1 Element Coordinate System and Degrees of Freedom

The locations of both the start joint A and end joint B are defined in the GCS. Let (X_{ga}, Y_{ga}, Z_{ga}) be the coordinates of the start joint A, and (X_{gb}, Y_{gb}, Z_{gb}) be the coordinates of the end joint B. As shown in Equation 5.7, the distance between joint A and joint B is given by L. The unit vector, $\vec{V}_x$, from the start joint to the end joint is

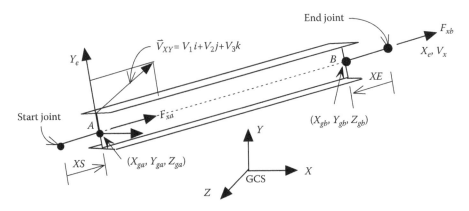

FIGURE 5.8 Brace element.

given in Equation 5.8. Let the ECS be denoted (X_e, Y_e, Z_e). The vector $\vec{V}_x$ defines the orientation of the brace element's X_e axis and can be written in matrix form

$$\{V_e\} = \vec{V}_x = \begin{bmatrix} C_{11} & C_{12} & C_{13} \end{bmatrix} \begin{bmatrix} \vec{i} \\ \vec{j} \\ \vec{k} \end{bmatrix} = [C_e] \begin{bmatrix} \vec{i} \\ \vec{j} \\ \vec{k} \end{bmatrix} \tag{5.56}$$

where $[C_e]$ is the direction cosine matrix of the ECS. The element has two dofs as shown in Figure 5.8. The element forces and displacements in the ECS are

$$\{F_e\} = \{F_{xa}, F_{xb}\}^T \tag{5.57}$$

$$\{\delta_e\} = \{\delta_{xa}, \delta_{xb}\}^T \tag{5.58}$$

5.4.2 ELEMENT STIFFNESS MATRIX IN ECS

The stiffness matrix of a bracing element corresponding to ECS can be expressed as follows:

$$\{F_e\} = \begin{Bmatrix} F_{xa} \\ F_{xa} \end{Bmatrix} = k_{br} \begin{bmatrix} 1 & -1 \\ -1 & 1 \end{bmatrix} \begin{Bmatrix} \delta_{xa} \\ \delta_{xb} \end{Bmatrix} \tag{5.59}$$

where k_{br} represents the axial stiffness coefficient obtained from the brace member hysteresis model described in Section 3.4.12.

5.4.3 ELEMENT STIFFNESS MATRIX IN GDOF

The procedures are similar to those used for the 3D-BEAM and IE3DBEAM elements in Section 5.2.3 and are therefore briefly presented herein. The transformation between the global forces and forces in the ECS is given by

$$\{F_e\} = \begin{bmatrix} [C_e] & 0 \\ 0 & [C_e] \end{bmatrix}\{F_{GCS}\} = [\bar{C}_e]\{F_{GCS}\} \tag{5.60}$$

where

$\{F_e\}$ contains the forces in the ECS
$\{F_{GCS}\}$ is the force vector in GCS
$[\bar{C}_e]$ is the direction cosine matrix of the ECS

Thus, rotating the element forces, $\{F_e\}$, to global forces, $\{F_{GCS}\}$, is achieved by

$$\{F_{GCS}\} = [\bar{C}_e]^T\{F_e\} \tag{5.61}$$

and rotating the global forces to joint forces, $\{F_j\}$, is achieved by

$$\{F_j\} = \begin{bmatrix} [C_j]_A & 0 \\ 0 & [C_j]_B \end{bmatrix}\{F_{GCS}\} = [\bar{C}_j]\{F_{GCS}\} \tag{5.62}$$

Substituting Equation 5.61 into Equation 5.62 leads to

$$\{F_j\} = [\bar{C}_j][\bar{C}_e]^T\{F_e\} \tag{5.63}$$

The constraint transformation for the forces from each slave joint to the master joint yields

$$\begin{Bmatrix} F_{jmx} \\ F_{jmy} \\ F_{jmz} \\ M_{jmx} \\ M_{jmy} \\ M_{jmz} \end{Bmatrix} = \begin{bmatrix} 1 & 0 & 0 \\ 0 & 1 & 0 \\ 0 & 0 & 1 \\ 0 & -Z_{ms} & Y_{ms} \\ Z_{ms} & 0 & -X_{ms} \\ -Y_{ms} & X_{ms} & 0 \end{bmatrix} \begin{Bmatrix} F_{jsx} \\ F_{jsy} \\ F_{jsz} \end{Bmatrix} \tag{5.64}$$

or

$$\{F_{jm}\} = [\bar{T}_{ms}]\{F_{js}\} \tag{5.65}$$

where $\{F_{jm}\}$ and $\{F_{js}\}$ represent the forces acting on the master and slave joints, respectively. The distances X_{ms}, Y_{ms}, and Z_{ms} are in the master joint's JCS and can be calculated using Equation 5.5. The formulation of the stiffness matrix expressed in Gdof, $[k_e]_G$, follows exactly as shown in Equations 5.38 through 5.41.

5.5   PLATE ELEMENT

The plate element consists of four joints as shown in Figure 3.20 and reproduced here as Figure 5.9. The ECS X_e axis goes from joint 3 toward joint 4. The orientation of

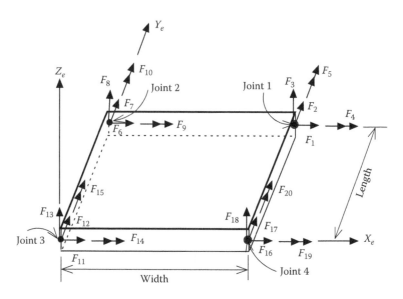

FIGURE 5.9 Plate element.

the ECS Y_e axis goes from joint 3 toward joint 2. The ECS Z_e axis is perpendicular to X_e and Y_e axes, oriented according to the right-hand rule. The plate element is elastic, and nonlinear behavior of the plate element is not considered.

5.5.1 ELEMENT COORDINATE SYSTEM AND DEGREES OF FREEDOM

There are three unit vectors, $\vec{V}_x$, $\vec{V}_y$, and $\vec{V}_z$, which define the ECS, denoted X_e, Y_e, Z_e, with the origin at the joint 3. The three unit vectors that define the orientation of the ECS are written in matrix form as follows:

$$\{V_e\} = \begin{bmatrix} \vec{V}_x \\ \vec{V}_y \\ \vec{V}_z \end{bmatrix} = \begin{bmatrix} C_{11} & C_{12} & C_{13} \\ C_{21} & C_{22} & C_{23} \\ C_{31} & C_{32} & C_{33} \end{bmatrix} \begin{bmatrix} \vec{i} \\ \vec{j} \\ \vec{k} \end{bmatrix} = [C_e] \begin{bmatrix} \vec{i} \\ \vec{j} \\ \vec{k} \end{bmatrix} \tag{5.66}$$

where $[C_e]$ is the direction cosine matrix for the ECS. The element has 20 dofs as shown in Figure 5.9. In matrix form, these local forces and displacements in the ECS are

$$\{F_e\} = \{F_1, F_2, F_3, \ldots, F_{20}\}^T \tag{5.67}$$

$$\{\delta_e\} = \{\delta_1, \delta_2, \delta_3, \ldots, \delta_{20}\}^T \tag{5.68}$$

5.5.2 ELEMENT STIFFNESS MATRIX IN ECS

The stiffness matrix of a plate element corresponding to the ECS can be expressed as follows:

$$\{F_e\} = [k_e]\{\delta_e\} \tag{5.69}$$

in which $[k_e]$ is given in Appendix J.

5.5.3 ELEMENT STIFFNESS MATRIX IN GDOF

The transformation of dofs from the ECS to Gdofs consists of two steps. First, the dofs at each of the four joints are rotated from the ECS to JCS at joints 1–4. Second, the constraint transformation moves dofs from each of the slave joints to the master joint, if constrained dofs are considered.

The transformation between the global forces and forces in an ECS is given by

$$\{F_e\}_{20\times1} = \begin{bmatrix} [C_e'] & 0 & 0 & 0 \\ 0 & [C_e'] & 0 & 0 \\ 0 & 0 & [C_e'] & 0 \\ 0 & 0 & 0 & [C_e'] \end{bmatrix}_{20\times24} \{F_{GCS}\}_{24\times1} = [\bar{C}_e]\{F_{GCS}\} \tag{5.70}$$

in which

$$[C_e'] = \begin{bmatrix} [C_e]_{3\times3} & 0 \\ 0 & [C_e^*]_{2\times3} \end{bmatrix} \tag{5.71}$$

where

$$[C_e^*] = \begin{bmatrix} C_{11} & C_{12} & C_{13} \\ C_{21} & C_{22} & C_{23} \end{bmatrix} \tag{5.72}$$

Thus, $\{F_{GCS}\}$ is achieved by

$$\{F_{GCS}\} = [\bar{C}_e]^T\{F_e\} \tag{5.73}$$

Rotating the global forces to joint forces, $\{F_j\}$, is achieved by

$$\{F_j\} = \begin{bmatrix} [C_j]_1 & 0 & 0 & 0 \\ 0 & [C_j]_2 & 0 & 0 \\ 0 & 0 & [C_j]_3 & 0 \\ 0 & 0 & 0 & [C_j]_4 \end{bmatrix} \{F_{GCS}\} = [\bar{C}_j]\{F_{GCS}\} \tag{5.74}$$

in which

$$[C_j]_i = \begin{bmatrix} [C_j] & 0 \\ 0 & [C_j] \end{bmatrix} \tag{5.75}$$

with $[C_j]$ as defined in Equation 5.1.

Substituting Equation 5.73 into Equation 5.74 leads to

$$\{F_j\} = [\bar{C}_j][\bar{C}_e]^T\{F_e\} \tag{5.76}$$

If constrained dofs are considered, the constraint transformation from the slave joints to the master joint(s) can be expressed as follows:

$$[F_{jm}] = \begin{bmatrix} [T_{ms}]_1 & 0 & 0 & 0 \\ 0 & [T_{ms}]_2 & 0 & 0 \\ 0 & 0 & [T_{ms}]_3 & 0 \\ 0 & 0 & 0 & [T_{ms}]_4 \end{bmatrix} \{F_j\} = [\bar{T}_{ms}]\{F_j\} \tag{5.77}$$

If a joint i of a plate element is not constrained, the corresponding $[T_{ms}]_i$ is an identity matrix. The formulation of the stiffness matrix expressed in Gdof, $[k_e]_G$, follows exactly as shown in Equations 5.38 through 5.41.

5.6   UNBALANCED FORCES

5.6.1   UNBALANCED ELEMENT FORCES

As described in the material library in Chapter 3, material nonlinearity is simulated by different hysteresis models. During pushover analysis, the relationship between a material's stress (force) and strain (displacement) is not linear. Hence, the internal element forces may not be in equilibrium with the forces acting on a joint at the end of a given incremental load step. As a result, at each incremental load step, there may be unbalanced forces at the member ends. If these unbalanced forces are not eliminated or reduced, the analysis will converge to an inaccurate response. There are several ways to correct this phenomenon. The simplest approach is to reduce the size of the incremental load step. However, many steps are required, and this approach leads to an excessive computation time. Another technique is to locate the point where an element stiffness will be changed (say element bending stiffness changes at the point of yield moment, M_p) and then reanalyze the structure by adjusting the magnitude of the load increment vector such that the element moment reaches M_p. For a structure with many members, several elements may experience the stiffness change in a single load step, and this approach can lead to an excessive solution time. A third technique is to calculate the magnitude of these unbalanced forces and apply them as joint loads to the structure in the next incremental load step. This approach is adopted in INSTRUCT. It is fairly simple to execute and yet provides reasonable approximation of the actual nonlinear response.

Consider an IE3DBEAM element using a bilinear bending hysteresis (IA_BILN) model with a moment–rotation curve at element joint A shown in Figure 5.10. Assuming that point $\bar{A}$ is on the loading curve and has a rotation and moment of θ_{a0} and M_{a0}, an incremental rotation, $\Delta\theta_a$, is applied to the member, and the member tangent stiffness at the current load step is used to determine the end moment, which

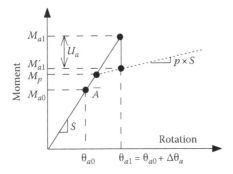

FIGURE 5.10 Member unbalance moment U_a

has a value of M_{a1}. However, the moment M_{a1} is apparently greater than the actual internal moment M'_{a1}. The internal moment, M'_{a1}, at rotation θ_{a1} is calculated by

$$M'_{a1} = M_{a1} - [(M_{a1} - M_p)(1-p)] \tag{5.78}$$

The moment acting on the element joint A, M_{a1}, exceeds the internal moment M'_{a1}, and the unbalanced moment at joint A is

$$U_a = M_{a1} - M'_{a1} \tag{5.79}$$

Similarly, the unbalanced moment at joint B is

$$U_b = M_{b1} - M'_{b1} \tag{5.80}$$

The unbalanced force for axial load can be obtained using the same approach mentioned above.

5.6.2 GLOBAL UNBALANCED JOINT FORCES

For the IE3DBEAM element, the unbalanced forces are determined by bending and axial hysteresis models. For the bracing element, the unbalanced force is determined by the bracing member hysteresis model only. The element unbalanced forces are then applied to the Gdof by

$$\begin{bmatrix} \{U_{Ame}\}_{6\times1} \\ \{U_{Bme}\}_{6\times1} \end{bmatrix} = [A][\{F_e\} - \{F'_e\}] \tag{5.81}$$

where
   $\{U_{Ame}\}_{6\times1}$ is the unbalanced force vector from element e, acting on the joint A
   $\{F'_e\}$ is the actual element internal force vector based on the hysteresis models
   $\{F_e\}$ is the calculated element force vector from the structural pushover analysis

The global unbalanced joint force vector, $\{U\}$, as shown in Equation 3.1, is assembled by

$$U(li) = U(li) + U_{jme}(i), \quad i = 1 \text{ to } 6$$

$$j = \text{joint } A \text{ or } B$$

$$e = 1 \text{ to } NELEM \tag{5.82}$$

$$li = Lm(i)_e$$

where
    e is the element number
    $NELEM$ is the total number of elements in the structure

Recall that $\{Lm\}_e$ in Equation 5.40 is a vector containing the Gdof numbers at element e's joints. $Lm(i)_e$ is the global degree of freedom number li corresponding to the ith degree of freedom of element e.

5.6.3 ASSEMBLY OF THE GLOBAL STRUCTURAL AND GEOMETRIC STIFFNESS

The structural stiffness matrix is assembled by the direct element method, where the element's stiffness is mapped into the Gdof. The global structural stiffness $[K]$ is given by

$$K(li,lj) = K(li,lj) + k_e(i,j)_G, \quad \text{for } i = 1 \text{ to } NELDOF$$

$$j = 1 \text{ to } NELDOF$$

$$e = 1 \text{ to } NELEM \tag{5.83}$$

$$li = Lm(i)_e$$

$$lj = Lm(j)_e$$

where
    $NELDOF$ is the number of dofs for element e
    $k_e(i,j)_G$ is the ijth term of the element stiffness matrix, $[k_e]_G$, as shown in Equation
      5.41 for element e

Similarly, the element's geometric stiffness is mapped into the Gdof. The structural global geometric stiffness $[G]$ is given by

$$G(li,lj) = G(li,lj) + G_e(i,j)_G, \quad \text{for } i = 1 \text{ to } NELDOF$$

$$j = 1 \text{ to } NELDOF$$

$$e = 1 \text{ to } NELEM \tag{5.84}$$

$$li = Lm(i)_e$$

$$lj = Lm(j)_e$$

where $G_e(i,j)_G$ is the ijth term of the element geometric stiffness matrix $[G_e]_G$, as shown in Equation 5.49, for element e.

Example 5.1

Find the global stiffness matrix corresponding to the Gdofs 1–6 shown in Figure 5.11a, based on the procedure described in Sections 5.2.3 and 5.6.3. In the figure, the direction of the JCS, (X_j, Y_j, Z_j), is same as that of the GCS, (X_g, Y_g, Z_g). The ECS, (X_e, Y_e, Z_e), for each 3D-BEAM member is also shown in Figure 5.11b.

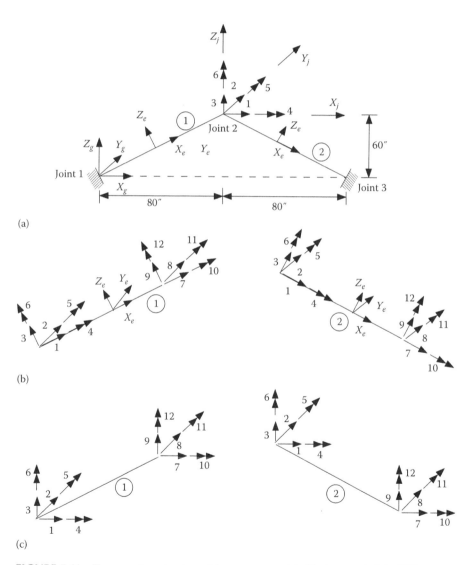

FIGURE 5.11 Two-member structure: (a) structure model; (b) element dofs in ECS (X_e, Y_e, Z_e); (c) element dofs in JCS (X_j, Y_j, Z_j).

The element stiffness matrix, $[k_e]$, for the 3D-BEAM element can be calculated using Equation 5.26. For demonstration purposes, the material properties of 3D-BEAM are assumed to be: $E=1000$ ksi; $G=1000$ ksi; $AX=AY=AZ=1$ in.^{2}; $J=1$ in.^{4}; $IY=2$ in.^{4}; and $IZ=4$ in.^{4}

Solution

1. Find the direction cosine matrix, $[C_e]$, between ECS and GCS:
   From Figure 5.11, the direction cosine matrices for members 1 and 2 are

$$[C_e]_1 = \begin{bmatrix} 0.8 & 0 & 0.6 \\ 0 & 1 & 0 \\ -0.6 & 0 & 0.8 \end{bmatrix} \text{ for member 1} \tag{5.85}$$

and

$$[C_e]_2 = \begin{bmatrix} 0.8 & 0 & -0.6 \\ 0 & 1 & 0 \\ 0.6 & 0 & 0.8 \end{bmatrix} \text{ for member 2} \tag{5.86}$$

2. Find the direction cosine matrix, $[C_j]$, between JCS and GCS:
   Since the directions of JCS and GCS are the same, $[C_j] = \begin{bmatrix} 1 & 0 & 0 \\ 0 & 1 & 0 \\ 0 & 0 & 1 \end{bmatrix}$.

3. Find the transformation matrix, $[A]$, for each element:
   Since there is no rigid body constraint, the transformation matrix, $[\bar{T}_m]$, shown in Equation 5.37, is an identity matrix. Therefore, $[A]$, for member 1 can be obtained based on Equation 5.38 as follows:

$$[A]_1 = [\bar{T}_m][\bar{C}_j][\bar{C}_e]_1^T = \begin{bmatrix} [C_e]_1^T & 0 & 0 & 0 \\ 0 & [C_e]_1^T & 0 & 0 \\ 0 & 0 & [C_e]_1^T & 0 \\ 0 & 0 & 0 & [C_e]_1^T \end{bmatrix} \tag{5.87}$$

Similarly, $[A]$, for member 2 is

$$[A]_2 = [\bar{T}_m][\bar{C}_j][\bar{C}_e]_2^T = \begin{bmatrix} [C_e]_2^T & 0 & 0 & 0 \\ 0 & [C_e]_2^T & 0 & 0 \\ 0 & 0 & [C_e]_2^T & 0 \\ 0 & 0 & 0 & [C_e]_2^T \end{bmatrix} \tag{5.88}$$

4. Find the element stiffness matrix in the JCS direction (i.e., in the Gdof directions):
   The element stiffness matrix, $[k_e]$, in the ECS for both elements can be obtained from Equation 5.26 as follows:

$$[k_e]_1 = [k_e]_2 = \begin{bmatrix} 10 & 0 & 0 & 0 & 0 & 0 & -10 & 0 & 0 & 0 & 0 & 0 \\ & 0.048 & 0 & 0 & 0 & 2.4 & 0 & -0.048 & 0 & 0 & 0 & 2.4 \\ & & 0.024 & 0 & -1.2 & 0 & 0 & 0 & -0.024 & 0 & -1.2 & 0 \\ & & & 10 & 0 & 0 & 0 & 0 & 0 & -10 & 0 & 0 \\ & & & & 80 & 0 & 0 & 0 & 1.2 & 0 & 40 & 0 \\ & & & & & 160 & 0 & -2.4 & 0 & 0 & 0 & 80 \\ & & & & & & 10 & 0 & 0 & 0 & 0 & 0 \\ & \text{Symm.} & & & & & & 0.048 & 0 & 0 & 0 & -2.4 \\ & & & & & & & & 0.024 & 0 & 1.2 & 0 \\ & & & & & & & & & 10 & 0 & 0 \\ & & & & & & & & & & 80 & 0 \\ & & & & & & & & & & & 160 \end{bmatrix}$$

(5.89)

From Equation 5.41, the element stiffness matrix in the JCS direction for element 1 is

$$[k_1]_G = [A]_1[k_e]_1[A]_1^T = \begin{bmatrix} 6.41 & 0 & 4.79 & 0 & 0.72 & 0 & -6.41 & 0 & -4.79 & 0 & 0.72 & 0 \\ & 0.048 & 0 & -1.44 & 0 & 1.92 & 0 & -0.048 & 0 & -1.44 & 0 & 1.92 \\ & & 3.62 & 0 & -0.96 & 0 & -4.79 & 0 & -3.62 & 0 & -0.96 & 0 \\ & & & 64 & 0 & -72 & 0 & 1.44 & 0 & 22.4 & 0 & -43.2 \\ & & & & 80 & 0 & -0.72 & 0 & 0.96 & 0 & 40 & 0 \\ & & & & & 106 & 0 & -1.92 & 0 & -43.2 & 0 & 47.6 \\ & & & & & & 6.41 & 0 & 4.79 & 0 & -0.72 & 0 \\ & \text{Symm.} & & & & & & 0.048 & 0 & 1.44 & 0 & -1.92 \\ & & & & & & & & 3.62 & 0 & 0.96 & 0 \\ & & & & & & & & & 64 & 0 & -72 \\ & & & & & & & & & & 80 & 0 \\ & & & & & & & & & & & 106 \end{bmatrix}$$

(5.90)

Similarly, the element stiffness matrix in the JCS direction for element 2 is

$$[k_2]_G = [A]_2[k_e]_2[A]_2^T = \begin{bmatrix} 6.41 & 0 & -4.79 & 0 & -0.72 & 0 & -6.41 & 0 & 4.79 & 0 & -0.72 & 0 \\ & 0.048 & 0 & 1.44 & 0 & 1.92 & 0 & -0.048 & 0 & 1.44 & 0 & 1.92 \\ & & 3.62 & 0 & -0.96 & 0 & 4.79 & 0 & -3.62 & 0 & -0.96 & 0 \\ & & & 64 & 0 & 72 & 0 & -1.44 & 0 & 22.4 & 0 & 43.2 \\ & & & & 80 & 0 & 0.72 & 0 & 0.96 & 0 & 40 & 0 \\ & & & & & 106 & 0 & -1.92 & 0 & 43.2 & 0 & 47.6 \\ & & & & & & 6.41 & 0 & -4.79 & 0 & 0.72 & 0 \\ & \text{Symm.} & & & & & & 0.048 & 0 & -1.44 & 0 & -1.92 \\ & & & & & & & & 3.62 & 0 & 0.96 & 0 \\ & & & & & & & & & 64 & 0 & 72 \\ & & & & & & & & & & 80 & 0 \\ & & & & & & & & & & & 106 \end{bmatrix}$$

(5.91)

5. Find the global stiffness matrix corresponding to Gdof 1–6:

In Equation 5.83, the Gdof numbers for element 1 are $Lm(i)_e = (7,8,9,10,11,12,1, 2,3,4,5,6)$, corresponding to element dofs of $i = 1, 2, 3, 4, 5, 6, 7, 8, 9, 10, 11$, and 12 in the JCS (see Figure 5.11c), respectively. Similarly, the Gdof numbers for element 2 are $Lm(i)_e = (1,2,3,4,5,6,13,14,15,16,17,18)$, corresponding to element dofs of $i = 1, 2, 3, 4, 5, 6, 7, 8, 9, 10, 11$, and 12 in the JCS, respectively. Note that Gdofs 7, 8, 9, 10, 11, 12, 13, 14, 15, 16, 17, and 18 are restrained dof as described in Section 5.1.5.

From Equations 5.90 and 5.91, and $Lm(i)_e$ for members 1 and 2, the global stiffness matrix corresponding to Gdofs 1–6 is obtained as follows:

$$[K] = \begin{bmatrix} 12.82 & 0 & 0 & 0 & -1.44 & 0 \\ & 0.096 & 0 & 2.88 & 0 & 0 \\ & & 7.24 & 0 & 0 & 0 \\ & & & 128 & 0 & 0 \\ & Symm. & & & 160 & 0 \\ & & & & & 212 \end{bmatrix} \quad (5.92)$$

6 Input Data for INSTRUCT Program

As described in Chapter 3, INSTRUCT is divided into several blocks. The input data of each block is described below. The blocks may be executed in any order, except as noted. Multiple solutions of the same structure can be performed by using multiple solution blocks. Multiple structures may be analyzed by redefining the structure with the STRUCT block. The readers can download the INSTRUCT executable file from the Web site at http://www.crcpress.com/product/isbn/9781439837634.

Block	Description
STRUCT	Defines the structure to be analyzed. Joints, materials, elements, mass, and damping are defined.
SOL01	Elastic static solution of the structure with multiple load cases. SOL01 must be preceded by the block STRUCT.
SOL04	Incremental static (pushover) solution of a nonlinear structure. SOL04 must be preceded by the block STRUCT.

Secondary Blocks	
BUG	Sets flags to print out detailed information
READ	Reads results written to an output file during SOL04 and prints the results
NOECHO	Inhibits the input echo
DUMP	Prints out the contents of the memory
RELEASE	Releases the memory used for the previous solution
STOP	Terminates execution of the program

NOTES ON INPUT

1. Input is free format, unless otherwise noted.
2. Input variables beginning with I–N are integers and should not contain a decimal point.
3. Input variables beginning with A–H and O–Z are real and may contain a decimal point.
4. Logical variables are identified in the input description and have the value .TRUE. or .FALSE.
5. Character variables are identified in the input description and are enclosed in single quotes, except as noted.
6. The input data is read from unit 05, except as noted.
7. The output is printed on unit 06, except as noted.

8. Units 10 through 47 and units 51 through 60 are reserved for plot files (i.e., temp10.out through temp47.out and temp51.out through temp60.out), which include the output responses at selected joints, elements, or degrees of freedom (dofs).

9. Input cards are identified by a box. If that card is repeated, the entire box is repeated.

input card image

The data for one card may be input on one or more lines in the data file, provided that all of the character variables are on the first line.

10. Consistent units are used throughout the program. Thus, input in inches, kips, seconds yields output in inches, inch-kips, etc. Mixing units will yield unpredictable results. Units are indicated parenthetically where appropriate.

Example

	Note
'BUG=K'	(1)
'STRUCT'	(2)
.	
.	
. Structure input is omitted	
.	
'SOL04'	(3)
.	
. Solution input is omitted	
.	
'READ UNIT=21'	(4)
'STOP'	(5)

(1) The bug option is set. This prints out the hysteresis model data for each load step.

(2) The structure is defined.

(3) A cyclic static incremental solution of the structure is performed.

(4) Data from the plot file 'temp21.out' on unit 21 is written to the output file on unit 06.

(5) The program is terminated.

6.1 STRUCTURE—DEFINE THE STRUCTURAL MODEL

These cards define the structural model to be analyzed. The following cards are each input once.

'STRUCT'
'TITLE'

STRUCT   Signifies that the structural model is to be input. Character variable, enclosed in single quotes.

TITLE   User input title, 80 characters maximum.

6.1.1 JOINTS AND DEGREES OF FREEDOM

These cards are used to define the coordinates of the joints, joint restraints, constraints, and the dof to be condensed out. The dof numbers are assigned by the program and printed in the output. The following card is input once.

NJOINT	NCOS	NSUPT	NCOND	NCONST	SCALE

NJOINT   The number of joints defined by the joint coordinate cards.
NCOS   The number of joint direction cosine cards input.
NSUPT   The number of joint restraint cards input.
NCOND   The number of joint condensation cards input.
NCONST   The number of joint constraint cards input.
SCALE   Scale factor that the joint coordinates are to be multiplied by. For example, if SCALE = 12, the user inputs the joint coordinates in feet, and the structure is defined in inches

   a. *Joint coordinates:* These cards are used to define the coordinates of the joints, in the global coordinate system, GCS, and identify the direction cosine of the joint. The total number of joints defined in this section is less than or equal to NJOINT. The second card is only used when the preceding card has a value of IGEN that is greater than zero. These cards are repeated until (1) NJOINT joints have been defined or (2) an input or generated joint ID number is less than or equal to zero.

ID	X	Y	Z	ICOS	IGEN
Δ ID	Δ X	Δ Y	Δ Z		

ID   The joint identification number. ID numbers can be input in any convenient order and need not be consecutive. However, the band width of the structural stiffness matrix is dependent on the joint ID numbers. An ID ≤ 0 terminates the input of the joint coordinates.
X   The GCS X-coordinate of the joint (length).
Y   The GCS Y-coordinate of the joint (length).
Z   The GCS Z-coordinate of the joint (length).
ICOS   The joint's direction cosine number.
IGEN   The number of additional joints to be generated from this joint.
Δ ID   The increment between the generated ID number and the previous joint's ID number.
     A generated ID ≤ 0 terminates the input of the joint coordinates.

Δ X   The increment between the generated joint's GCS X-coordinate and the previous joint's X-coordinate (length).

Δ Y   The increment between the generated joint's GCS Y-coordinate and the previous joint's Y-coordinate (length).

Δ Z   The increment between the generated joint's GCS Z-coordinate and the previous joint's Z-coordinate (length).

Example: NJOINT = 7

	Note
10 0. 0. 0. 1 2	(1,2)
10 0. 0. 3.	(2)
1 4. 3. −1. 2 0	(3)
−1 0 0 0 0 0	(4)

(1) Joint 10 has the coordinates (0,0,0) and uses direction cosine #1.

(2) Two joints are generated from joint 10: Joint 20 (0,0,3) and Joint 30 (0,0,6).

(3) Joint 1 has the coordinates (4,3,−1) and uses direction cosine #2.

(4) Input of the joint coordinates is terminated.

b. *Joint direction cosines*: These cards are used to input the joint direction cosines, which in turn define the joint coordinate system, JCS. The joint direction cosines are numbered from 1 to NCOS.

This card is repeated NCOS times.

Vxi	Vxj	Vxk	Vyi	Vyj	Vyk

Vxi   The projection on the GCS X-axis of a unit vector parallel to the JCS X-axis.

Vxj   The projection on the GCS Y-axis of a unit vector parallel to the JCS X-axis.

Vxk   The projection on the GCS Z-axis of a unit vector parallel to the JCS X-axis.

Vyi   The projection on the GCS X-axis of a unit vector parallel to the JCS Y-axis.

Vyj   The projection on the GCS Y-axis of a unit vector parallel to the JCS Y-axis.

Vyk   The projection on the GCS Z-axis of a unit vector parallel to the JCS Y-axis.

c. *Joint restraints*: These cards are used to define the joint restraints. This card is repeated NSUPT times.

ID	ITX	ITY	ITZ	IRX	IRY	IRZ	IGEN	Δ ID

ID   The joint identification number. An ID of zero indicates that all the joints are restrained by this card.

ITX   Restraint flag for translation in the JCS X-direction.

ITY   Restraint flag for translation in the JCS Y-direction.

ITZ   Restraint flag for translation in the JCS Z-direction.

IRX   Restraint flag for rotation about the JCS X-axis.

IRY   Restraint flag for rotation about the JCS Y-axis.

IRZ   Restraint flag for rotation about the JCS Z-axis.

IGEN The number of additional joints, with the same restraints, to be generated from this joint.

Δ ID   The increment between the generated ID number and the last ID number.

Valid joint restraint flags are

0	Free or unrestrained dof.
1	Restrained dof.
2	Restrained dof. A restraint flag of 2 forces the program to assign the dof a higher number. This option can be used to reduce the bandwidth of the stiffness matrix.

Example: NSUPT=2

	Note
0 0 0 0 0 0 1 0 0	(1)
1 1 1 1 1 1 1 0 0	(2)

(1) The rotation about the JCS Z-axis of all joints is restrained.

(2) Joint 1 has all six dofs restrained.

d. *Joint condensation*: These cards are used to identify which dofs are condensed out. This card is repeated NCOND times and is omitted if NCOND equals zero.

ID	ITX	ITY	ITZ	IRX	IRY	IRZ	IGEN	Δ ID

ID   The joint identification number. An ID of zero indicates that all of the joints are affected by this card.

ITX   Condensation flag for translation in the JCS X-direction.

ITY   Condensation flag for translation in the JCS Y-direction.

ITZ   Condensation flag for translation in the JCS Z-direction.

IRX   Condensation flag for rotation about the JCS X-axis.

IRY   Condensation flag for rotation about the JCS Y-axis.

IRZ   Condensation flag for rotation about the JCS Z-axis.

IGEN The number of additional joints, with the same condensation, to be generated from this joint.

Δ ID   Increment between the generated ID number and the last ID number.

Valid condensation flags are

0	dof is not condensed out.
1	dof is condensed out. Condensation of a restrained dof is ignored.

Example: NCOND = 1
30 0 0 0 1 1 1 0 0

The rotations of joint 30 are condensed out. If the Z-axis rotation has been previously restrained, only the X- and Y-axes rotations are condensed out.

e. *Joint constraints*: These cards are used to identify which dofs are constrained. This card is repeated NCONST times and omitted if NCONST equals zero.

ITYPE	MASTER	ISLAVE	IGEN	Δ ID

ITYPE   The type of constraint.
            0 Rigid body constraint. A rigid body constraint transfers all six joint dofs from the slave to the master joint.
            1 XY-planar constraint. An XY-planar constraint transfers the joint's JCS X- and Y-axes translational dof and the joint's JCS Z-axis rotational dof from the slave to the master joint.
MASTER   The joint identification number of the master joint.
ISLAVE   The joint identification number of the slave joint.
IGEN   The number of additional slave joints, constrained to the same master joint to be generated.
Δ ID   The increment between the generated ID number and the previous ID number.

Note: Both the slave and master joints must have the same joint direction cosine number (ICOS).

Example: NCONST=1
1 10 20 0 0

Joint 20 is constrained in the JCS XY-plane to Joint 10.

6.1.2   MATERIALS AND HYSTERESIS MODELS

These cards are used to input the material and hysteresis model information. The first card is input once. The second card is repeated NMAT times.

NMAT			
TYPE	VALUE1	VALUE2	...

NMAT   Number of material input.
TYPE   Material type. Valid types are discussed below. Character variable, enclosed in single quotes.
VALUEi   Input required by a given material type. The values for each TYPE are discussed below (real or integer).
Note

A given material may not be compatible with all the elements. For example, the 3D-BEAM material cannot be used with the PLATE element. Compatible materials for each element are specified under element input or Table 3.2.

a. TYPE = '3D-BEAM'. Material data for an elastic 3D-beam (MAT01 subroutine)

'3D-BEAM'	E	G	AX	AY	AZ	J	IY	IZ

E Young's modulus (force/length^2).
G Shear modulus (force/length^2).
AX Cross-sectional area (length^2).
AY Y-axis shear area (length^2).
AZ Z-axis shear area (length^2).
J Torsional moment of inertia (length^4).
IY Moment of inertia about the element coordinate system (ECS) Y-axis (length^4).
IZ Moment of inertia about the ECS Z-axis (length^4).

b. TYPE = 'BILINEAR or ELSPLS'. Material data for bilinear spring model (MAT07)

'BILINEAR'	E	PY	SIE	μf	βDI

E Elastic stiffness (force/length).
PY Yield load (force).
SIE Inelastic stiffness (force/length).
μf Failure ductility.
βDI Parameter for damage index. (Note: Current version of the program does not calculate it. Dummy variable. The damage index is a parameter (Park and Ang, 1985), showing the damage condition of a structure. A damage index greater than one indicates structure is fully damaged and collapsed.)

b2. TYPE = 'TAKEDA'. Material data for TAKEDA hysteresis model that was developed to model bending deformation in reinforced concrete members (MAT06).

'TAKEDA'	EI	PC	DC	PY	DY	PU	DU	βDI

EI Initial bending stiffness of the member (force * length^2/rad)
PC Cracking moment (force * length)
DC Cracking rotation, for a unit length member (radian/length)
PY Yield moment (force * length)
DY Yield rotation, for a unit length member (radian/length)
PU Ultimate moment (force * length)
DU Ultimate rotation, for a unit length member (radian/length)
βDI Parameter for damage index (dummy variable)

b3. TYPE = 'GAP'. Material data for gap or restrainer model (MAT15B) (see Figure 3.5).

'GAP'	KC	KT	KTIE	DC	DX	PPY	IOPTION

KC	Compression stiffness when $D \le DC$ (i.e., gap closed)
KT	Restrainer elastic stiffness
KTIE	Restrainer post-yield stiffness
DC	Displacement at which gap closed
DX	Displacement at which restrainer engaged
PPY	Yield displacement of restrainer
IOPTION	1: Inelastic restrainer.
	2: Gap without restrainer.
	3: Special gap: Spring stiffness = 0 if D = DC.
	Spring stiffness = KC if D ≠ DC.

c. TYPE= 'HINGE'. Material data for plastic hinge length method (MAT19).

'HINGE'	RHA		VA	RATIO	μf	βDI	STIELE	PRMAX	ELAS
ICHOICE1 (if ELAS = 1 or 3)									
A0	A1	A2	A3 (if ICHOICE1 = 0 and ELAS = 1 or 3)						
M (if ICHOICE1 = 1 and ELAS = 1 or 3)									
X1	Y1								
X2	Y2								
...	...	(if ICHOICE1 = 1 and ELAS = 1 or 3)							
Xm	Ym								
ICHOICE2 (if ELAS = 2 or 3)									
B0	B1	B2	B3 (if ICHOICE2 = 0 and ELAS = 2 or 3)						
M (if ICHOICE2 = 1 and ELAS = 2 or 3)									
X1	Y1	FMODE1							
X2	Y2	FMODE2							
...	...	......	(if ICHOICE2 = 1 and ELAS = 2 or 3)						
Xm	Ym	FMODEm							
ICHOICE3 (use ICHOICE3 for concrete column only)									
VCI	VCF		VS	ALFA	(if ICHOICE3 = 1)				

RHA	Yield rotation at VA of plastic hinge segment.
VA	Plastic moment, M_n, of plastic hinge segment.
RATIO	Ratio of post-yield stiffness to elastic stiffness of plastic hinge segment.
μf	Failure ductility of plastic hinge segment.
βDI	Parameter for damage index of plastic hinge segment (dummy variable).
STIELE	Elastic bending stiffness, EI, for elastic member portion.

PRMAX	Plastic rotation capacity, θ_p, of a member (if PRMAX<0, member plastic rotation capacity is not checked).
ELAS	Index for nonlinear analysis:
	0 Axial load and moment interaction curve is not considered.
	1 Nominal moment, M_n, in ECS Y-direction is adjusted based on interaction axial load–moment (P–M) curve. The axial load–plastic rotation capacity interaction (P–PRC) is not considered.
	2 P–PRC is considered, but P–M interaction is not considered.
	3 Both P–M and P–PRC interactions are considered.
ICHOICE1	0: A0−A3 are input by user.
	1: The program will calculate A0−A3 based on user input interaction data points.
A0−A3	If ELAS=1 or 3, input coefficients used for the nominal moment–axial load interaction curve are
	$M = A0 + A1 * P + A2 * P^2 + A3 * P^3$ in which M and P are nominal moment and axial load, respectively. If A2=A3=0, the maximum moment along the interaction curve is VA. This is mainly for steel member with linear interaction curve.
M	If ICHOICE1 = 1, total number of $P-M$ data points input by user.
$X_i Y_i$	The ith data point values (X_i represents the ith axial load and Y_i represents the ith nominal moment).
ICHOICE2	0: B0−B3 are input by user.
	1: The program will calculate B0−B3 based on user input interaction data points.
B0−B3	If ELAS=2 or 3, input coefficients used for the axial–plastic rotation capacity interaction (P–PRC) interaction curve are
	$PRC = B0 + B1 * P + B2 * P^2 + B3 * P^3$ in which PRC and P are plastic rotation capacity and axial load, respectively.
M	If ICHOICE2 = 1, total number of $P-PRC$ data points input by user.
$X_i Y_i$	The ith data point values (X_i represents the ith axial load, and Y_i represents the ith plastic rotation capacity).
FMODEi	Failure mode corresponding to ($X_i Y_i$). Character variable, enclosed in single quotes.
	FMODEi could be one of the following:
	'CONCRETE': Concrete compression failure.
	'FRACTURE': Longitudinal steel tensile fracture.
	'BUCKLING' Longitudinal steel buckling.
	'FATIGUE': Longitudinal steel low-cycle fatigue.
	'SPLICE': Longitudinal steel lap-splice failure.
ICHOICE3	0: Element shear capacity is not considered.
	1: Element shear capacity is considered.
VCI	Initial concrete shear capacity.
VCF	Concrete shear capacity after ductility $\mu \geq 15$.
VS	Shear strength due to transverse rebars.
ALFA	Angle for calculating shear strength due to compression axial load (in degree) (see Figure 4.18).

Notes

1. If user inputs RATIO=0, the program uses RATIO=0.00001 for computation to prevent numerical overflow.
2. For ELAS=1 and RATIO > 0: The P–M interaction curve is only used for determining the nominal moment M_n in ECS Y-direction. Once the moment reaches M_n according to the P–M interaction curve, it will not be adjusted along the interaction curve if axial load varies.
3. For ELAS=1 and RATIO=0: The P–M interaction curve is used for determining the nominal moment M_n in ECS Y-direction. Once the moment reaches M_n according to the P–M interaction curve, it will also be adjusted along the interaction curve if axial load changes.
4. ELAS=1, 2, and 3 are not applicable in ECS Z-direction.
5. When member ductility > μf, it is assumed that failure occurs and moment is reduced to zero.
6. Currently damage index is not calculated.

   d. TYPE='IA_BILN'. Material data for bilinear model (MAT10)

'IA_BILN'	ELAS	SP	E	TI	TMP	μf	βDI	PRMAX
ICHOICE1 (if ELAS=4 or 6)								
A0 A1 A2 A3 (if ICHOICE1=0 and ELAS=4 or 6)								
M (if ICHOICE1=1 and ELAS=4 or 6)								
X1 Y1								
X2 Y2								
... ... (if ICHOICE1=1 and ELAS=4 or 6)								
Xm Ym								
ICHOICE2 (if ELAS=5 or 6)								
B0 B1 B2 B3 (if ICHOICE2=0 and ELAS=5 or 6)								
M (if ICHOICE2=1 and ELAS=5 or 6)								
X1 Y1 FMODE1								
X2 Y2 FMODE2								
... ... (if ICHOICE2=1 and ELAS=5 or 6)								
Xm Ym FMODEm								
ICHOICE3 (use ICHOICE3 if ELAS ≠ 0) (use ICHOICE3 for concrete column only)								
VCI VCF VS ALFA (if ICHOICE3=1)								

ELAS   Index for elastic or nonlinear analysis:
    0   Elastic material property.
    1   For bilinear material property.
    2   Dummy.
    3   Dummy.
    4   Nominal moment, M_n, is adjusted based on interaction axial load–moment (P–M) curve. ELAS=4 is not applicable to nominal moment in ECS Z-direction.

	5 P–PRC is considered, but P–M interaction is not considered.
	6 Both P–M and P–PRC interactions are considered.
SP	Post-yield hardening ratio.
E	Elastic modulus.
TI	Section area, torsional rigidity, or moment of inertia.
TMP	Axial yield load, torsional yield load, or plastic moment.
μf	Failure ductility (if μf< 1, failure ductility is not checked).
βDI	Parameter for damage index (not available currently).
PRMAX	Plastic rotation capacity, θ_p, of a member (if PRMAX<0, member plastic rotation capacity is not checked).
ICHOICE1	0: A0 – A3 are input by user.
	1. The program will calculate A0 – A3 based on user input interaction data points.
A0 – A3	If ELAS=4 or 6, coefficients used in the nominal moment–axial load interaction curve are $M = A0 + A1 * P + A2 * P^2 + A3 * P^3$ in which M and P are nominal moment and axial loads, respectively. If A2=A3=0, the maximum moment along the interaction curve is TMP. This is mainly for steel member with linear interaction curve.
M	If ICHOICE1 = 1, total number of $P-M$ data points input by user.
$X_i \, Y_i$	The ith data point values (X_i represents the ith axial load, and Y_i represents the ith nominal moment)
ICHOICE2	0: B0 – B3 are input by user.
	1: The program will calculate B0 – B3 based on user input interaction data points.
B0 – B3	If ELAS=5 or 6, input coefficients used for the axial–plastic rotation capacity interaction (P–PRC) curve are $PRC = B0 + B1 * P + B2 * P^2 + B3 * P^3$ in which PRC and P are plastic rotation capacity and axial load, respectively.
M	If ICHOICE2=1, total number of $P-PRC$ data points input by user.
$X_i \, Y_i$	The ith data point values (X_i represents the ith axial load, and Y_i represents the ith plastic rotation capacity).
FMODEi	Failure mode corresponding to ($X_i \, Y_i$). Character variable, enclosed in single quotes.
	FMODEi could be one of the following:
	'CONCRETE': Concrete compression failure.
	'FRACTURE': Longitudinal steel tensile fracture.
	'BUCKLING': Longitudinal steel buckling.
	'FATIGUE': Longitudinal steel low-cycle fatigue.
	'SPLICE': Longitudinal steel lap-splice failure.
ICHOICE3	0: Element shear capacity is not considered.
	1: Element shear capacity is considered.
VCI	Initial concrete shear capacity.
VCF	Concrete shear capacity after ductility $\mu \geq 15$.
VS	Shear strength due to transverse rebars.

ALFA   Angle for calculating shear strength due to compression axial load (in
              degree).
Notes

1. For ELAS = 4 and SP > 0: The P–M interaction curve is only used for determining the nominal moment, M_n, in ECS Y-direction. Once the moment
   reaches M_n according to the P–M interaction curve, it will not be adjusted
   along the interaction curve if axial load varies.
2. For ELAS = 4 and SP = 0: The P–M interaction curve is used for determining the nominal moment M_n in ECS Y-direction. Once the moment reaches
   M_n according to the P–M interaction curve, it will also be adjusted along the
   interaction curve if axial load changes.
3. When member ductility > μf, it is assumed that failure occurs and the
   moment is reduced to zero.
4. Currently damage index is not calculated.
5. ELAS = 4, 5, and 6 are not applicable in ECS Z-direction.

e. TYPE = 'STABILITY1'. Material data are mainly for finite-segment steel
   member (MAT12).

'STABILITY1' NSEG YS EM LIBN HH UU WW ZZ INEB INEH ST
IREV1 IREV2 IREV3 IREV4 IECOP SMALL RATIX0 RATIY0 TOTA
IAUTO IMATER RATIO3 IR G QRNEE ISTIF IELAS PCMAX

NSEG   Number of segments considered. Maximum number of NSEG is 32.
YS   Material yield stress (force/length²).
EM   Elastic modulus.
LIBN   Cross section library number (see Figures 6.1 through 6.6):
              1   Box section.
              2   Tube (one layer) section.
              3   Rectangular section.
              4   Wide-flange section.
              5   Not available.
              6   Tube (two layers) section.
              7   Equal leg angle section.
HH   Height for LIBN = 1, 3, 4.
              Radius of tube for LIBN = 2, 6.
              Angle leg length for LIBN = 7.
UU   Width for LIBN = 1, 3, 4.
              Dummy variable for LIBN = 2, 6.
              Dummy variable for LIBN = 7.
WW U_0 direction eccentricity from section reference coordinate origin to
              applied load location (see Figure 6.6, for example).
ZZ V_0 direction eccentricity from section reference coordinate origin to
              applied load location (see Figure 6.6, for example).

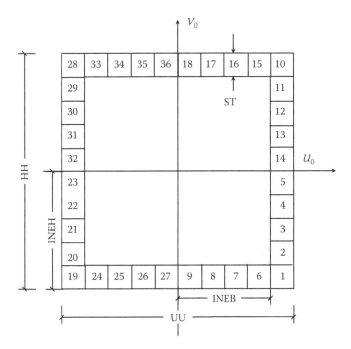

FIGURE 6.1   Box section, INEB = 4 and INEH = 5 per quarter.

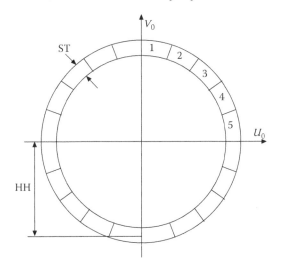

FIGURE 6.2   Tube section (one layer), INEB = 5 per quarter.

INEB   Total number of section elements in half width for LIBN = 1, 3.
           Web thickness for LIBN = 4.
           Total number of section elements in one quarter of a circle for
           LIBN = 2, 6.
           Dummy variable for LIBN = 7.

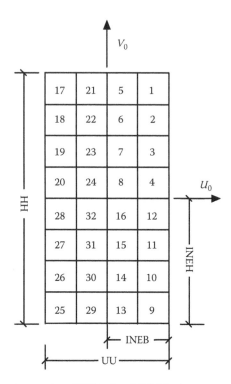

FIGURE 6.3   Rectangular section, INEB = 2 and INEH = 4.

INEH	Total number of section elements in half height for LIBN = 1, 3.
	Dummy variable for LIBN = 2, 4, 6, and 7.
ST	Thickness for LIBN = 1, 2, 6, and 7; flange thickness for LIBN = 4 (length).
IREV1	IREV1 = 0 for LIBN = 1.
	Dummy variable for LIBN = 2, 3, and 6.
	Number of rows in flange's U_0 direction for LIBN = 4.
	Number of columns in segment U_0 direction for LIBN = 7.
IREV2	Dummy variable for LIBN = 1, 2, 3, and 6.
	Number of columns in flange's U_0 direction for LIBN = 4.
	Number of rows in segment U_0 direction for LIBN = 7.
IREV3	Dummy variable for LIBN = 1, 2, 3, and 6.
	Number of columns in web's V_0 direction for LIBN = 4.
	Number of rows in segment V_0 direction for LIBN = 7.
IREV4	Dummy variable for LIBN = 1, 2, 3, and 6.
	Number of rows in web's V_0 direction for LIBN = 4.
	Number of columns in segment V direction for LIBN = 7.
IECOP	0 Eccentricity is not considered.
	1 Eccentricity is considered.
	Note: If IECOP = 0, WW and ZZ are ignored by program.

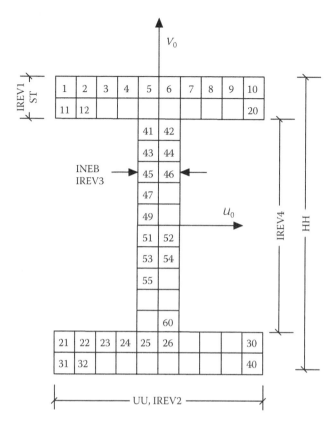

FIGURE 6.4   Wide-flange section, IREV1 = 2, IREV2 = 10, IREV3 = 2, and IREV4 = 10.

SMALL   Length of segments at two ends and center right and left sides. The length of a segment other than end segments or center right and left side segments is equal to member total length divided by (NSEG-4). If SMALL ≤ 0, the length of every segment = member total length divided by NSEG.

RATIX0   Initial imperfection ratio in element coordinate U_0 direction.

RATIY0   Initial imperfection ratio in element coordinate V_0 direction.

TOTA   Section gross area (length^2).

IAUTO   1   Element stiffness parameter, SP, is calculated (see Appendix D).
        0   Element SP is not calculated.

IMATER   0   For bilinear stress–strain model.
         2   For Ramberg–Osgood stress–strain model.
         3   For elastic case.

RATIO3   Finite-segment element strain-hardening ratio for IMATER = 0 or 3; $\bar{b}$ for IMATER = 2 (see Figure 3.7).

IR   Parameter, R, for Ramberg–Osgood stress–strain model.

G   Shear modulus.

QRNEE   Dummy variable.

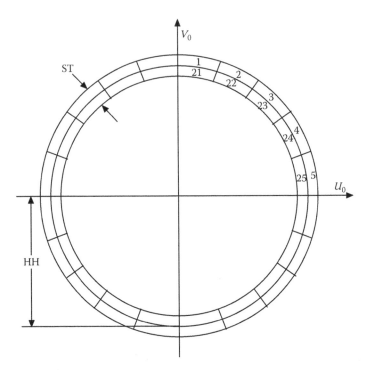

FIGURE 6.5   Tube section (two layers), INEB = 5 for each layer per quarter.

ISTIF   Segment stiffness formulation index:
            0   Exact approach (Chen and Atsuta, 1977).
            1   Approximated approach (see Appendix B).
IELAS   IELAS = 0: use elastic material properties.
            IELAS = 1: use nonlinear material properties with iterations for axial
            load (see Appendix C).
            IELAS = 2: use nonlinear material properties without iterations for axial
            load (see Appendix C).
PCMAX   Maximum allowable plastic curvature defined by user. If $PCMAX \leq 0$,
            the maximum plastic curvature is defined by INSTRUCT as five times
            of the curvature at nominal moment, M_y (see Figure 6.7).

Notes

1. µf and βDI are not considered.
2. When IAUTO = 1, element SP is calculated by the program. If an element's
   $SP \leq 0$, it is in the unstable condition (for example, element buckled), and
   the program output will show a message that the element is in the unstable
   condition.
3. PCMAX is the curvature between maximum allowable curvature and the
   curvature corresponding to nominal moment (i.e., yield moment), M_y. M_y is
   defined as the moment at which compression yield stresses occurred in a
   steel cross section (i.e., $M_n = M_y$).

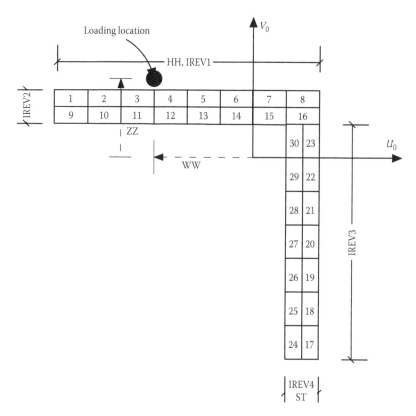

FIGURE 6.6 Equal-leg angle section, IREV1 = 8, IREV2 = 2, IREV3 = 7, and IREV4 = 2.

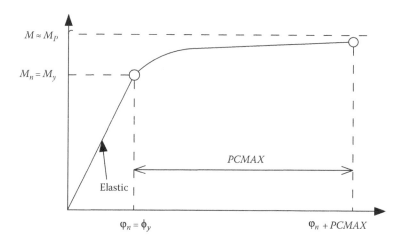

FIGURE 6.7 Moment–curvature curve for steel member.

f. TYPE = 'R/CONCRETE1'. Material data for finite-segment R/C concrete
member (MAT17)

'R/CONCRETE1' NSEG YS EM LIBN HH UU WW ZZ INEB INEH ST IREV1 IREV2 IREV3 IREV4 IECOP SMALL RATIX0 RATIY0 TOTA IAUTO IMATER RATIO3 IR G QRNEE ISTIF NLBAR SMAT(31) NLBARB SMAT(33) NLBARS SMAT(35) YSHOOP IELAS PCMAX ECU ESU RHOOPDIA RHOOPSPA STRPERIOD SPLICELT

NSEG	Number of segments considered. Maximum number of NSEG is 32.
YS	Steel rebar yield stress (psi).
EM	Elastic modulus of steel rebar (psi).
LIBN	Cross section library number (see Figures 6.8 through 6.10):
	1. Circular concrete section with steel ring.
	2. Circular concrete section with steel rebars.
	3. Rectangular concrete section.
HH	Width for LIBN = 3 along V_0 (i.e., Z_e) direction (in.).
	Diameter of column for LIBN = 1 and 2 (in.).
UU	Width for LIBN = 3 along U_0 (i.e., Y_e) direction (in.).
	Concrete cover (from surface of column to the edge of longitudinal rebar) for LIBN = 1 and 2 (in.).
WW	U_0 direction eccentricity from section reference coordinate origin to applied load location (in.).
ZZ	V_0 direction eccentricity from section reference coordinate origin to applied load location (in.).
INEB	Total number of section elements in half width for LIBN = 3 in U_0 direction (exclude cover) (in.).
	Total number of section elements in one quarter of circle for each layer LIBN = 1, 2.
INEH	Total number of section elements in half height for LIBN = 3 in V_0 direction (exclude cover) (in.).
	Total number of layers including cover layer for LIBN = 1 2.
ST	Concrete f_c' at 28 days (psi).
IREV1	Cross-sectional diameter (in inch) of hoop or spiral for LIBN = 1, 2.
	Volumetric ratio of transverse reinforcement in the U_0 direction (i.e., ρ_{U0}) for LIBN = 3.
IREV2	Spacing of hoop (in inch) or spiral for LIBN = 1, 2. Volumetric ratio of transverse reinforcement in the V_0 direction (i.e., ρ_{V0}) for LIBN = 3.
IREV3	Effective confinement coefficient, K_e. $K_e = 0.95$ for circular section; 0.75 for rectangular section; and 0.6 for rectangular wall section.
IREV4	Steel layer thickness for LIBN = 1 (in.).
	Diameter of longitudinal rebar for LIBN = 2 (in.).
	Concrete cover thickness for LIBN = 3 (in.).
IECOP	0: Eccentricity is not considered.
	1: Eccentricity is considered.

	Note: If IECOP=0, WW and ZZ are ignored by program.
SMALL	Length of segments at two ends and center right and left sides (in.). The length of a segment other than end segments or center right and left side segments is equal to member total length divided by (NSEG-4). If SMALL≤0, the length of every segment=member total length divided by NSEG.
RATIX0	Initial imperfection ratio in element coordinate U_0 direction (in.).
RATIY0	Initial imperfection ratio in element coordinate V_0 direction (in.).
TOTA	Section gross area (in.^{2}).
IAUTO	1 Element SP is calculated (see Appendix D).
	0 Element SP is not calculated.
IMATER	1: only for LIBN−1 (bilinear steel model and Mander's concrete model).
	2: only for LIBN=2 (bilinear steel model and Mander's concrete model).
	3: only for LIBN=3 (bilinear steel model and Mander's concrete model).
RATIO3	Finite-segment element steel strain-hardening ratio.
IR	0: concrete tension strength is not considered.
	1: concrete tension strength is considered.
G	Concrete shear modulus.
QRNEE	Dummy variable.
ISTIF	Segment stiffness formulation index:
	0: Exact approach.
	1: Approximated approach.
NLBAR	Total numbers of top bars for LIBN=3.
	Dummy variable for LIBN=1.
	Total numbers of long bars for LIBN=2.
SMAT(31)	Top bar diameter for LIBN=3 (in.). SMAT(31) is used for the buckling of longitudinal rebar analysis.
	Dummy variable for LIBN=1, 2.
NLBARB	Total numbers of bottom bars for LIBN=3.
	Dummy variable for LIBN=1, 2.
SMAT(33)	Bottom bar diameter for LIBN=3 (in.).
	Dummy variable for LIBN=1, 2.
NLBARS	Total number of side bar (each side) for LIBN=3.
	Dummy variable for LIBN=1, 2.
SMAT(35)	Side bar diameter for LIBN=3 (in.).
	Dummy variable for LIBN=1, 2.
YSHOOP	Yield stress of hoop (psi).
IELAS	IELAS=0: use elastic material properties.
	IELAS=1: use nonlinear material properties with iterations for axial load (see Appendix C).
	IELAS=2: use nonlinear material properties without iterations for axial load (see Appendix C).

PCMAX   Maximum allowable plastic curvature. If $PCMAX \leq 0$, maximum plastic curvature is determined when the concrete strain reaches its ultimate compression strain, ε_{cu}.

ECU   User-defined ultimate concrete compression strain, ε_{cu}. If ECU ≤ 0, the ε_{cu} is determined by the program per Equation 3.24. This option is useful for the seismic retrofit of column using either steel jacket or composite material jacket. For example, the ultimate concrete compression strain of steel-jacketed column can be input by user and expressed as $\varepsilon_{cu} = 0.004 + 1.4\rho_{sj}f_{yj}\varepsilon_{sm}/f'_{cc}$ where $\rho_{sj} = 4t_j/D =$ volumetric ratio of confining steel; $t_j =$ jacket thickness; $D =$ diameter of steel jacket; $\varepsilon_{sm} =$ jacket strain at maximum stress $= 0.15$ for grade 40 and 0.12 for grade 60; and $f_{yj} =$ yield stress of jacket. For the composite jacket, the ultimate concrete compression strain can be input by user and be expressed as $\varepsilon_{cu} = 0.004 + 2.5\rho_{sj}f_{uj}\varepsilon_{uj}/f'_{cc}$ in which $\rho_{sj} = 4t_j/D$ for circular column and $\rho_{sj} = 2t_j[(b+h)/bh]$ for rectangular column, where $\rho_{sj} = \rho_X + \rho_Y$; f_{uj} and ε_{uj} are the ultimate stress and strain of the jacket material, and b and h are the section dimensions of the column.

ESU   User-defined ultimate steel strain of longitudinal rebar. If ESU ≤ 0, the ultimate steel strain of 0.09 is used by the program per Chapter 3.

RHOOPDIA   Hoop diameter for rectangular section in LIBN $= 3$.
RHOOPSPA   Hoop spacing for rectangular section in LIBN $= 3$.
STRPERIOD   Structural period. If STRPERIOD > 0 and column concrete is confined, low-cycle fatigue is checked. If STRPERIOD ≤ 0, low-cycle fatigue is not checked.
SPLICELT   Lap-splice length in the plastic hinge region. If SPLICELT > 0, splice failure is checked. If SPLICELT ≤ 0, splice failure is not checked.

Notes

1. μf and βDI are not considered.
2. The units used for R/CONCRETE1 shall be in "pound" and "inch" because the concrete confined model in the program is based on these units.
3. When IAUTO $= 1$, element SP is calculated by the program. If an element's $SP \leq 0$, it is in the unstable condition and the program output will show a message that the element is in the unstable condition.

g. TYPE $=$ 'MOMCURVA1'. Material data for finite-segment moment–curvature element (MAT18).

```
'MOMCURVA1' NSEG YS EM LIBN HH UU WW ZZ PCMAX
CURNM ST IREV1 IREV2 IREV3 IREV4 IECOP SMALL RATIX0
RATIY0 TOTA IAUTO IMATER RATIO3 IR G QRNEE ISTIF
IELAS
```

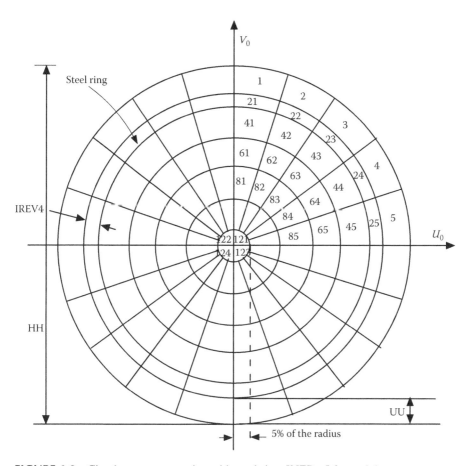

FIGURE 6.8 Circular concrete section with steel ring, INEB = 5 for each layer per quarter and INEH = 6.

AXIAL(1) AXIAL(2) … AXIAL(IREV2)		
For AXIAL(1): MOM(1)	MOM(2) … MOM(IREV1)	
For AXIAL(1): CUR(1)	CUR(2) … CUR(IREV1)	
For AXIAL(2): MOM(1)	MOM(2) … MOM(IREV1)	
For AXIAL(2): CUR(1)	CUR(2) … CUR(IREV1)	
…		
For AXIAL(IREV2): MOM(1)	MOM(2) … MOM(IREV1)	
For AXIAL(IREV2): CUR(1)	CUR(2) … CUR(IREV1)	

NSEG   Number of segments considered. Maximum number of NSEG is 32.
YS   Concrete shear modulus, G (force/length^2).
EM   Concrete elastic modulus.
LIBN   1: general cross section.

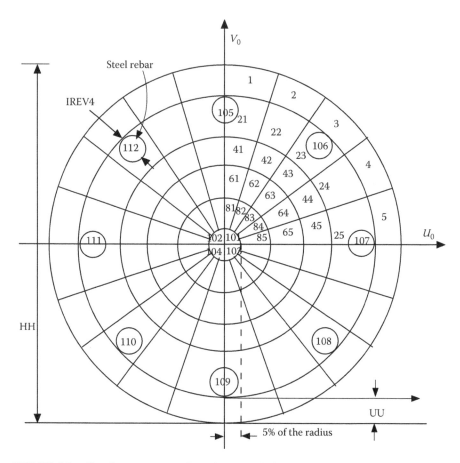

FIGURE 6.9 Circular concrete section with steel reinforcement, INEB = 5 for each layer per quarter, INEH = 5, and NLBAR = 8.

HH	Cross-sectional elastic moment of inertia, I_V, in the segment's local coordinate V_0 direction.
UU	Cross-sectional polar moment of inertia, J.
WW	U_0 direction eccentricity from section reference coordinate origin to applied load location.
ZZ	V_0 direction eccentricity from section reference coordinate origin to applied load location.
PCMAX	Maximum allowable plastic curvature. If $PCMAX \leq 0$, plastic curvature is unlimited. This is only for IREV2 = 1 (i.e., only one set of moment–curvature data points is input). If IREV2 > 1, the program ignores PCMAX and treats the last curvature control point, CUR(IREV1), as the maximum curvature capacity.
CURNM	Curvature corresponding to nominal moment, M_n. This is only for IREV2 = 1 (i.e., only one set of moment–curvature data points is input). If IREV2 > 1, the program ignores CURNM.

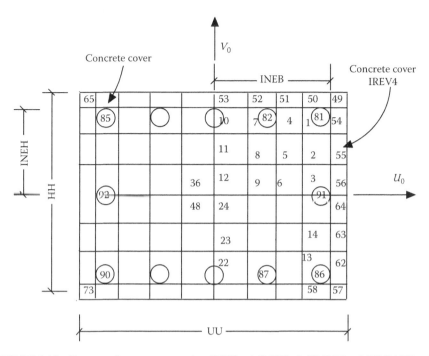

FIGURE 6.10 Rectangular concrete section, INEB = 4, INEH = 3, NLBAR = 5, NLBARB = 5, and NLBARS = 1.

ST   Dummy variable.

IREV1   Total number of control points along the $M-\phi$ backbone curve (exclude origin).

IREV2   Total number of axial load cases considered.

IREV3   0: Moment–curvature relationship is not adjusted for axial load effect. Moment–curvature relationship is based on AXIAL(1).
          1: Moment–curvature relationship is adjusted due to axial load.

IREV4   Dummy variable.

IECOP   0: Eccentricity is not considered.
          1: Eccentricity is considered.
          Note: If IECOP = 0, WW and ZZ are ignored by program.

SMALL   Length of segments at two ends and center right and left sides. The length of a segment other than end segments or center right and left side segments is equal to member total length divided by (NSEG-4). If SMALL ≤ 0, the length of every segment = member total length divided by NSEG.

RATIX0   Initial imperfection ratio in element coordinate U_0 direction.

RATIY0   Initial imperfection ratio in element coordinate V_0 direction.

TOTA   Section gross area (length²).

IAUTO   1   Element SP is calculated (see Appendix D).
          0   Element SP is not calculated.

IMATER	1: Moment–curvature relationship is not influenced by shear or splice failures.
	2: Moment–curvature relationship is influenced by shear or splice failures (currently not available).
RATIO3	Dummy variable.
IR	0: concrete tension strength is not considered (dummy variable).
	1: concrete tension strength is considered (dummy variable).
G	Concrete shear modulus (dummy variable).
QRNEE	Dummy variable.
ISTIF	Segment stiffness formulation index:
	0: Exact approach.
	1: Approximated approach.
IELAS	IELAS = 0: use elastic material properties.
	IELAS = 1: use nonlinear material properties.
AXIAL(n)	Column axial load for load case n, where n = 1 to IREV2. Input axial load values in ascending order. Positive in compression and negative in tension.
MOM(m)	mth moment control point, where m = 1 to IREV1.
CUR(m)	mth curvature control point, where m = 1 to IREV1.

Notes

1. μf and βDI are not considered.
2. The moment–curvature control points (MOM, CUR) are for the calculation of segment's sectional moment of inertia, I_U, in the segment local coordinate U_0 direction.
3. I_V is constant and in elastic.
4. When IAUTO = 1, element SP is calculated by the program. If an element's $SP \leq 0$, it is in the unstable condition, and the program output will show a message that the element is in the unstable condition.

h. TYPE = 'PLATE'. Material data for plate element (MAT16).

'PLATE'	E	NU	THK

E	Young's modulus (force/length²).
NU	Poisson's ratio of plate material.
THK	Thickness of plate (length).

i. TYPE = 'POINT'. Material data for Point Element (MAT20).

'POINT'		
S(1,1)	S(2,2)	S(3,3)
S(4,4)	S(5,5)	S(6,6)
S(1,2)	S(1,3)	S(1,4)
S(1,5)	S(1,6)	S(2,3)

S(2,4)	S(2,5)	S(2,6)
S(3,4)	S(3,5)	S(3,6)
S(4,5)	S(4,6)	S(5,6)

Note

1. When using point element (see Figure 3.21) to simulate pile–soil interaction, in general, the nonzero terms in the stiffness matrix are S(1,1), S(2,2), S(3,3), S(4,4), S(5,5), S(6,6), S(2,6), and S(3,5). Since the stiffness matrix is symmetric, the program will generate S(6,2) and S(5,3).

j. TYPE = 'BRACE'. Material data for bracing element (MAT08).

'BRACE'	E	A	R	YS	SHAPE	FDMAX	βDI

E   Elastic modulus (force/length²).
A   Section area (length²).
R   Radius of gyration (length).
YS   Yielding stress (force/length²).
SHAPE   Member cross section shape.
         1 For box or equal-leg angle section.
         2 For I-shape section.
FDMAX   Failure ductility due to elongation (if FDMAX < 0, failure ductility is not considered).
βDI   Parameter for damage index (dummy variable).

6.1.3 Geometric Stiffness Data

This card is used to determine the type of geometric stiffness used in the analysis. This card is input once.

KGLOAD	KGTYPE	KGFORM	KGCOND

KGLOAD   The type of axial force used to calculate the geometric stiffness.
          0 The geometric stiffness is omitted.
          1 The axial force is equal to the input force, magnified by the ground acceleration in the global Z direction, if applicable.
          2 The internal element force of the previous load step is used to generate the geometric stiffness.
KGTYPE   The type of geometric stiffness formulation.
          0 The geometric stiffness is omitted.
          1 A "lumped parameter" formulation is used for the element geometric stiffness (see Equation 5.48a).
          2 A "consistent parameter" formulation is used for the element geometric stiffness (see Equation 5.48b). If a consistent parameter formulation for an individual element is not available, the lumped

parameter formulation is used. Refer to individual element specifications for applicability.

KGFORM   Form of the geometric stiffness used.
           0   The geometric stiffness is omitted.
           1   The geometric stiffness is subtracted from the structural stiffness.
           2   Separate structural stiffness and geometric stiffness matrices are
              formed.

KGCOND   A logical flag. If KGCOND = .TRUE., the geometric stiffness matrix is
           condensed when applicable. Logical variable.

6.1.4 ELEMENT DATA

These cards are used to define the elements. Elements are numbered by the program
in the order they are input from 1 to NELMT. Element numbers are used to identify
elements in the output. The first card is input once. Second and third cards (if used)
are repeated until (1) NELMT elements are input or (2) a TYPE = 'END' is encountered. The third card follows each second card with a value of IGEN > 1.

NELMT					
TYPE	NAME	VALUE1	VALUE2 ...		IGEN
	ΔVALUE1	ΔVALUE2 ...			

NELMT   Number of elements input.
TYPE   Element type. Valid types are described below. Character variable,
           enclosed in single quotes.
NAME   A user-defined name. Character variable, enclosed in single quotes.
VALUEi   Input required by a given element type. Values for each TYPE are dis-
           cussed below.
IGEN   Number of additional elements to be generated from the element.
ΔVALUEi   Incremental value used to generate subsequent elements.

$$\text{VALUEi}_{\text{generated}} = \text{VALUEi}_{\text{previous}} + \Delta \text{VALUEi}$$

   a. TYPE = '3D-BEAM'. This card is used to define element data for elastic
   prismatic element (ELE01).

'3D-BEAM' NAME MAT JOINTI JOINTJ V1 V2 V3 XS XE PKG IRELT IGEN

NAME   A user-defined name. Character variable, enclosed in single quotes.
MAT   Material number. This number must correspond to the material type
           '3D-BEAM'.
JOINTI   Start joint ID number.
JOINTJ   End joint ID number.
V1   Projection on the GCS X-axis of a vector in the element's local XY-plane.
           This vector defines the orientation of element's local Y-axis.

V2   Projection on the GCS Y-axis of a vector in the element's local XY-plane.

V3   Projection on the GCS Z-axis of a vector in the element's local XY-plane.

XS   Offset distance (i.e., rigid zone) from the start joint to the beginning of the element. Positive in the direction of the element's local X-axis.

XE   Offset distance from the end joint to the end of the element. Negative in the direction of the element's local X-axis.

PKG   Axial load used to calculate the geometric stiffness, if KGLOAD = 1. Positive is compression.

IRELT   A six-digit release code that is used to release the rotational dof at both ends of the element. A nonzero value of the ith digit signifies a released dof.

Digit

1: Release the moment about the element's X-axis at the start joint.
2: Release the moment about the element's Y-axis at the start joint.
3: Release the moment about the element's Z-axis at the start joint.
4: Release the moment about the element's X-axis at the end joint.
5: Release the moment about the element's Y-axis at the end joint.
6: Release the moment about the element's Z-axis at the end joint.

IGEN   Element generation parameter. See discussion under "Element Data."
  Note: Both KGTYPE = 1 and 2 are available.

  b. TYPE = 'IE3DBEAM'. This card is used to define element data for inelastic 3D-Beam element (ELE09).

'IE3DBEAM' NAME MATMYA MATMYB MATMZA MATMZB MATMXA MATFXA JOINTI JOINTJ V1 V2 V3 XS XE PKG
ENDI PTII HCI BCI HBI BBI PRESI ENDJ PTIJ HCJ BCJ HBJ BBJ PRESJ IGEN

NAME   A user-defined name. Character variable, enclosed in single quotes.

MATMYA   Bending hysteresis material number at the start joint in the element's local Y-axis.

MATMYB   Bending hysteresis material number at the end joint in the element's local Y-axis.

MATMZA   Bending hysteresis material number at the start joint in the element's local Z-axis.

MATMZB   Bending hysteresis material number at the end joint in the element's local Z-axis.

MATMXA   Element torsional hysteresis material number.

MATFXA   Element axial hysteresis material number.

JOINTI	Start joint ID number.
JOINTJ	End joint ID number.
V1	Projection on the GCS X-axis of a vector in the element's local XY-plane. This vector defines the orientation of element's local Y-axis.
V2	Projection on the GCS Y-axis of a vector in the element's local XY-plane.
V3	Projection on the GCS Z-axis of a vector in the element's local XY-plane.
XS	Offset distance from the start joint to the beginning of the element. Positive in the direction of the element's local X-axis.
XE	Offset distance from the end joint to the end of the element. Negative in the direction of the element's local X-axis.
PKG	Axial load used to calculate the geometric stiffness, if KGLOAD = 1. Positive is compression.
ENDI	0: Joint shear at joint i is not checked.
	1: Joint shear at joint i is checked.
PTII	Initial principle stress at joint i.
HCI	Column width (or diameter of circular column) (see Figure 4.20).
BCI	Column depth (= 0 for circular column) (see Figure 4.20).
HBI	Beam depth at joint i (see Figure 4.20).
BBI	Beam width at joint i (see Figure 4.20).
PRESI	Prestress force of beam at joint i.
ENDJ	0: Joint shear at joint j is not checked.
	1: Joint shear at joint j is checked.
PTIJ	Initial principle stress at joint j.
HCJ	Column width (or diameter of circular column).
BCJ	Column depth (= 0 for circular column).
HBJ	Beam depth at joint j (see Figure 4.20).
BBJ	Beam width at joint j (see Figure 4.20).
PRESJ	Prestress force of beam at joint j.
IGEN	Element generation parameter. See discussion under "Element Data."

Notes

1. Bending material specified for IE3DBEAM element may consist of IA-BILN, HINGE, or TAKEDA.
2. Bending hysteresis material number at the start joint in the element's local Y-axis should be the same as that at the end joint in the element's Y-axis.
3. Bending hysteresis material number at the start joint in the element's local Z-axis should be the same as that at the end joint in the element's Z-axis.
4. IA-BILN model with ELAS = 0 or 1 is specified for the torsional and axial materials.
5. Axial load–moment interaction (i.e., IA-BILN model with ELAS = 4, 5, and 6; HINGE model with ELAS = 1, 2, and 3) only applies to moments in the ECS Y-direction.

6. Only KGTYPE=1 (lump parameter formulation) is available for this element.

7. The program outputs the yielding condition of element by a six-digit code called "STBFAG."

STBFAG could be:

100000: axial yield load occurred.

010000: torsion yield load occurred.

001000: yield moment occurred at 'START' joint in ECS Y-direction.

000100: yield moment occurred at 'END' joint in ECS Y-direction.

000010: yield moment occurred at 'START' joint in ECS Z-direction.

000001: yield moment occurred at 'END' joint in ECS Z direction.

In addition, the program outputs the failure mode of the element by a variable called "FLP."

FLP could be

0: Elastic condition.

1: Axial yield occurs.

2: Torsion yield occurs.

3: Norminal moment M_n occurs at End A Y-direction.

4: Norminal moment M_n occurs at End B Y-direction.

5: Norminal moment M_n occurs at End A Z-direction.

6: Norminal moment M_n occurs at End B Z-direction.

7: Plastic rotation capacity reaches at End A Y-direction due to concrete compression failure.

8: Plastic rotation capacity reaches at End A Y-direction due to longitudinal steel reinforcement tensile fracture.

9: Plastic rotation capacity reaches at End A Y-direction due to longitudinal steel reinforcement buckling.

10: Plastic rotation capacity reaches at End A Y-direction due to longitudinal steel reinforcement low-cycle fatigue.

11: Plastic rotation capacity reaches at End A Y-direction due to longitudinal steel reinforcement lap-splice failure.

12: Plastic rotation capacity reaches at End A Y-direction due to shear failure.

13: Plastic rotation capacity reaches at End A Y-direction due to possible joint shear crack.

14: Plastic rotation capacity reaches at End B Y-direction due to concrete compression failure.

15: Plastic rotation capacity reaches at End B Y-direction due to longitudinal steel reinforcement tensile fracture.

16: Plastic rotation capacity reaches at End B Y-direction due to longitudinal steel reinforcement buckling.

17: Plastic rotation capacity reaches at End B Y-direction due to longitudinal steel reinforcement low-cycle fatigue.

18: Plastic rotation capacity reaches at End B Y-direction due to longitudinal steel reinforcement lap-splice failure.

19: Plastic rotation capacity reaches at End B Y-direction due to shear failure.

20: Plastic rotation capacity reaches at End B Y-direction due to possible joint shear crack.

c. TYPE = 'STABILITY'. This card is used to define element data for Finite Segment–Finite String element (ELE12).

| 'STABILITY' NAME MAT JOINTI JOINTJ V1 V2 V3 XS XE IGEN |

NAME   A user-defined name. Character variable, enclosed in single quotes.

MAT   Material number.

JOINTI Start joint ID number.

JOINTJ End joint ID number.

V1   The projection on the GCS X-axis of a vector in the member XY-plane. This vector defines the orientation of the element's local Y-axis.

V2   The projection on the GCS Y-axis of a vector in the member XY-plane.

V3   The projection on the GCS Z-axis of a vector in the member XY-plane.

XS   The offset distance from the start joint to the beginning of the element. Positive in the direction of the element's local X-axis (length).

XE   The offset distance from the end joint to the end of the element. Negative in the direction of the element's local X-axis (length).

IGEN   Element generation parameter. See discussion under "Element Data."

Notes

1. The STABILITY1, R/CONCRETE1, or MOMCURVA1 material model is specified for this element.

2. Each STABILITY element shall have its own assigned material number. One material number can only be assigned to one STABILITY element. Material numbers are defined in the Material data block.

3. For 'STABILITY' element using R/CONCRETE1 or MOMCURVA1 material model, the program outputs the yielding condition of element by a variable called "FLP."

FLP for 'R/CONCRETE1' material could be

0: Concrete extreme fiber strain is less than 0.004 and steel rebar stress is below yield stress.

1: The first tensile rebar reaches yield.

2: After yielding of the first rebar and before reaching nominal moment, M_n.

3: Concrete extreme fiber strain reaches 0.004, at which nominal moment, M_n, is defined.

3.5: Tension rebar's strain reaches 0.015, at which nominal moment, M_n, is defined.

4: Concrete extreme fiber strain is greater than 0.004 or tension rebar is greater than 0.015.

5: Concrete extreme fiber strain reaches its ultimate compression strain, ε_{cu}.

5.1: After 5.

6: Longitudinal rebar buckled.

6.1: After 6.

7: Longitudinal steel reaches its ultimate strain of 0.09.

8: Concrete average compression strain reaches its ultimate compression strain, ε_{cu}.

10: Plastic curvature reaches maximum allowable curvature, PCMAX.

10.1: Plastic curvature is beyond PCMAX.

11: Longitudinal rebar lap-splice failure occurred.

12: Longitudinal rebar low-cycle fatigue occurred.

FLP for 'MOMCURVA1' material could be

3: Curvature reaches nominal curvature, CURNM.

4: Curvature is between CURNM and PCMAX.

10: Plastic curvature reaches maximum allowable curvature, PCMAX.

4. For 'STABILITY' element using STABILITY1 material model, the program outputs the yielding condition of element by a variable called "FLP." FLP could be

0: Steel extreme fiber strain is less than yield strain.

1: The compression yield stress occurred, the yield moment, M_y, is defined.

2: After M_y.

5: Steel member global buckling occurred.

5.1: After 5.0.

10: Plastic curvature reaches the maximum plastic curvature, PCMAX.

5. The program outputs the stability condition of 'STABILITY' element by a variable SP (stiffness parameter). If SP ≤ 0, the element is unstable. If SP > 0, the element is stable.

d. TYPE = 'SPRING'. This card is used to define element data for the one-dimensional spring (ELE02).

'SPRING' NAME MAT JOINTI JOINTJ KTYPE XLEN V1 V2 V3 XS XE IGEN

NAME   A user-defined name. Character variable, enclosed in single quotes.

MAT   Material number.

JOINTI   Start joint ID number.

JOINTJ   End joint ID number.

KTYPE   Type of spring.

      1   Axial spring.

      2   Shear spring in the element's local Y-axis.

3	Shear spring in the element's local Z-axis.
4	Torsional spring.
5	Rotational spring about the element's local Y-axis.
6	Rotational spring about the element's local Z-axis.
XLEN	Length of the spring used to calculate the stiffness.
V1	Projection on the GCS X-axis of a vector in the spring's local XY-plane. This vector defines the orientation of element's local Y-axis.
V2	Projection on the GCS Y-axis of a vector in the spring's local XY-plane.
V3	Projection on the GCS Z-axis of a vector in the spring's local XY-plane.
XS	Offset distance from the start joint to the beginning of the spring. Positive in the direction of the element's local X-axis.
XE	Offset distance from the end joint to the end of the spring. Negative in the direction of the element's local X-axis.
IGEN	Element generation parameter. See discussion under "Element Data."

Notes

1. Material specified for the spring element may consist of any of the following: 3D-BEAM, BILINEAR, TAKEDA, or GAP.
2. Spring uses the axial stiffness from the 3D-BEAM material.
3. If the distance between the start and end joints is zero, the spring is oriented such that the ECS is parallel to the start joint's JCS.
4. Geometric stiffness matrix is not calculated for SPRING element.

e. TYPE = 'PLATE'. This card is used to define element data for the plate element (ELE16).

'PLATE'	NAME	MAT	J1	J2	J3	J4	IGEN

NAME	A user-defined name. Character variable, enclosed in single quotes.
MAT	Material number. This number corresponds to the material type 'PLATE'.
J1	Joint 1 ID number (see Figure 3.20).
J2	Joint 2 ID number.
J3	Joint 3 ID number.
J4	Joint 4 ID number.
IGEN	Element generation parameter. See discussion under "Element Data."

Example

	Note
5	(1)
'3D-BEAM' 'A1' 1 10 20 0. 0. 1. 0. 0. 100000 2	(2)
0 10 10 0. 0. 0. 0. 0. 000000	(3)
'SPRING' 's1' 2 10 11 1 0. 0. 0. 1. 0. 0. 0	(4)
'PLATE' 'p1' 3 20 10 1 2 0	(5)

(1) Five elements are to be input.

(2) A 3D-BEAM element is input between joint 10 and 20 with moment released about the element's X_e axis at the start joint.

(3) Two more 3D-BEAM elements are generated from the first element.

(4) An axial spring is input between joints 10 and 11.

(5) A plate element is input, which is connected at joints 20, 10, 1, and 2.

(6) Geometric stiffness matrix is not calculated for PLATE element.

  f. TYPE = 'POINT'. This card is used to define element data for the point element (ELE20).

'POINT' NAME MAT JOINT VY1 VY2 VY3 VX1 VX2 VX3 IGEN

NAME   A user-defined name. Character variable, enclosed in single quotes.

MAT   Material number. This number corresponds to the material type 'POINT'.

JOINT   Joint that element is located.

VY1   Projection on the GCS X-axis of the element's Y-axis.

VY2   Projection on the GCS Y-axis of the element's Y-axis.

VY3   Projection on the GCS Z-axis of the element's Y-axis.

VX1   Projection on the GCS X-axis of the element's X-axis.

VX2   Projection on the GCS Y-axis of the element's X-axis.

VX3   Projection on the GCS Z-axis of the element's X-axis.

IGEN   Element generation parameter. See discussion under "Element Data."

Note: Geometric stiffness matrix is not calculated for this element.

  g. TYPE = 'BRACE'. This card is used to define the element data for the bracing member (ELE08).

'BRACE' NAME MAT JOINTI JOINTJ KTYPE SLK V1 V2 V3 XS XE IGEN

NAME   A user-defined name. Character variable, enclosed in single quotes.

MAT   Material number. This number corresponds to the material type 'BRACE'.

JOINTI Start joint ID number.

JOINTJ End joint ID number.

KTYPE   Equal to 1.

SLK   K factor for slenderness ratio.

V1   The projection on the GCS X-axis of a vector in the member XY-plane. This vector defines the orientation of the element's local Y-axis.

V2   The projection on the GCS Y-axis of a vector in the member XY-plane.

V3   The projection on the GCS Z-axis of a vector in the member XY-plane.

XS   The offset distance from the start joint to the beginning of the brace element. Positive in the direction of the element's local X-axis (length).

XE   The offset distance from the end joint to the end of the brace element. Negative in the direction of the element's local X-axis (length).

IGEN   Element generation parameter. See discussion under "Element Data."

Notes

1. The stiffness of bracing member is based on the 'BRACE' material only.
2. The cross section of the bracing member could be box, angle, or wide flange section.

6.1.5 MASS

These cards are used to input lumped masses at the joints. The first card is input once. The second card is repeated INMASS times or until a joint ID ≤ 0 is encountered. If INMASS is less than one, or FMASS is zero, omit the second card.

INMASS FMASS MCOND											
ID PX PY PZ RXX RYY RZZ RXY RXZ RYZ IGEN Δ ID											

INMASS   The number of mass cards to be read.

FMASS   Mass flag.
 0   The mass matrix is omitted.
 1   The mass matrix due to concentrated joint masses is formed.

MCOND   A logical flag. If MCOND = .TRUE., the mass matrix is condensed when applicable.
 Logical variable.

ID   Identification number of the joint.

PX   Translational mass in the joint's JCS X-direction (mass).

PY   Translational mass in the joint's JCS Y-direction (mass).

PZ   Translational mass in the joint's JCS Z-direction (mass).

RXX   Rotational mass moment of inertia about the joint's JCS X-axis (mass * length^2).

RYY   Rotational mass moment of inertia about the joint's JCS Y-axis (mass * length^2).

RZZ   Rotational mass moment of inertia about the joint's JCS Z-axis (mass * length^2).

RXY   Rotational mass product of inertia about the joint's JCS XY-axis (mass * length^2).

RXZ   Rotational mass product of inertia about the joint's JCS XZ-axis (mass * length^2).

RYZ   Rotational mass product of inertia about the joint's JCS YZ-axis (mass * length^2).

IGEN   Number of joints with identical mass, to be generated.

Δ ID   Increment of joint ID number for generated values.

Example: INMASS = 1, FMASS = 1,
20 8. 8. 8. 0 0 0 0 0 0 0 0

Joint 20 has a translational mass of 8.0 in the X-, Y-, and Z-directions.

6.1.6 DAMPING

This card is used to input proportional damping data.

ALPHA	BETA

ALPHA   Proportional damping coefficient for mass.
BETA   Proportional damping coefficient for stiffness.

6.2 SOL01—ELASTIC STATIC SOLUTION

The following solution is an elastic solution with multiple load cases. The following cards are input once.

'SOL01'	
TITLE	
NLOAD	MAXELD

SOL01   Signifies solution #1. Character variable, enclosed in single quotes.
TITLE   User input title, 80 characters maximum. Character variable, enclosed
               in single quotes.
NLOAD   Number of load cases.
MAXELD   Maximum number of element loads.

Notes

1. Condensation increases the band width of the stiffness matrix for SOL01
   without any other benefits. Condensation is not recommended for SOL01.
2. A separate geometric stiffness (KGFORM = 2) is not used by SOL01. The
   geometric stiffness may be included by using KGFORM = 1.
3. If the geometric stiffness is included, the element axial loads must be input,
   KGLOAD = 1.

6.2.1 JOINT LOADS

These cards are used to apply loads to joints. Loads applied to restrained joints are considered as displacements, support settlement, or displacement-control solutions. Loads applied to constrained joints are transferred to their "master" joints. If the "master" joint is restrained, loads transferred to restrained dof are considered as displacements. Joint loads are additive; applying two loads to the same joint results in the sum of the joint loads being considered. The following card is repeated until the value of DIR is 'END'.

LOAD	ID	IGEN	Δ ID	DIR	VALUE

LOAD   Number of the load case that the load is applied to.
ID   Joint ID number that the load is applied to.
IGEN   Number of identical joint loads to be generated.
Δ ID   Increment of joint ID number for generated values.
DIR   Direction of load in the JCS. Valid directions and the units of VALUE are given below. Character variable, enclosed in single quotes.

'FX'   Applied force in the joint's JCS X-direction (force).
'FY'   Applied force in the joint's JCS Y-direction (force).
'FZ'   Applied force in the joint's JCS Z-direction (force).
'MX'   Applied moment about the joint's JCS X-axis (force * length).
'MY'   Applied moment about the joint's JCS Y-axis (force * length).
'MZ'   Applied moment about the joint's JCS Z-axis (force * length).
'END'   Terminate the input of element loads.

VALUE   Magnitude of the applied load.

Example

1 2 2 1 'FZ' –3.00	(1)
2 7 0 0 'MX' 33.5	(2)
0 0 0 0 'END' 0	(3)

(1) Joints 2, 3, and 4 have an applied force of –3.00 in the Z-direction, load case 1.
(2) Joints 7 has an applied moment of 33.50 in the X-direction, load case 2.
(3) Joint loading input is terminated.

6.2.2 ELEMENT LOADS

These cards are used to apply element loads to the '3D-BEAM', 'IE3DBEAM', or 'PLATE' element. The loads are applied to the portion of the beam between points A and B of Figure 5.4 in the ECS. The loads are transferred by the program to the start and end joints. The following card is only included if MAXELD>0. The card is repeated MAXELD times or until the value of TYPE is 'END'.

LOAD	IELE	IGEN	Δ IELE	TYPE	DIR	VALUE1, VALUE2, ...

LOAD   Load case number.
IELE   Element number.
IGEN   Number of similar element loads to be generated.
Δ IELE   Increment of element number for generated values.
TYPE   Type of load. Valid types are described in detail below. Character variable, enclosed in single quotes.

'CONC'   Concentrated load applied in direction DIR. VALUE1 is the magnitude of the load (force or force * length). VALUE2 is the ratio of the distance to the load, divided by the flexible length of the member. The distance is measured from the beginning of the flexible length at the member's start end. VALUE2 is between 0 and 1. Only two values are input. 'CONC' is not available for 'PLATE' element.

'UNIF' Uniform load applied in direction DIR. VALUE1 is the magnitude of the load. Only one value is input (force/length for '3D-BEAM' and 'IE3DBEAM' elements and force/length^2 for 'PLATE' element).

'FEM' Input the fixed end forces on the ends of the member. DIR is not used and may be set to any value. VALUE1 to VALUE12 are required.

VALUE1 Fixed end axial force, at point A of Figure 5.4, in the ECS X-direction (force).

VALUE2 Fixed end shear, at point A of Figure 5.4, in the ECS Y-direction (force).

VALUE3 Fixed end shear, at point A of Figure 5.4, in the ECS Z-direction (force).

VALUE4 Fixed end torsion, at point A of Figure 5.4, about the ECS X-axis (force * length).

VALUE5 Fixed end moment, at point A of Figure 5.4, about the ECS Y-axis (force * length).

VALUE6 Fixed end moment, at point A of Figure 5.4, about the ECS Z-axis (force * length).

VALUE7 Fixed end axial force, at point B of Figure 5.4, in the ECS X-direction (force).

VALUE8 Fixed end shear, at point B of Figure 5.4, in the ECS Y-direction (force).

VALUE9 Fixed end shear, at point B of Figure 5.4, in the ECS Z-direction (force).

VALUE10 Fixed end torsion, at point B of Figure 5.4, about the ECS X-axis (force * length).

VALUE11 Fixed end moment, at point B of Figure 5.4, about the ECS Y-axis (force * length).

VALUE12 Fixed end moment, at point B of Figure 5.4, about the ECS Z-axis (force * length).

'END' Terminate the input of element loads.

DIR Direction of load in ECS. Valid directions are given below. Character variable is enclosed in single quotes.

'FX' Axial load is applied.

'FY' Force is applied in the local element's Y-direction.

'FZ' Force is applied in the local element's Z-direction.

'MX' Torque is applied.

'MY' Moment is applied about the local element's Y-axis.

'MZ' Moment is applied about the local element's Z-axis.

VALUEi Values used to calculate the element loads.

Notes

1. Multiple loads may be put on a single element.
2. 'UNIF' 'MY' and 'UNIF' 'MZ' are not available for '3D-BEAM' and 'IE3DBEAM' elements.
3. Only 'UNIF' 'FZ' is available for 'PLATE' element.
4. Element load TYPE = 'FEM' is not modified to reflect member end releases.

Example: MAXELD = 1,
1 13 0 0 'UNIF' 'FZ' –3.00

Element 13 has a uniform load of –3.00 applied in the element's local Z-direction, for load case 1.

6.3   SOL04—INCREMENTAL STATIC (PUSHOVER) SOLUTION

The following solution is used to calculate the static cyclic response of nonlinear structures. The following cards are input once.

'SOL04'				
TITLE				
MAXELD	IPRINT	IWRITE	UNBAL	SPLIMIT

SOL04   Signifies solution #4. Character variable, enclosed in single quotes.

TITLE   User input title, 80 characters maximum. Character variable, enclosed in single quotes.

MAXELD Maximum number of element loads.

IPRINT   Step increment for printed output.

IWRITE   Step increment for plot data written to output files.

UNBAL   A logical flag. If UNBAL = .TRUE., the unbalanced loads from the preceding step is added to the current load. Logical variable.

SPLIMIT   Stiffness parameter limit. The SP is calculated at each load step. The unbalanced forces will be added to the next load step if ABS(SP) is greater than SPLIMIT. Otherwise, the unbalanced forces are not added to the next load step, and the simple Euler incremental method is performed until ABS(SP) is greater than SPLIMIT (normally SPLIMIT = 10^{-2}–10^{-5}). If SPLIMIT = 0, the unbalanced forces are adjusted at every step regardless of SP value.

1. The load case for this solution is always input as one.
2. A separate geometric stiffness (KGFORM = 2) is not used by SOL04. The geometric stiffness may be included by using KGFORM = 1.

6.3.1   OUTPUT DATA TO PLOT FILES

These cards control the data that are written to separate plot files. This data is used to print reports and plot data. The card is repeated until the value of TYPE is 'END'.

TYPE	NUMB	IUNIT	IGEN	Δ NUMB	Δ IUNIT

TYPE   Type of data to be written to plot file. Valid types are given below. Character variable, enclosed in single quotes.

        'DOF'   Data is printed for a dof. The dof ID number, as assigned by the program, is used.

        'JOINT FX'   Data is printed for the dof corresponding to the joint's JCS X translation.

        'JOINT FY'   Data is printed for the dof corresponding to the joint's JCS Y translation.

        'JOINT FZ'   Data is printed for the dof corresponding to the joint's JCS Z translation.

        'JOINT MX'   Data is printed for the dof corresponding to the joint's JCS X rotation.

        'JOINT MY'   Data is printed for the dof corresponding to the joint's JCS Y rotation.

        'JOINT MZ'   Data is printed for the dof corresponding to the joint's JCS Z rotation.

        'ELE'   Element data is printed.

        'END'   Terminate the input of element loads.

NUMB   Element, dof, or joint number.

IUNIT   Plot file unit number.

IGEN   Number of similar data groups to be printed.

Δ NUMB   Incremental element dof or joint number for generation.

Δ IUNIT   Incremental plot file unit number for generation.

Example	
	Note
'DOF' 5 10 0 0 0	(1)
'ELE' 32 11 1 1 1	(2)
'END' 0 0 0 0 0	(3)

(1) dof #5 plot data is written to plot file 'temp10.out'.

(2) Element 32's data is written to file 'temp11.out', and element 33's data is written to file 'temp12.out'.

(3) End of plot file generation.

(4) For a structure with element(s) using R/CONCRETE1 material type, the data written from 'JOINT XX' plot file to output file (unit 06) will show the occurrence of limit state(s) for the structure. The possible limit states are

Limit State 1: Yield moment occurred, a symbol of *1* will be shown in the output file.

Limit State 2: Nominal moment occurred, a symbol of *2* will be shown in the output file.

Limit State 3: Ultimate moment occurred, a symbol of *3* will be shown in the output file.

Limit State 4: Buckling of longitudinal rebar occurred, a symbol of *4* will be shown in the output file.

Limit State 5: Lap-splice failure occurred, a symbol of *5* will be shown in the output file.

Limit State 6: Low-cycle fatigue occurred, a symbol of *6* will be shown in the output file.

6.3.2   JOINT LOADS

These cards are used to apply loads to joints. These cards are identical to the 'Joint Load' cards in block SOL01. Refer to SOL01 for a detailed description. The following card is repeated until the value of DIR is 'END'.

LOAD	ID	IGEN	Δ ID	DIR	VALUE

6.3.3   ELEMENT LOADS

These cards are used to apply element loads to the '3D-BEAM', 'IE3DBEAM', or 'PLATE' element. These cards are identical to the 'Element Load' cards in block SOL01. Refer to SOL01 for a detailed description. The following card is only included if MAXELD > 0. The card is repeated MAXELD times or until the value of TYPE is 'END'.

LOAD	IELE	IGEN	Δ IELE	TYPE	DIR	VALUE1,	VALUE2, ...

6.3.4   LOAD FACTORS

These cards contain the load factors that are used to generate incremental static loads. Each load step is subdivided into N small load steps. The card is repeated until the value of STEP is 'END'.

STEP	FACTOR	PMAX	DMAX	N

STEP   User input step name, 80 characters maximum. Character variable, enclosed in single quotes. If STEP is 'END', the current solution is terminated.

FACTOR   Load factor. The applied load at the end of the step is FACTOR * (applied joint and element loads).

PMAX   Dummy variable, use PMAX = 0.

DMAX   Dummy variable, use DMAX = 0.

N   Number of load steps between the previous and current factor. N is only used if both PMAX and DMAX are equal to zero.

Example

		Note
'LOAD'	10. 0 0 100	(1)
'UNLOAD'	0.0 0. 0. 20	(2)
'END'	0 0. 0. 0	(3)

Two load steps are inputted. They are 'LOAD' and 'UNLOAD.'

(1) The structure is loaded to 10 times the joint and element loads with 100 equal size load steps

(2) The structure is then unloaded with 20 equal size load steps.

(3) SOL04 is terminated.

6.4   BUG—SET BUG OPTIONS

This card is used to set the bug options, which print out the intermediate results listed below. The entire statement is a character variable and is enclosed in single quotes.

'BUG = options'

Option	Description
A	Print element displacements.
	Print loads applied to dofs.
B	Not used.
C	Not used.
D	Print joint, element, and dynamic loading data.
E	Print numerical integration data for linear and average acceleration methods.
F	Print the element's structural and geometric stiffness. Print the global mass, structural stiffness, geometric stiffness, loads, and displacements.
G	Print the condensed global mass, structural stiffness, and geometric stiffness matrices.
H	Print the element transformations, structural and geometric stiffness, etc.
I	Print contents of memory for elements.
J	Print a skyline map for matrices.
K	Print the material data for each load step (e.g., cross-sectional elements of a finite segment).
L	Print the energy balance.
M	Print the skyline data for matrices.
N	Print the contents of memory when DUMP is called.

Notes

1. Any number of options may be specified at one time.
2. Options specified with the last bug statement are the only options active.

6.5   READ—READ PLOT FILES

This card is used to read plot data written to output file (unit 06). The entire statement is a character variable and is enclosed in single quotes.

'READ	INC=I	UNIT=NO'

Where NO is the unit number of the plot file that contains the plot data, and I is the increment of the steps printed out. Multiple UNIT=NO statements may exist on each read card.

6.6   NOECHO—INHIBIT INPUT ECHO

This card is used to inhibit the input echo. Character variable, enclosed in single quotes.

'NOECHO'

6.7   DUMP—PRINT MEMORY

This card is used to print the addresses of the data in memory. If 'BUG=N' was previously specified, 'DUMP' also prints the nonzero values in the linear array. Character variable, enclosed in single quotes.

'DUMP'

6.8   RELEASE—RELEASE MEMORY

This card is used to release or 'free up' memory used for previous solutions. Global displacements, velocities, etc., are reset to zero. The entire statement is a character variable and is enclosed in single quotes.

'RELEASE	OPTION'

If OPTION = 'ELEMENT', the element forces, displacements, and hysteresis models are also reset to their initial values.

Example

	Note
'STRUCT'	(1)
.	
. Structure input is omitted.	
.	
'SOL01'	(2)
.	
. Solution input is omitted.	
.	
'RELEASE'	(3)

'SOL04'	(4)
Solution input is omitted.	
'STOP'	(5)

(1) The structure is defined.

(2) A static solution is performed.

(3) The static solution is released. The memory required for load, displacement, stiffness, etc., is released. The element forces are not released because the ELEMENT statement was omitted from the RELEASE card.

(4) Releasing the memory after the static solution allows the same memory to be used for SOL04. If the memory had not been released, then the total memory required would be the sum of the memory required for SOL01 and SOL04.

(5) Terminate the program.

6.9   STOP—TERMINATE EXECUTION

This card is used to terminate execution of the program. The statement is a character variable and is enclosed in single quotes.

'STOP'

7 Numerical Examples

Several numerical examples are included in this chapter for the illustration of how to use INSTRUCT to perform moment–curvature analyses of structural members and pushover analyses of bridge concrete and steel bents and other structures. The examples are as follows:

Example 1: Moment–Curvature Analysis
Example 2: Single-Column Bent
Example 3: Steel Member Plastic Analysis
Example 4: Two-Column Bent (Displacement Control)
    PHL Method
    PM Method
    FSFS Method
    FSMC Method
Example 5: Two-Column Bent (Force Control)
Example 6: Concrete Column with Rectangular Section
Example 7: Three-Column Bent with Different Elements
Example 8: Four-Column Bent
Example 9: Steel Pile Cap Bent
Example 10: Steel Cross Frame Analysis
Example 11: Concrete Column with Shear Failure
Example 12: Concrete Beam–Column Joint Failure
Example 13: Cyclic Response of a Cantilever Beam

7.1 STRUCTURAL LIMIT STATE INDICATORS

INSTRUCT provides several structural limit state indicators in the output. Some indicators are shown in the "structural joint" output at certain pushover load steps, depending on the nonlinear condition of the structure. Table 7.1 lists the possible limit state indicators shown at the structural joint force–displacement output.

For example, during the pushover analysis of a multiple-column bent, the structural limit state at which the first column reaches its M_n occurs at load step "X." Then, *2* will be shown in the "structural joint" output corresponding to load step "X." As shown in Table 7.1, ultimate limit state of a structure may be controlled by limit state indicator *3*, *4*, *5*, or *6*. Normally, *3* controls the ultimate structure-displacement capacity, if a structure is designed based on the current American Association of State Highway and Transportation Officials (AASHTO) design specifications. However, for old bridges, structural limit states may be controlled by *4*, *5*, or *6*.

TABLE 7.1

Limit State Indicators at Structural Joint Output

Limit State Indicators	Occurrence at the First Member
1	Yield moment, M_y, as defined in Section 4.2
2	Nominal moment, M_n, as defined in Section 4.2
3	Ultimate moment, M_u, due to failure modes 1, 2, or 4 (i.e., unconfined concrete compression failure, confined concrete compression failure, or tensile fracture of longitudinal rebar, respectively), as described in Section 4.6
4	Ultimate moment, M_u, due to failure mode 3 (i.e., buckling of longitudinal rebar), as described in Section 4.6
5	Ultimate moment, M_u, due to failure mode 6 (i.e., failure in lap-splice of longitudinal rebar), as described in Section 4.6
6	Ultimate moment, M_u, due to failure mode 5 (i.e., low cycle fatigue of longitudinal rebar), as described in Section 4.6

Note that INSTRUCT performs pushover analysis until the user-defined target pushover displacement at a joint is reached. For example, if the user assigned a target pushover displacement of 5 in. at structure joint "Y," the structural limit state *3* occurs at a pushover displacement of 3 in. The program will not stop at pushover displacement of 3 in., but will continue to push the structure to the displacement of 5 in. However, the INSTRUCT output will show the indicator *3* at the joint "Y" corresponding to pushover displacement of 3 in.

7.2   MEMBER YIELD INDICATORS

INSTRUCT output also provides yield conditions of individual members by using an indicator called FLP (flag for limit state of plasticity). FLPs are shown in the "element" output and described in Chapter 6 for "STABILITY" and "IE3DBEAM" elements.

7.3   NUMERICAL EXAMPLES

7.3.1   Example 1: Moment–Curvature Analysis

Moment–curvature curve of a reinforced concrete circular section shown in Figure 7.1 is generated. The details of the section are as follows: diameter = 48″, 20-#10 longitudinal bars, $f_c' = 4$ ksi, $f_y = 60$ ksi, spiral = #5 @3.25″, concrete cover = 2.6″ and applied column axial dead loads = 765 kip. The post-yield modulus of the steel stress–strain curve is assumed to be 1% of the elastic modulus. The finite segment–finite string (FSFS) method is used to (1) generate the moment–curvature curve for axial dead load $P = 765$ kip and (2) generate column axial load–nominal

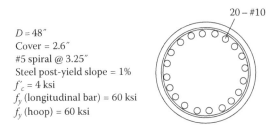

D = 48″
Cover = 2.6″
#5 spiral @ 3.25″
Steel post-yield slope = 1%
f'_c = 4 ksi
f_y (longitudinal bar) = 60 ksi
f_y (hoop) = 60 ksi

20 – #10

FIGURE 7.1 Cross section details.

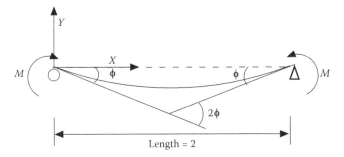

FIGURE 7.2 Structural model.

moment interaction curve using P=−1000, 0, 765, 2000, 3000, and 5000 (kip). The ultimate concrete compression strain, ε_{cu}, is based on Equation 3.24, which is $\varepsilon_{cu} = 0.004 + 1.4\rho_s f_{yh}\varepsilon_{su}/f'_{cc}$.

The structural model based on Section 2.5 for the moment–curvature analysis is shown in Figure 7.2. A "STABILITY" element with only one segment is used. The "R/CONCRETE1" material is considered. The length of the member is chosen to be 2, and so the end rotation of the member represents the curvature of the cross section. The same magnitudes of rotations are applied at both ends of the element by incremental displacement control.

1. Output (EX1_MC.out, Moment–Curvature Curve, P=765 kip)
   The output results are shown below. Note that the symbols, %7, %8, and %9, in the output file indicate the limit states 1, 2, and 3, respectively. Limit state 1 occurs when moment reaches yield moment, M_y; limit state 2 occurs when moment reaches nominal moment, M_n; and limit state 3 occurs when moment reaches ultimate moment, M_u. Limit states 1, 2, and 3 (with symbols *1*, *2*, and *3*, respectively) are also shown in the user-defined joint's force–displacement plot file written to the output. The moment–curvature curves are shown in Figures 7.3 and 7.4.

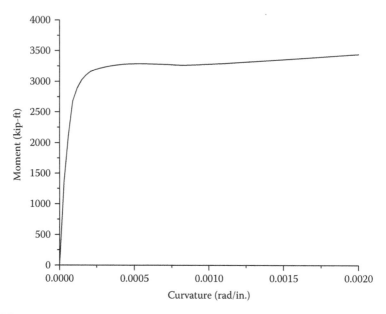

FIGURE 7.3 Moment–curvature curve (with 765 kip dead load).

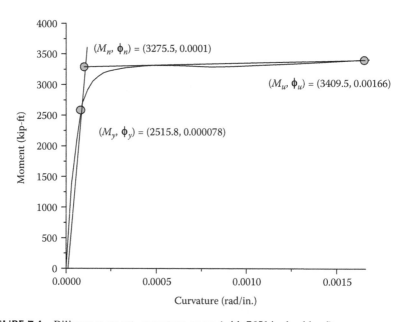

FIGURE 7.4 Bilinear moment–curvature curve (with 765 kip dead load).

```
1   ECHO OF INPUT DATA

LINE  ....|.. 10....|.. 20....|.. 30....|.. 40....|.. 50....|.. 60....|.. 70....|.. 80....|.. 90....|..100....|..110
   1: 'STRUCTURE DEFINITION - R/C Circular Sec. 20-#10 rebars with 48 in. Dia.'
   2: 'Moment-curvature analysis; Axial load P=765000 lbs'
   3: 2   1   2   0   1
   4: 1   0.0   0.0   1   0
   5: 2   2.0   0.0   1   0
   6: 1   0   0   1   0   | Direction Cosine
   7: 1   0   1   1   1   0   0
   8: 2   1   1   1   1   0   0
   9: 1   | Number of Material
  10: 'R/CONCRETE1   MAT#1'   1   60000   29000000   2   48.   2.625   0   0   10   6   4000
  11: 0.625   3.25   0.95   1.27   1809.6   0   2   0.01   0   1200780.   0.01   1
  12: 20   0   0   0   60000.   1   -1   -1   0   -1   -1
  13: 0   0   .FALSE.   | Geometric stiffness
  14: 1   | Number of Element
  15: 'STABILITY'   'R/C CIRCULAR SECTION ELE.'   1   1   2   0   1   0   0   0
  16: 0   0   .FALSE.   | Mass
  17: 0   0   | Damp
  18: 'SOL01 Elastic Static Analysis'
  19: 'APPLY AXIAL LOAD AT JOINT 1'
  20: 1   0
  21: 1   0   0   'FX'   765000.0   | Joint Load
  22: 0   0   'END'   0   | Joint Load
  23: 'SOL04 Inelastic Incremental Pushover Analysis'
  24: 'incremental disp. control'
  25: 0   50000   10   .TRUE.   0
  26: 'JOINT MZ'   1   13   0   0
  27: 'END'   0   0   0   0
  28: 1   0   0   'MZ'   0.003   | Joint Load
  29: 2   0   0   'MZ'   -0.003   | Joint Load
  30: 0   0   0   'END'   0   | Joint Load
  31: 'disp. from 0 to 0.003'   1   0   0   1000
  32: 'END'   0   0   0   0
  33: 'READ   UNIT=13'
  34: 'STOP'

1 STRUCTURE.....: Moment-curvature analysis; Axial load P=765000 lbs
SOLUTION......:
```

```
*** PROGRAM INSTRUCT ***    (VERSION 1.0)

%1S%

NODE COORDINATES AND DEGREES OF FREEDOM

TOTAL NUMBER OF DEGREES OF FREEDOM........:    12
NUMBER OF DEGREES OF FREEDOM CONDENSED OUT.:     0
NUMBER OF FREE DEGREES OF FREEDOM.........:     1
NUMBER OF RESTRAINED DEGREES OF FREEDOM....:    11

NODE COS#   X-COORD    Y-COORD    Z-COORD    FX     FY     FZ     MX     MY     MZ
  1   1     0.0000     0.0000     0.0000     1      2-R    3-R    4-R    5-R    6-R
  2   1     0.0000     2.0000     0.0000     7-R    8-R    9-R   10-R   11-R   12-R
     NOTE: R - RESTRAINED DEGREE OF FREEDOM
           C - CONSTRAINED DEGREE OF FREEDOM

%1E%

%2S%

DIRECTION COSINES ...

COS( 1)  VX: 1.00000 I +0.00000 J +0.00000 K   VY: 0.00000 I +1.00000 J +0.00000 K   VZ: 0.00000 I +0.00000 J +1.00000 K
%2E%

%3S%
1 STRUCTURE.....: Moment-curvature analysis; Axial load P=765000 lbs          TIME: 08:48:29, DATE: 31-AUG-09
  SOLUTION......:
```

```
R/CONCRETE1 ELEMENT
===================

MAT. NO.     =   1    R/C CIRCULAR SECTION
NSEG         =   1          YS          =  6.000E+04
EM           =  2.900E+07   LIBN        =  2
COL. DIA     =  48.0        COL. THK    =  2.63
ECCX0        =  0.00        ECCY0       =  0.00
ELE #/QT.LAY=    10         NO.OF LAYERS=  6
F"C          =  4.000E+03   DIA. OF HOOP=  0.625
S            =  3.25        KE          =  0.950
L.BAR DIA.   =  1.27        IECOP       =  1
NUMP         =     264      SMALL       =  0.00
RATIX0       =  0.00        RATIY0      =  0.00
TOTA         =  1.810E+03   IAUTO       =  0
IMATER       =  2           RATIO3      =  1.000E-02
IR           =  0           G           =  1.201E+06
QRNEE        =  1.000E-02   ISTIF       =  1
YSHOOP       =  6.000E+04   IELAS       =  1.00
PCMAX        =  -1.00       NLBAR       =  20
ECU          =  -1.00       ESU         =  -1.00
STR. PERIOD  =  -1.00       SPLICE LEN. =  -1.00

%3E%

%4S%
1 STRUCTURE.....: Moment-curvature analysis; Axial load P=765000 lbs
  SOLUTION......:

MAT. NO.     =   1
INITIAL CONCRETE SHEAR CAPACITY (VCI in lb)   =  3.204E+05
FINAL CONCRETE SHEAR CAPACITY (VCF in lb)     =  5.493E+04
TRANS.STEEL SHEAR CAPACITY,U0 DIR (VSu in lb) =  6.684E+05
TRANS.STEEL SHEAR CAPACITY,V0 DIR (VSv in lb) =  6.684E+05
```

TIME: 08:48:29, DATE: 31-AUG-09

```
ELEMENT 12, STABILITY ELEMENT

R/C CIRCULAR SE     #   MATL   START   END   LENGTH   --------- Y-AXIS ---------   START DIST   END DIST
%4E%                1    1       1      2    2.000    0.00000 I +1.00000 J +0.00000 K   0.000    0.000
   .
   .
   .
1 STRUCTURE.....: Moment-curvature analysis; Axial load P=765000 lbs      TIME: 08:48:29, DATE: 31-AUG-09
  SOLUTION......: APPLY AXIAL LOAD AT JOINT 1                              TIME: 08:48:29, DATE: 31-AUG-09

SOLUTION #1, STATIC - ELASTIC ANALYSIS
======================================

NUMBER OF LOAD CASES .............  1
APPLIED JOINT LOADS
====================

LOAD CASE:  1  JOINT:   1  DIRECTION: FX   DOF(S)  1        MAGNITUDE:   765000.
1 STRUCTURE.....: Moment-curvature analysis; Axial load P=765000 lbs      TIME: 08:48:29, DATE: 31-AUG-09
  SOLUTION......: APPLY AXIAL LOAD AT JOINT 1                              TIME: 08:48:29, DATE: 31-AUG-09

GCS DISPLACEMENTS, LOADING #   1
================================

NODE     DX              DY          DZ          RX          RY          RZ

  1   2.1079969E-04    0.00000     0.00000     0.00000     0.00000     0.00000
  2   0.00000          0.00000     0.00000     0.00000     0.00000     0.00000

1 STRUCTURE.....: Moment-curvature analysis; Axial load P=765000 lbs      TIME: 08:48:29, DATE: 31-AUG-09
  SOLUTION......: APPLY AXIAL LOAD AT JOINT 1                              TIME: 08:48:29, DATE: 31-AUG-09
```

GCS RESTRAINT REACTIONS,   LOADING #   1
==

NODE	FX	FY	FZ	MX	MY	MZ
1	0.000000	0.000000	0.000000	0.000000	0.000000	0.000000
2	-765000.0	0.000000	0.000000	0.000000	0.000000	0.000000

SUMMATION -765000.0   0.000000   0.000000   0.000000   0.000000   0.000000
----

1 STRUCTURE.....: Moment-curvature analysis; Axial load P=765000 lbs   TIME: 08:48:29, DATE: 31-AUG-09
SOLUTION......: APPLY AXIAL LOAD AT JOINT 1   TIME: 08:48:29, DATE: 31-AUG-09

STABILITY ELEMENT FORCES...

ELEMENT	LOAD		NODE	AXIAL	FY	FZ	TORSION	MY	MZ	FLP,SP
1	1	DISPL	1	2.107969E-04	0.00000	0.00000	0.00000	0.00000	0.00000	FLP: 0.0
		DISPL	2	0.00000	0.00000	0.00000	0.00000	0.000C0	0.00000	
1	1	FORCE	1	765000.	0.00000	0.00000	0.00000	0.000C0	0.00000	SP: 0.0
		FORCE	2	-765000.	0.00000	0.00000	0.00000	0.000C0	0.00000	

1 STRUCTURE.....: Moment-curvature analysis; Axial load P=765000 lbs   TIME: 08:48:29, DATE: 31-AUG-09
SOLUTION......: incremental disp. control   TIME: 08:48:29, DATE: 31-AUG-09
SOLUTION #4, STATIC NONLINEAR SOLUTION
===
INTERVAL FOR PRINTING DATA...........50000
INTERVAL FOR WRITING DATA TO FILE...... 10

UNBALANCED JOINT FORCES ARE ADDED TO THE NEXT CYCLE
UNBALANCED FORCES ARE NOT ADDED WHEN THE ABS(SP) IS LESS THAN SPLIMIT OF   0.00000

```
DATA WRITTEN TO FILES
=========================

DEGREE OF FREEDOM  #    6 IS WRITTEN TO UNIT #   13   JOINT:    1   DIRECTION: MZ
APPLIED JOINT LOADS
=========================

LOAD CASE:  1  JOINT:  1  DIRECTION: MZ  DOF(S)   6      MAGNITUDE:  3.000000E-03  ...JOINT DISPLACEMENT...
LOAD CASE:  1  JOINT:  1  DIRECTION: MZ  DOF(S)  12      MAGNITUDE: -3.000000E-03  ...JOINT DISPLACEMENT...

****  STABILITY ELEMENT       1REACHES FIRST YIELD, WHICH OCCURRED AT THE FOLLOWING STAGE:

STABILITY ELEMENT FORCES...
    disp. from STEP:   26 FACTOR:  1.000
ELEMENT LOAD  NODE   AXIAL          FY         FZ       TORSION      MY           MZ         FLP,SP

    1    1  DISPL   2.107969E-04   0.00000    0.00000    0.00000    0.00000    7.800001E-05
         2  DISPL   0.00000        0.00000    0.00000    0.00000    0.00000   -7.800001E-05  FLP:  1.0

    1    1  FORCE   765000.        0.00000    0.00000    0.00000    0.328492   3.025547E+07
         2  FORCE  -765000.        0.00000    0.00000    0.00000   -0.328492  -3.025547E+07  SP:   0.0
%7 % LIMIT STATE POINT 1 DEFINED DUE TO ELEMENT NO    1 AT STEP=   26

****  STABILITY ELEMENT       1 SEGMENT NO.    1 REACHES NOMINAL MOMENT AT CURVATURE OF CURNM =   0.333000E-03

****  STABILITY ELEMENT       1 EXTREME FIBER REACHES COMPRESSION STRAIN OF 0.004, AT WHICH
NOMINAL MOMENT, MN, IS DEFINED AT THE FOLLOWING STAGE:

STABILITY ELEMENT FORCES...
    disp. from STEP:  111 FACTOR:  1.000
ELEMENT LOAD  NODE   AXIAL          FY         FZ       TORSION      MY           MZ         FLP,SP

    1    1  DISPL   2.107969E-04   0.00000    0.00000    0.00000    0.00000    3.329996E-04
         2  DISPL   0.00000        0.00000    0.00000    0.00000    0.00000   -3.329996E-04  FLP:  3.0

    1    1  FORCE   765000.        0.00000    0.00000    0.00000    1.44113    3.963655E+07
         2  FORCE  -765000.        0.00000    0.00000    0.00000   -1.44113  -3.963655E+07  SP:   0.0
%8 % LIMIT STATE POINT 2 DEFINED DUE TO ELEMENT NO    1 AT STEP=  111
****%MEMBER   1ELE.    1SEG.    60STRAIN=  -0.159741E-01AT ISTEP   554
```

```
WHICH EXCEEDS ECU of    0.159607E-01
STRESS =     -4753.18
STRAIN =    -0.159741E-01
FCCP   =      5502.39
ECC    =     0.575599E-02

**** STABILITY ELEMENT    1 SEGMENT NO.    1 REACHES ULTIMATE CONCRETE COMPRESSION STRAIN.
PLASTIC CURVATURE CURPLS =    0.132901E-02
CUR : TOTAL CURVATURE =    0.166201E-02
COMPRESSION DEPTH TO N.A. IN U0 DIR. =    12.8653

COMPRESSION DEPTH TO N.A. IN V0 DIR. =    0.00000

***** STABILITY ELEMENT    1 EXTREME FIBER REACHES ULTIMATE COMPRESSION STRAIN,        PROGRAM CONTINUE
***** THE ELEMENT FORCES disp. from STEP:  554 FACTOR:  1.000    ARE:

   1  2  DISPL   1    2.107969E-04    0.00000    0.00000    0.00000    1.659008E-03
          DISPL   2    0.00000         0.00000    0.00000    0.00000   -1.659008E-03   FLP:   5.0

   1  2  FORCE   1    765000.         0.00000    0.00000    5.26463    4.124242E+07
          FORCE   2   -765000.        0.00000    0.00000   -5.26463   -4.124242E+07   SP:   0.0
%9 % LIMIT STATE POINT 3 DEFINED DUE TO ELEMENT NO    1 AT STEP=   554

**** STABILITY ELEMENT    1 SEGMENT NO.    1 LONG. REBAR REACHES ULTIMATE TENSION STRAIN.
PLASTIC CURVATURE CURPLS =    0.254402E-02
CUR : TOTAL CURVATURE =    0.287702E-02
COMPRESSION DEPTH TO N.A. IN U0 DIR. =    13.3807

COMPRESSION DEPTH TO N.A. IN V0 DIR. =    0.00000

***** STABILITY ELEMENT    1 LONG. REBARREACHES ULTIMATE TENSION STRAIN OF 0.09,      PROGRAM CONTINUE
***** THE ELEMENT FORCES disp. from STEP:  959 FACTOR:  1.000    ARE:

   1  2  DISPL   1    2.107969E-04    0.00000    0.00000    0.00000    2.874017E-03
          DISPL   2    0.00000         0.00000    0.00000    0.00000   -2.874017E-03   FLP:   7.0

   1  2  FORCE   1    765000.         0.00000    0.00000   -5.46955    4.349969E+07
          FORCE   2   -765000.        0.00000    0.00000    5.46955   -4.349969E+07   SP:   0.0
```

TIME: 08:48:29, DATE: 31-AUG-09
TIME: 08:48:29, DATE: 31-AUG-09

1 STRUCTURE.....: Moment-curvature analysis; Axial load P=765000 lbs
  SOLUTION......: incremental disp. control

%5S%

DEGREE OF FREEDOM #   6 IS READ FROM UNIT #   13   JOINT #   1, DIRECTION: MZ

STEP	TIME	LOAD	DISPLACEMENT	VELOCITY	ACCELERATION
0	0.0000	0.0000	0.0000	0.0000	0.0000
10	0.0000	1.64879E+07	3.00000E-05	0.0000	0.0000
20	0.0000	2.53391E+07	6.00000E-05	0.0000	0.0000
1 30	0.0000	3.24775E+07	9.00000E-05	0.0000	0.0000
2 120	0.0000	3.97430E+07	3.60000E-04	0.0000	0.0000
130	0.0000	3.98538E+07	3.89999E-04	0.0000	0.0000
3 560	0.0000	4.12866E+07	1.68001E-03	0.0000	0.0000
570	0.0000	4.13497E+07	1.71001E-03	0.0000	0.0000
980	0.0000	4.36019E+07	2.94002E-03	0.0000	0.0000
990	0.0000	4.36441E+07	2.97002E-03	0.0000	0.0000
1000	0.0000	4.36825E+07	3.00002E-03	0.0000	0.0000

%5E%

```
*-------------------------------------------*
*--- MEMORY UTILIZATION .........   ----*
*--- IZ=  7681,    MEM= 0.015%     ----*
*-------------------------------------------*
*--- ELAPSED CPU TIME    0.00 SEC  ----*
*--- TOTAL CPU TIME      0.59 SEC  ----*
*-------------------------------------------*
```

2. Generate P–M Interaction Curve

Similar to the above moment–curvature analysis for $P = 765$ kip, the moment–curvature plots corresponding to different column axial dead loads are shown in Figure 7.5 by using the FSFS method.

For each axial load case, the nominal moment M_n is shown in the INSTRUCT output file (with %8 symbol). Once all the nominal moments corresponding to different axial loads are known, the P–M interaction curve can be plotted as in Figure 7.6.

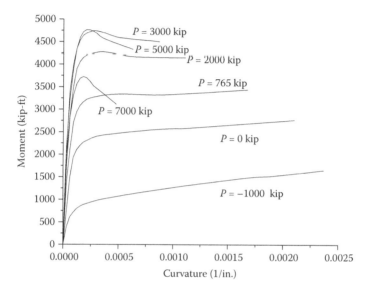

FIGURE 7.5 Moment–curvature curves.

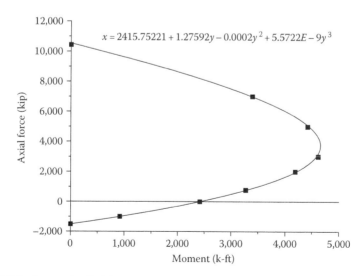

FIGURE 7.6 Interaction P–M curve.

7.3.2   EXAMPLE 2: SINGLE-COLUMN BENT

This example compares the numerical result with a full-scale column test result. The full-scale column test was conducted by the National Institute of Standards and Technology (NIST) (Stone and Cheok, 1989). The height of the column was 30′ measured from the top of the footing to the top of the column. The diameter of the column was 60″. The plastic hinge length based on Equation 4.5 is 46.3″. The material properties of the column are shown in Figure 7.7. As shown in Figure 7.8, the column is modeled as a "STABILITY" element with eight segments. The length of the first segment near the foundation is equal to the plastic hinge length of 46.3″. The

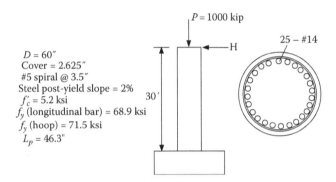

FIGURE 7.7   NIST 30′ full-scale column.

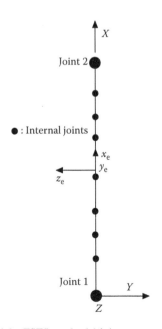

FIGURE 7.8   Structural models: FSFS method (eight segments).

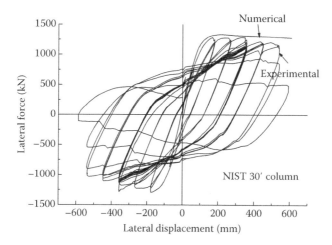

FIGURE 7.9   Experimental and numerical comparison.

column is pushed by the incremental displacement control at the top of the column until the total displacement at the top of the column is equal to 24″.

The pushover curve is shown in Figure 7.9. It can be seen that the lateral force–lateral displacement curve generated by the FSFS method is in agreement with the test results when lateral displacement is between 0 and 350 mm. The output results show that the concrete ultimate compression strain, ε_{cu}, is reached (i.e., limit state 3) at a pushover displacement of 505 mm. Once ε_{cu} is developed, fracture of the transverse reinforcement may occur, and the concrete is no longer in the confined condition.

Output (EX2_NIST3B62.out file)

The output results are shown below. The symbols, %7, %8, and %9, in the output file indicate the limit states 1, 2, and 3, respectively. Limit state 1 occurs when the moment reaches yield moment, M_y; limit state 2 occurs when the moment reaches nominal moment, M_n; and limit state 3 occurs when the moment reaches ultimate moment, M_u.

```
1     ECHO OF INPUT DATA

LINE  ....|.. 10....|.. 20....|.. 30....|.. 40....|.. 50....|.. 60....|.. 70....|.. 80....|.. 90....|..100....|..110
   1: 'STRUCTURE DEFINITION - NIST 30' COLUMN'
   2: 'TEST FULL-SCALE R/C COLUMN, use one ele & 8 segments'
   3: 2   1   2   0   0   1
   4: 1   0.00   0.00   .00   1   0
   5: 2   360.0   0.00   .00   1   0   | Direction Cosine
   6: 1   0   0   0   1   0   1   0   0
   7: 1   1   1   1   1   1   1   0   0
   8: 2   0   1   1   1   1   0   0   0
   9: 1   | Number of Material
  10: 'R/CONCRETE1   MAT#1'   8   68900   27438000   2   60.   2.625   0   0   10   6   5200.
  11: 0.625   3.5   0.95   1.69   0   46.3   0.   0.   2827.43   0   2   0.02   1580895.   0.01   1
  12: 25   0   0   0   0   71500.   1   -1   0   -1   0   -1   -1
  13: 0   0   0   .FALSE.   | Geometric stiffness
  14: 1   | Number of Element
  15: 'STABILITY'   'R/C CIRCULAR MEMBER 1'   1   1   2   0   0   1   0   0   0
  16: 0   0   .FALSE.   | Mass
  17: 0   0   0   | Damp
  18: 'SOL01 Elastic Static Analysis'
  19: 'APPLY AXIAL LOAD AT JOINT 2'
  20: 1   0
  21: 1   2   0   0   'FX'   -1000000.   | Joint Load
  22: 0   0   0   0   'END'   0   | Joint Load
  23: 'SOL04 Inelastic Incremental Pushover Analysis'
  24: 'INCREMEMTAL DISPLACEMENT CONTROL AT JOINT 2'
  25: 0   50000   2   .TRUE.   0
  26: 'JOINT FY'   2   11   0   0   0
  27: 'ELE  '   1   12   0   0   0
  28: 'END'   0   0   0   0   0
  29: 1   2   0   0   'FY'   24   | Joint Load
  30: 0   0   0   0   'END'   0   | Joint Load
  31: 'DISP. FROM   0   TO   24'   1   0   0   800
  32: 'END'   0   0   0   0
  33: 'READ   UNIT=11   UNIT=12'
  34: 'STOP'
```

. . .

```
****  STABILITY ELEMENT          1REACHES FIRST YIELD, WHICH OCCURRED AT THE FOLLOWING STAGE:

      STABILITY ELEMENT     FORCES...
        DISP. FROM STEP: 1C2 FACTOR: 1.000
      ELEMENT LOAD    NODE     AXIAL           FY              FZ         TORSION        MY            MZ        FLP,SP

          1    1    DISPL   1   0.00000        0.00000         0.00000    0.00000        0.00000       0.00000
                    DISPL   2  -4.235713E-02   0.00000        -3.06000    0.00000        1.209287E-02  0.00000    FLP:    1.0

          1    1    FORCE   1   998511.        1.378863E-02    228886.    0.00000       -8.545890E+07  4.57937
                    FORCE   2  -1.000030E+06  -1.378863E-02   -228886.    0.00000        1.63044       0.384536   SP:     0.0

  %7 % LIMIT STATE POINT 1 DEFINED DUE TO ELEMENT NO   1 AT STEP= 102

****  STABILITY ELEMENT          1 SEGMENT NO.       1 REACHES NOMINAL MOMENT AT CURVATURE OF CURNM =   0.268345E-03

****  STABILITY ELEMENT          1 EXTREME FIBER REACHES COMPRESSION STRAIN OF 0.004, AT WHICH
      NOMINAL MOMENT, MN, IS DEFINED AT THE FOLLOWING STAGE:

      STABILITY ELEMENT     FORCES...
        DISP. FROM STEP: 247 FACTOR: 1.000
      ELEMENT LOAD    NODE     AXIAL           FY              FZ         TORSION        MY            MZ        FLP,SP

          1    1    DISPL   1   0.00000        0.00000         0.00000    0.00000        0.00000       0.00000
                    DISPL   2  -0.110737       0.00000        -7.41002    0.00000        2.675829E-02  0.00000    FLP:    3.0

          1    1    FORCE   1   995383.        1.239864E-02    299526.    0.00000       -1.152396E+08  3.94567
                    FORCE   2  -1.000029E+06  -1.239864E-02   -299526.    0.00000       -17.3179       0.517838   SP:     0.0

  %8 % LIMIT STATE POINT 2 DEFINED DUE TO ELEMENT NO   1 AT STEP= 247
  ***%MEMBER   1SEG.   1ELE.   61STRAIN=  -0.127191E-01AT ISTEP   665
      WHICH EXCEEDS ECU of  0.126985E-01
      STRESS =  -5271.18
      STRAIN =  -0.127191E-01
      FCCP   =   6557.84
      ECC    =   0.461123E-02

****  STABILITY ELEMENT          1 SEGMENT NO.       1 REACHES ULTIMATE CONCRETE COMPRESSION STRAIN.
      PLASTIC CURVATURE CURPLS =   0.730140E-03
      CUR : TOTAL CURVATURE =   0.998485E-03
      COMPRESSION DEPTH TO N.A. IN U0 DIR. =    0.00000

      COMPRESSION DEPTH TO N.A. IN V0 DIR. =   16.2017
```

```
**** STABILITY ELEMENT      1 EXTREME FIBER REACHES ULTIMATE COMPRESSION STRAIN,          PROGRAM CONTINUE
***** THE ELEMENT FORCES DISP. FROM STEP:  665 FACTOR:  1.000    ARE:

        1   2   DISPL        0.00000        0.00000        0.00000        0.00000        0.00000
                DISPL       -0.607292      -19.9200        0.00000        6.49325BE-02   0.00000        FLP:  5.0

        1       FORCE     987204.        2.922991E-02    292462.        0.00000      -1.252070E+08   10.1835
        2       FORCE   -1.000028E+06   -2.922991E-02   -292462.        0.00000      -37.0481         0.339241     SP:  0.0
%9 % LIMIT STATE POINT 3 DEFINED DUE TO ELEMENT NO   1 AT STEP=  665
```

DEGREE OF FREEDOM #   9 IS READ FROM UNIT # 11   JOINT #   2, DIRECTION: FY

STEP	TIME	LOAD	DISPLACEMENT	VELOCITY	ACCELERATION
0	0.0000	0.0000	0.0000	0.0000	0.0000
2	0.0000	11768.	6.00000E-02	0.0000	0.0000
4	0.0000	23532.	0.12000	0.0000	0.0000
100	0.0000	2.25491E+05	3.0000	0.0000	0.0000
1 102	0.0000	2.28887E+05	3.0600	0.0000	0.0000
104	0.0000	2.32136E+05	3.1200	0.0000	0.0000
246	0.0000	2.99435E+05	7.3800	0.0000	0.0000
2 248	0.0000	2.99600E+05	7.4400	0.0000	0.0000
250	0.0000	2.99763E+05	7.5000	0.0000	0.0000
664	0.0000	2.92458E+05	19.920	0.0000	0.0000
3 666	0.0000	2.92412E+05	19.980	0.0000	0.0000
668	0.0000	2.92369E+05	20.040	0.0000	0.0000
798	0.0000	2.86694E+05	23.940	0.0000	0.0000
800	0.0000	2.88623E+05	24.000	0.0000	0.0000

%5E%

TIME: 08:59:24, DATE: 31-AUG-09
TIME: 08:59:24, DATE: 31-AUG-09

```
1 STRUCTURE....: TEST FULL-SCALE R/C COLUMN, use one ele & 8 segments
  SOLUTION......: INCREMEMTAL DISPLACEMENT CONTROL AT JOINT 2

%5S%
   ELEMENT #    1 IS READ FROM UNIT #   12

   STABILITY ELEMENT FORCES...
   STEP    TIME       NODE      AXIAL           FY              FZ              TORSION      MY              MZ            FLP,SP
```

STEP	TIME	NODE	AXIAL	FY	FZ	TORSION	MY	MZ	FLP,SP
0	0.0000	1	0.100000E+07	0.111844E-02	-0.807263E-03	0.00000	1.33950	0.140470	
		2	-0.100000E+07	-0.111844E-02	0.807263E-03	0.00000	-1.04890	0.262160	SP: 0.0
		DISP	-0.273549E-01	0.00000	0.00000	0.00000	0.00000	0.00000	FLP: 0.0
2	0.0000	1	0.100000E+07	-0.340459E-03	11777.0	0.00000	-0.429976E+07	-0.149020	
		2	-0.100000E+07	0.340459E-03	-11777.0	0.00000	-10.3760	0.264571E-01	SP: 0.0
		DISP	-0.273578E-01	0.00000	-0.600000E-01	0.00000	0.250191E-03	0.00000	FLP: 0.0
102	0.0000	1	998511.	0.137886E-01	228886.	0.00000	-0.854589E+08	4.57940	
		2	-0.100003E+07	-0.137886E-01	-228886.	0.00000	1.63040	0.384540	SP: 0.0
		DISP	-0.423571E-01	0.00000	-3.06000	0.00000	0.120929E-01	0.00000	FLP: 2.0
104	0.0000	1	998460.	0.399111E-02	232143.	0.00000	-0.866916E+08	1.46510	
		2	-0.100003E+07	-0.399111E-02	-232143.	0.00000	15.2960	-0.282924E-01	SP: 0.0
		DISP	-0.429546E-01	0.00000	-3.12000	0.00000	0.123327E-01	0.00000	FLP: 2.0
248	0.0000	1	995366.	0.184611E-01	299608.	0.00000	-0.115299E+09	6.55870	
		2	-0.100003E+07	-0.184611E-01	-299608.	0.00000	-15.1610	0.872474E-01	SP: 0.0
		DISP	-0.111390	0.00000	-7.44000	0.00000	0.268520E-01	0.00000	FLP: 4.0
250	0.0000	1	995332.	0.123476E-01	299771.	0.00000	-0.115418E+09	4.16940	
		2	-0.100003E+07	-0.123476E-01	-299771.	0.00000	-26.1080	0.275740	SP: 0.0
		DISP	-0.112710	0.00000	-7.50000	0.00000	0.270393E-01	0.00000	FLP: 4.0

```
664   0.0000

        1     987204.        0.292299E-01    292462.    0.00000   -0.125207E+09   10.1840      SP:    0.0
        2    -0.100003E+07  -0.292299E-01   -292462.    0.00000   -37.0480         0.339240
 DISP         0.00000        0.00000         0.00000    0.00000    0.00000         0.00000
             -0.607290                      -19.9200               0.649326E-01    0.00000     FLP:   4.0

666   0.0000

        1     987173.        0.310389E-01    292419.    0.00000   -0.125252E+09   11.0430      SP:    0.0
        2    -0.100003E+07  -0.310389E-01   -292419.    0.00000   -37.8750         0.130930
 DISP         0.00000        0.00000         0.00000    0.00000    0.00000         0.00000
             -0.610780                      -19.9800               0.651177E-01    0.00000     FLP:   5.1

800   0.0000

        1     984053.        0.909737E-01    288630.    0.00000   -0.127907E+09   32.7050      SP:    0.0
        2    -0.100003E+07  -0.909737E-01   -288630.    0.00000   -38.1730         0.452513E-01
 DISP         0.00000        0.00000         0.00000    0.00000    0.00000         0.00000
             -0.868400                      -24.0000               0.775572E-01    0.00000     FLP:   5.1
```

. . .   . . .

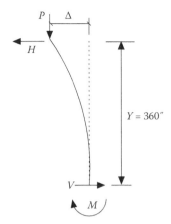

FIGURE 7.10 Equilibrium check.

Equilibrium check at Step = 800 (Figure 7.10):
From output plot file units 11 and 12, $P = 1{,}000\,\text{kip}$; $H = 288.6\,\text{kip}$; $\Delta = 24\,\text{in.}$; $M = 127{,}907$ (k-in.); $V = 288.6\,\text{kip}$.

$V = H$ (ok)

$$\bar{M} = P\Delta + HY = 127{,}896 \approx M \text{ (ok)}$$

7.3.3 EXAMPLE 3: STEEL MEMBER PLASTIC ANALYSIS

This example is to compare the numerical solution with that in Example 8.6 of McGuire's text book (McGuire et al., 2000). The structural model is shown in Figure 7.11. The bilinear interaction axial load–moment (PM) interaction method is used to determine the plastic limit load P at which the failure mechanism occurs. The following P–M interaction curve for the W-shape sections is used here (ASCE, 1989):

$$M = 1.18\left(1 - \frac{P}{P_y}\right)M_p \leq M_p \tag{7.1}$$

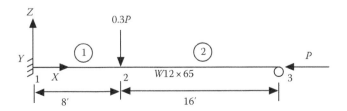

FIGURE 7.11 Structure model.

Solution

For $W12 \times 65$, $A = 19.1$ (in.²); $I_y = 533$ (in.⁴); $Z_y = 96.8$ (in.³); $E = 29,000$ (ksi); $\sigma_y = 50$ (ksi); $P_y = 19.1 \times 50 = 955$ (kip); $M_p = 96.8 \times 50 = 4,840$ (in.-kip).

From Equation 7.1

$$\frac{P}{P_y} + \frac{M}{1.18M_p} = 1$$

$$\Rightarrow \frac{P}{955} + \frac{0.85M}{4840} = 1$$

$$\Rightarrow M = 5694.11 - 5.9624P$$

Therefore, $A_0 = 5694.11$; $A_1 = -5.9624$; $A_2 = 0$; $A_3 = 0$. $A_0 - A_3$ were input into the program.

The moment–curvature curve is assumed to be an elastoplastic curve. Therefore, the moment–rotation curve is also elastoplastic (i.e., post-yield hardening ratio, SP = 0). The moment–rotation relations for members 1 and 2 are shown in Figure 7.12.

As mentioned in Chapter 4, the program does not consider material isotropic hardening or kinematic hardening (i.e., the interaction P–M yield curve will not move outward). Therefore, for the fully plastic condition, the force (M, P) points are always on the interaction curve. Performing pushover analysis with force control at joints 2 and 3, the pushover curve at joint 2 is shown in Figure 7.13.

Figure 7.14 shows the member's axial load–moment interactions at joints 1 and 2 while pushover load P increases. It can be seen that the force (M, P) points are always on the interaction curve when the fully plastic condition develops at joint 1. The output results are in favorable agreement with those shown in McGuire's textbook.

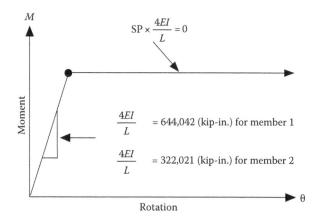

FIGURE 7.12   Elastoplastic moment–rotation curves for Members 1 and 2.

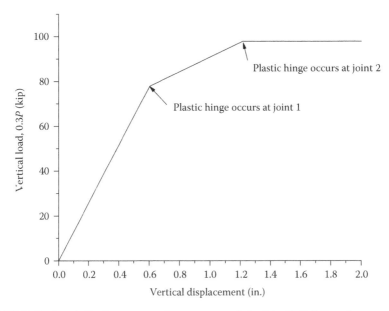

FIGURE 7.13 Load–displacement pushover curve at Joint 2 in GCS Z direction.

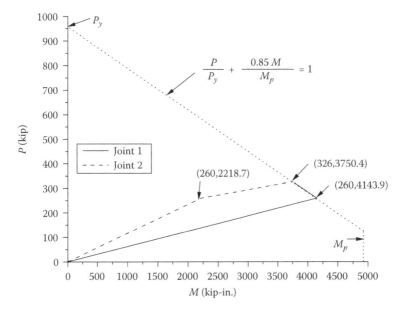

FIGURE 7.14 Member forces and interaction diagram.

Input Data (EX3_Test_PM.dat)

```
1   ECHO OF INPUT DATA

LINE  ....|..10....|..20....|..30....|..40....|..50....|..60....|..70....|..80....|..90....|..100....|..110
  1:  'STRUCTURE DEFINITION-TEST PM Method'
  2:  'EXAMPLE: Use Example 8.6 of William McGuire's Text Book'
  3:  3 1 3 0 0 1.   NNODE,NCOS , NSUPT,NCOND,NCONST    SCALE
  4:      1   0.00    0.00    .00 1 0
  5:      2  96.0     0.00    .00 1 0
  6:      3 288.0    00.0     .00 1 0
  7:      1 0 0   0 1 0 | DIRECTION COSINE
  8:      1 1 1 1 1 1 1  0 0
  9:      2 0 1 0 1 0 1  0 0
 10:      3 0 1 1 1 0 1  0 0
 11:              6 |NMAT
 12:  'IA_BILN MAT#1: MYA ' 4 0.00001 29000.0 533.0 4840.0 -1 0 -1.
 13:  0
 14:  5694.1123 -5.96242 0.   0.
 15:  0
 16:  'IA_BILN MAT#1: MYA ' 4 0.00001 29000.0 533.0 4840.0 -1 0 -1.
 17:  0
 18:  5694.1123 -5.96242 0.   0.
 19:  0
 20:  'IA_BILN MAT#3: MZA ' 4 0.0 0.0 29000.0 533.0 4840.0 -1 0 -1.
 21:  'IA_BILN MAT#4: MZA ' 4 0.0 0.0 29000.0 533.0 4840.0 -1 0 -1.
 22:  'IA_BILN MAT#5: MXA ' 4 0.0 0.0 13000.0 533.0 4840.0 -1 0 -1.
 23:  'IA_BILN MAT#6: FXA ' 4 0.0 0.0 29000.0 19.1  955.0 -1 0 -1.
 24:  0 0 0 .FALSE.          |KG: AXL, FORM, ASSY
 25:      2  NELEM
 26:  'IE3DBEAM'  'W12x65 STEEL MEMBER 1' 1 2 3 4 5 6 1 2 0 1 0 0 0 0.
 27:  0 0 0 0 0 0 0 0 0 0 0 0 0 0
 28:  'IE3DBEAM'  'W12x65 STEEL MEMBER 2' 1 2 3 4 5 6 2 3 0 1 0 0 0 0.
 29:  0 0 0 0 0 0 0 0 0 0 0 0 0 0
 30:  0 0 .FALSE. | MASS
 31:  0 0 | DAMP
 32:  'SOL04 SOLUTION'
```

```
33:  'APPLY HORIZONTAL P AT JOINT 3 AND VERTICAL 0.3P AT JOINT 2'
34:  1 50000 1  .TRUE. 0.000001  | MAXELD IPRINT IWRITE UNBAL SPLIMIT
35:  'JOINT FZ'  2  11  0  0  0
36:  'ELE  ,  1  12  0  0  0
37:  'ELE  ,  2  13  0  0  0
38:  'END  ,  3  14  0  0  0
39:  1  3  0  0  'FX'  -327
40:  1  2  0  0  'FZ'  -98.1
41:  0  0  0  0  'END'  0    |JOINT LOAD
42:  0  0  0  0  'END' 'FY'  0  0  | ELEMENT LOAD
43:  'FORCE. FROM  0  TO  -327  '  1  0  0  327
44:  'END OF FORCE CONTROL  '  0  0  0  0
45:  'READ UNIT=11 UNIT=12 UNIT=13'
46:  'STOP'
```

7.3.4 EXAMPLE 4: TWO-COLUMN BENT (DISPLACEMENT CONTROL)

A two-column bent used in the MCEER/ATC Design Example No. 8 (MCEER/ATC, 2003) was chosen for the pushover analysis (see Figure 7.15). The column size and its cross-sectional details are the same as those in Example 1. They are diameter = 48″, 20-#10 longitudinal bars, $f_c' = 4$ ksi, $f_y = 60$ ksi, spiral = #5 @3.25″, concrete cover = 2.6″, and applied column axial dead loads = 765 kip. The post-yield modulus of the steel stress–strain curve is 1% of the elastic modulus. The column plastic hinge length is 33 in. from Equation 4.5. The foundation of the bent structure is assumed to be fixed in this example. The column shear failure mode and the cap beam–column joint shear failure mode are not considered in this example.

The structural model is shown in Figure 7.15. The superstructural bent cap beam is assumed to be an elastic 3D-BEAM member with the properties of $AX = 3888$ (in.²); $J = IY = IZ = 2.07 \times 10^8$ (in.⁴).

Perform pushover analyses by using (1) plastic hinge length (PHL), (2) PM, (3) FSFS, and (4) finite segment–moment curvature (FSMC) methods. Pushover

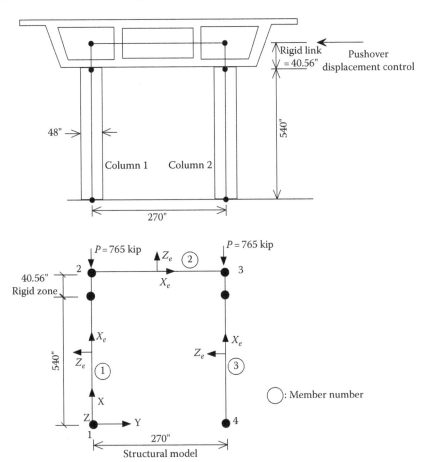

FIGURE 7.15 Two-column bent structural model.

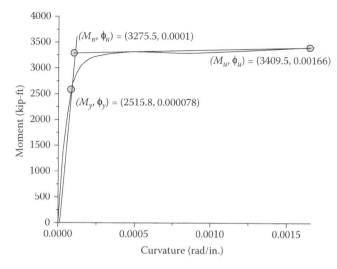

FIGURE 7.16 Bilinear moment–curvature curve (with 765 kip dead load).

displacement control is used at joint 3 in the negative global coordinate system (GCS) Y-direction until the displacement reaches 35″. P-δ effect is considered in the analysis. Find the displacement capacity of the bent from the pushover analysis. For the PHL and PM methods, the displacement capacity of the bent is defined when the first column reaches its plastic rotational capacity. The plastic rotation capacity of the column used here is 0.0515 (rad), which was calculated based on the bilinear moment–curvature curve in Figure 7.16 and plastic hinge length. The calculation of plastic rotation capacity is described below.

1. PHL Method

From Example 1, the bilinear moment curvature curve is shown as follows:

The moment–rotation relationship for PHL method is shown in Figure 7.17, which was calculated based on the idealized bilinear moment–curvature curve from Figure 7.16. This curve has a post-yield slope equal to 0.24% of the initial slope. The 0.24% post-yield slope can be calculated as follows:

Based on Equation 4.16, the rotation θ_n at moment equal to $M_n = 3275.5$ k-ft is

$$\theta_n = \phi_n \left[L_p - \frac{1}{L} * \frac{L_p^2}{2} \right] = 0.0001 \left(33.0 - \frac{1}{270} \frac{(33.0)^2}{2} \right) = 0.0031 \text{ (rad)}$$

From Equation 4.17, the plastic rotation capacity θ_H at ultimate moment of $M_u = 3409.5$ kip-ft is

$$\theta_H = (\phi_u - \phi_n)L_p = (0.00166 - 0.0001)(33.0) = 0.0515 \text{ (rad)}$$

The rotation θ_u at moment M_u is

$$\theta_u = \theta_n + \theta_H = 0.0031 + 0.0515 = 0.055 \text{ (rad)}$$

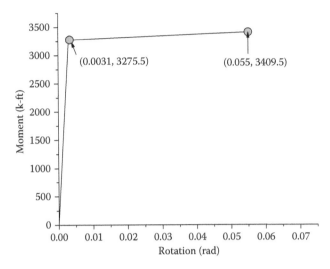

FIGURE 7.17   Moment–rotation relationships used in PHL method.

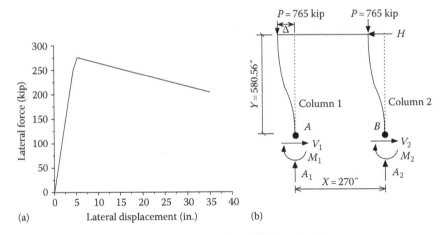

FIGURE 7.18   (a) Pushover curve at Joint 3; (b) equilibrium check.

The initial slope in the moment–rotation curve is $M_n/\theta_n = 3275.5/0.0031 = 1{,}056{,}612.9$ (k-ft/rad). The post slope is $(M_u - M_n)/(\theta_u - \theta_n) = (3409.5 - 3275.5)/(0.055 - 0.0031) = 2582$ (k-ft/rad). Therefore, the post slope is 0.24% of the initial slope.

The effect of column axial load and moment interaction is considered here, but the column axial load–plastic rotation capacity interaction is not considered. The P–M interaction curve with $A_0 = 28{,}989.6$, $A_1 = 15.31$, $A_2 = -0.0024$, and $A_3 = 6.684E - 8$ is shown in Figure 7.6 in Example 1. The output results are shown as follows (Figure 7.18a and b):

Output (EX4_PHL2.out file)

```
1  ECHO OF INPUT DATA

LINE  ....|....10....|....20....|....30....|....40....|....50....|....60....|....70....|....80....|....90....|...100....|...110
   1: 'STRUCTURE DEFINITION-TEST NCHRP-12-49 EXAMPLE 8'
   2: 'EXAMPLE: TWO-COLUMN BENT,use IE3DBEAM MAT19,plastic hinge length METHOD'
   3:  4 1 4 0 0 1.        NNODE,NCOS , NSUPT,NCOND,NCONST   SCALE
   4:  1    0.00    0.00    .00 1 0
   5:  2  580.56    0.00    .00 1 0
   6:  3  580.56  270.0     .00 1 0
   7:  4    0.00  270.0     .00 1 0
   8:  1 0 0    0 1 0  | DIRECTION COSINE
   9:  1 1 1 1 1 1 0 0
  10:  2 0 0 1 1 1 0 0
  11:  3 0 1 1 1 1 0 0
  12:  4 1 1 1 1 1 0 0
  13:  7  |NMAT
  14: 'HINGE  MAT#1: MYA '  0.0031  39306  0.0024  -1  0  393060000.  0.0515  1
  15: 0
  16: 28989.6  15.31  -0.0024  6.684E-8
  17: 0
  18: 'HINGE  MAT#2: MYB '  0.0031  39306  0.0024  -1  0  333060000.  0.0515  1
  19: 0
  20: 28989.6  15.31  -0.0024  6.684E-8
  21: 0
  22: 'IA_BILN MAT#3: MZA '  0  0.001  3605.  260576.  38400.  -1  0
  23: 'IA_BILN MAT#4: MZB '  0  0.001  3605.  260576.  38400.  -1  0
  24: 'IA_BILN MAT#5: MXA '  0  0.001  3605.  521152.  38400.  -1  0
  25: 'IA_BILN MAT#6: FXA '  0  0.001  3605.  1809.6   38400.  -1  0
  26: '3D-BEAM MAT#7'  3122.  1200.  3888.  0 0 207000000.  207000000.  207000000.
  27:  2 1 1 .TRUE.  |KG: AXL, FORM, ASSY
  28:  3  NELEM
  29: 'IE3DBEAM' 'R/C CIRCULAR MEMBER 1' 1 2 3 4 5 6  1 2  0 0 1  0 40.56  765.
  30:  0 0 0 0 0 0 0 0 0 0 0 0 0 0
  31: '3D-BEAM' 'MEMBER 2' 7
  32: 'IE3DBEAM' 'R/C CIRCULAR MEMBER 3' 1 2 3 4 5 6  4 3  0 0 1  0 40.56  765.
  33:  0 0 0 0 0 0 0 0 0 0 0 0
  34:  0 0 .FALSE. | MASS
```

```
35:        0  0   | DAMP
36:  'SOL01  SOLUTION'
37:  'APPLY AXIAL LOAD AT JOINT 2 AND JOINT 3'
38:  1 1      | NLOAD MAXELD
39:  1  2  0  0  'FX'  -765.  |JOINT DEAD LOAD
40:  1  3  0  0  'FX'  -765.  |JOINT DEAD LOAD
41:  0  0  0  0  'END'  0.
42:  0  0  0  0  'END'  'FZ'   0.    0.
43:  'SOL04 SOLUTION'
44:  'INCREMEMTAL DISPLACEMENT CONTROL AT JOINT 3'
45:  1 50000 2  .TRUE.  0 |MAXELD IPRINT IWRITE UNBAL SPLIMIT
46:  'JOINT FY'  3  11  0  0
47:  'ELE  `  1  12  0  0
48:  'ELE  `  3  13  0  0
49:  'JOINT FY'  1  24  0  0
50:  'END  `  2  15  0  0
51:  1  3  0  'FY'  -35
52:  0  0  0  0  'END'  0        |JOINT LOAD
53:  0  0  0  0  'END' 'FY'  0  0  | ELEMENT LOAD
54:  'DISP. FROM 0   TO  -35  `  1  0  0  800
55:  'END OF DISP. CONTROL  `  0  0  0
56:  'READ UNIT=11 UNIT=12 UNIT=13 UNIT=24'
```

PLASTIC HINGE LENGTH MOMENT-ROTATION MODEL
==

MAT.	HA	VA	RATIO	MAX DUC	BETA	STIELE	PRMAX
1	0.310000E-02	39306.0	0.240000E-02	-1.00000	0.00000	0.393060E+09	0.515000E-01

P-MY INTERACTION EQUATION:
$M(P)=A0 + A1*P + A2*P**2 + A3*P**3$
WHERE   A0=   28989.6
        A1=   15.3100
        A2=   -0.240000E-02
        A3=   0.668400E-07

```
2  0.310000E-02  39306.0     0.240000E-02-1.00000     0.00000     0.393060E+090.515000E-01

P-MY INTERACTION EQUATION:
M(P)=A0 + A1*P + A2*P**2 + A3*P**3
WHERE   A0=   28989.6
        A1=   15.3100
        A2=  -0.240000E-02
        A3=   0.668400E-07
```

1 STRUCTURE.....: EXAMPLE: TWO-COLUMN BENT,use IE3DBEAM MAT19,plastic hinge length METHOD   TIME: 10:10:58, DATE: 31-AUG-09
  SOLUTION.....:

BILINEAR INTERACTIVE MATERIAL PROPERTIES
==

MAT.	ELAS	SP	E	TI	MP	MAX DUC	BETA	REDU F	PRMAX
3	0.00000	0.100000E-02	3605.00	260576.	38400.0	0.00000	0.00000	0.00000	-1.00000
4	0.00000	0.100000E-02	3605.00	260576.	38400.0	0.00000	0.00000	0.00000	-1.00000
5	0.00000	0.100000E-02	3605.00	521152.	38400.0	0.00000	0.00000	0.00000	-1.00000
6	0.00000	0.100000E-02	3605.00	1809.60	38400.0	0.00000	0.00000	0.00000	-1.00000

1 STRUCTURE.....: EXAMPLE: TWO-COLUMN BENT,use IE3DBEAM MAT19,plastic hinge length METHOD   TIME: 10:10:58, DATE: 31-AUG-09
  SOLUTION.....:

3-D ELASTIC BEAM ELEMENT
=========================

MATL.	E	GAMMA	AX	AY	AZ	IX	IY	IZ
7	3.122E+03	1.200E+03	3.888E+03	0.00	0.00	0.00	2.070E+08	2.070E+08	2.070E+08

```
***** IA_BILN ELEMENT                     3 YIELD AT END A
****  THE ELEMENT FORCES DISP. FROM STEP:  103 FACTOR:  1.000      ARE:

  3   2   DISPL   4       0.00000        0.00000        0.00000      8.249833E-03    0.00000    FLP:  3.0
          DISPL   3      -3.844852E-02   0.00000        4.45491      8.062724E-03    0.00000

  3   2   FORCE   4     464.487          0.00000     -127.789      35590.2           0.00000    CHR: 001000
          FORCE   3    -464.487          0.00000      127.789      35485.0           0.00000
%8 % LIMIT STATE POINT 2 DEFINED DUE TO ELEMENT NO   3 AT STEP=  103


***** IA_BILN ELEMENT                     3 YIELD AT END B
****  THE ELEMENT FORCES DISP. FROM STEP:  104 FACTOR:  1.000      ARE:

  3   2   DISPL   4       0.00000        0.00000        0.00000      8.330754E-03    0.00000    FLP:  4.0
          DISPL   3      -3.827270E-02   0.00000        4.49861      8.142346E-03    0.00000

  3   2   FORCE   4     462.363          0.00000     -127.919      35593.8           0.00000    CHR: 001100
          FORCE   3    -462.363          0.00000      127.919      35562.1           0.00000


***** IA_BILN ELEMENT                     1 YIELD AT END A
****  THE ELEMENT FORCES DISP. FROM STEP:  124 FACTOR:  1.000      ARE:

  1   2   DISPL   1       0.00000        0.00000        0.00000      9.943586E-03    0.00000    FLP:  3.0
          DISPL   2      -9.080197E-02   0.00000        5.36953      9.736083E-03    0.00000

  1   2   FORCE   1    1096.96           0.00000     -147.995      42984.6           0.00000    CHR: 001000
          FORCE   2   -1096.96           0.00000      147.995      42822.8           0.00000


***** IA_BILN ELEMENT                     1 YIELD AT END B
****  THE ELEMENT FORCES DISP. FROM STEP:  125 FACTOR:  1.000      ARE:

  1   2   DISPL   1       0.00000        0.00000        0.00000      1.002457E-02    0.00000    FLP:  4.0
          DISPL   2      -9.085673E-02   0.00000        5.41327      9.816640E-03    0.00000

  1   2   FORCE   1    1097.62           0.00000     -148.218      42988.3           0.00000    CHR: 001100
          FORCE   2   -1097.62           0.00000      148.218      42991.1           0.00000
```

```
* ELEMENT    3  INELASTIC ROTATION IN MY  DIR. AT END A  =  0.5151E-01 >  PLASTIC ROTATION
CAPACITY OF   0.5150E-01AT STEP = 740
%9 % LIMIT STATE POINT 3 DEFINED DUE TO ELEMENT NO   3 AT STEP= 740
.
.
1 STRUCTURE.....: EXAMPLE: TWO-COLUMN BENT, use IE3DBEAM MAT19, plastic hinge length METHOD      TIME: 10:10:58, DATE: 31-AUG-09
SOLUTION......: INCREMEMTAL DISPLACEMENT CONTROL AT JOINT 3                                       TIME: 10:10:58, DATE: 31-AUG-09

  , LOADING #   0   PEAK DUCTILITIES

  IE3DBEAM MEMBER DUCTILITY:
              ELEM#     |MYA|      |MYB|      |MZA|      |MZB|      |MX |      |FX |
  DUCTILITY :   1     6.46363    6.52546    0.00000    0.00000    0.00000    0.00000
  YIELD DISP:   1     0.01002    0.00990    0.00000    0.00000    0.00000    0.00000

  IE3DBEAM MEMBER DUCTILITY:
              ELEM#     |MYA|      |MYB|      |MZA|      |MZB|      |MX |      |FX |
  DUCTILITY :   3     7.77826    7.85539    0.00000    0.00000    0.00000    0.00000
  YIELD DISP:   3     0.00833    0.00822    0.00000    0.00000    0.00000    0.00000

1 STRUCTURE.....: EXAMPLE: TWO-COLUMN BENT, use IE3DBEAM MAT19, plastic hinge length METHOD      TIME: 10:10:58, DATE: 31-AUG-09
SOLUTION......: INCREMEMTAL DISPLACEMENT CONTROL AT JOINT 3                                       TIME: 10:10:58, DATE: 31-AUG-09
%5S%

DEGREE OF FREEDOM #  15  IS READ FROM UNIT #  11   JOINT #   3, DIRECTION: FY

     STEP    TIME       LOAD          DISPLACEMENT      VELOCITY    ACCELERATION
       0    0.0000    -2.74601E-06      0.0000          0.0000        0.0000
       2    0.0000    -4.9246          -8.75000E-02     0.0000        0.0000
       4    0.0000    -9.8492          -0.17500         0.0000        0.0000
       .
       .
       .
     102    0.0000    -251.16          -4.4625          0.0000
*2*  104    0.0000    -253.36          -4.5500          0.0000
     106    0.0000    -255.73          -4.6375          0.0000
       .
       .
       .
```

```
        738    0.0000    -210.84    -32.288    0.0000    0.0000
*3*     740    0.0000    -210.63    -32.375    0.0000    0.0000
        742    0.0000    -210.42    -32.463    0.0000    0.0000
        .
        .
        .
        798    0.0000    -204.49    -34.913    0.0000    0.0000
        800    0.0000    -204.28    -35.000    0.0000    0.0000
%5E%
```

```
1 STRUCTURE.....: EXAMPLE: TWO-COLUMN BENT, use IE3DBEAM MAT19, plastic hinge length METHOD    TIME: 10:10:58, DATE: 31-AUG-09
  SOLUTION......: INCREMEMTAL DISPLACEMENT CONTROL AT JOINT 3                                   TIME: 10:10:58, DATE: 31-AUG-09
%5S%
```

```
ELEMENT #    1 IS READ FROM UNIT #   12
```

IE3D BEAM FORCES...

STEP	TIME	NODE	AXIAL	FY	FZ	TORSION	MY	MZ	STBFAG & FLP
0	0.0000	1	765.000	0.00000	-1.448560E-06	0.00000	2.794090E-04	0.00000	
		2	-765.000	0.00000	1.448560E-06	0.00000	5.076500E-04	0.00000	000000
		DISP	-6.332390E-02	0.00000	0.00000	0.00000	1.171610E-11	0.00000	
				0.00000	6.326700E-09	0.00000	1.684990E-10	0.00000	0.0
2	0.0000	1	770.890	0.00000	-2.46060	0.00000	700.680	0.00000	
		2	-770.890	0.00000	2.46060	0.00000	695.340	0.00000	000000
		DISP	-6.381160E-02	0.00000	0.00000	0.00000	1.616600E-04	0.00000	
				0.00000	8.729650E-02	0.00000	1.579920E-04	0.00000	0.0
					.				
					.				
					.				
800	0.0000	1	1099.10	0.00000	-94.1460	0.00000	44647.0	0.00000	
		2	-1099.10	0.00000	94.1460	0.00000	44649.0	0.00000	001100
		DISP	-9.098140E-02	0.00000	0.00000	0.00000	6.479510E-02	0.00000	
				0.00000	34.9890	0.00000	6.458690E-02	0.00000	4.0

```
%5E%
```

```
1 STRUCTURE.....: EXAMPLE: TWO-COLUMN BENT, use IE3DBEAM MAT19, plastic hinge length METHOD        TIME: 10:10:58, DATE: 31-AUG-09
  SOLUTION......: INCREMEMTAL DISPLACEMENT CONTROL AT JOINT 3                                        TIME: 10:10:58, DATE: 31-AUG-09
%5S%
  ELEMENT #    3 IS READ FROM UNIT #   13

  IE3D BEAM FORCES....
```

STEP	TIME	NODE	AXIAL	FY	FZ	TORSION	MY	MZ	STBFAG & FLP
0	0.0000	4	765.000	0.00000	-1.280530E-06	0.00000	2.470730E-04	0.00000	
		3	-765.000	0.00000	1.280530E-06	0.00000	4.487110E-04	0.00000	000000
		DISP	0.00000	0.00000	0.00000	0.00000	1.040050E-11	0.00000	
			-6.332390E-02	0.00000	5.617910E-09	0.00000	1.489120E-10	0.00000	0.0
2	0.0000	4	759.110	0.00000	-2.46410	0.00000	701.130	0.00000	
		3	-759.110	0.00000	2.46410	0.00000	695.790	0.00000	000000
		DISP	0.00000	0.00000	0.00000	0.00000	1.617620E-04	0.00000	
			-6.283620E-02	0.00000	8.735120E-02	0.00000	1.580330E-04	0.00000	0.0
800	0.0000	4	430.880	0.00000	-110.180	0.00000	37304.0	0.00000	
		3	-430.880	0.00000	110.180	0.00000	37271.0	0.00000	001100
		DISP	0.00000	0.00000	0.00000	0.00000	6.479880E-02	0.00000	
			-3.566640E-02	0.00000	34.9920	0.00000	6.455250E-02	0.00000	4.0

```
. . .
%5E%
```

The output result shows that the nominal moment occurred at member 3 at Step 103 (i.e., pushover displacement=4.5″), and the displacement capacity of the bent is reached at Step 740 (pushover displacement=32.3″) at which the plastic rotation of member 3 exceeds the plastic capacity of 0.0515 rad.

Equilibrium check at Step = 800:
From output plot file units 11, 12, and 13
$H=204.28$ kip; $\Delta=35''$
Column No. 1 end forces: $M_1=44,647$ k-in.; $V_1=94.146$ kip; $A_1=1,099.1$ kip
Column No. 2 end forces: $M_2=37,304$ k-in.; $V_2=110.18$ kip; $A_2=430.88$ kip

$$\sum A = 1530 \text{ (kip)} = 2P \text{ (ok)}$$

$$\sum V = 204.3 \text{ (kip)} \approx H \text{ (ok)}$$

$$\sum M_B = 0:$$

$$\bar{M}_1 + M_2 + A_1 * 270 - 580.56 * H - P\Delta - P*(270+\Delta) = 0$$

$$\rightarrow \bar{M}_1 = P*(270+2\Delta) + 580.56*(H) - 270*A_1 - M_2 = 44,662 \approx M_1 \text{ (ok)}$$

Similarly

$$\bar{M}_2 = -P*(270-2\Delta) + 580.56*(H) + 270*A_2 - M_1 = 37,288 \approx M_2 \text{ (ok)}$$

Output (EX4_PHL2B.out file)

If the column axial–plastic rotation capacity interaction is considered in the PHL method, the output results are shown as follows:

```
1   ECHO OF INPUT DATA

LINE   ....|.. 10....|.. 20....|.. 30....|.. 40....|.. 50....|.. 60....|.. 70....|.. 80....|.. 90....|..100....|..110
  1:  'STRUCTURE DEFINITION-TEST NCHRP-12-49 EXAMPLE 8'
  2:  'EXAMPLE: TWO-COLUMN BENT, use IE3DBEAM MAT19,plastic hinge length METHOD'
  3:                       NNODE,NCOS , NSUPT,NCOND,NCONST     SCALE
  4:      4  1 4 0 0  1.
  4:   1    0.00    0.00        .00 1 0
  5:   2  580.56    0.00        .00 1 0
  6:   3  580.56  270.0         .00 1 0
  7:   4    0.00  270.0         .00 1 0
  8:   1 0 0     0 1 0   | DIRECTION COSINE
  9:   1    1 1 1 1 1 1      0 0
 10:   2    0 0 1 1 1 0      0 0
 11:   3    0 1 1 1 1 0      0 0
 12:   4    1 1 1 1 1 1      0 0
 13:             7   |NMAT
 14:  'HINGE  MAT#1: MYA '  0.0031    39306   0.0024   -1   0   393060000.   0.0515   3
 15:  1   |ICHOICE
 16:  7   |M
 17:  0.
 18:  1.068400E+03        2.871880E+04
 19:  2.136810E+03        4.243420E+04
 20:  3.205210E+03        5.112170E+04
 21:  4.273610E+03        5.591830E+04
 22:  5.342020E+03        5.582170E+04
 23:  6.410420E+03        4.537660E+04
 24:  1   |ICHOICE
 25:  7   |M
 26:  0.        6.436713E-02
 27:  1.068400E+03        4.345796E-02
 28:  2.136810E+03        3.101939E-02
 29:  3.205210E+03        2.364369E-02
 30:  4.273610E+03        1.973302E-02
 31:  5.342020E+03        1.697527E-02
 32:  6.410420E+03        1.478998E-02
 33:  0   |ICHOICE3
 34:  'HINGE  MAT#2: MYB '  0.0031    39306   0.0024   -1   0   393060000.   0.0515   3
 35:  1   |ICHOICE
 36:  7   |M
 37:  0.        2.871880E+04
```

'FATIGUE'
'BUCKLING'
'SPLICE'
'CONCRETE'
'CONCRETE'
'CONCRETE'
'CONCRETE'

```
38:  1.068400E+03    4.243420E+04
39:  2.136810E+03    5.112170E+04
40:  3.205210E+03    5.591830E+04
41:  4.273610E+03    5.582170E+04
42:  5.342020E+03    5.144160E+04
43:  6.410420E+03    4.537660E+04
44:  1    |ICHOICE
45:  7    |M
46:  0.        6.436713E-02        'FATIGUE'
47:  1.068400E+03    4.345796E-02    'BUCKLING'
48:  2.136810E+03    3.101939E-02    'SPLICE'
49:  3.205210E+03    2.364369E-02    'CONCRETE'
50:  4.273610E+03    1.973302E-02    'CONCRETE'
51:  5.342020E+03    1.697527E-02    'CONCRETE'
52:  6.410420E+03    1.478998E-02    'CONCRETE'
53:  0    |ICHOICE3
54:  'IA_BILN MAT#3:  MZA '   0   0.001   3605.   260576.   38400.   -1   0   -1
55:  'IA_BILN MAT#4:  MZB '   0   0.001   3605.   260576.   38400.   -1   0   -1
56:  'IA_BILN MAT#5:  MXA '   0   0.001   3605.   521152.   38400.   -1   0   -1
     ECHO OF INPUT DATA

LINE   ....|...10....|...20....|...30....|...40....|...50....|...60....|...70....|...80....|...90....|..100....|..110
57:  'IA_BILN MAT#6:  FXA '   0   0.001   3605.   1809.6   38400.   -1   0   -1
58:  '3D-BEAM MAT#7'  3122.  1200.  3888.  0  207000000.  207000000.  207000000.
59:  2  1  1  .TRUE.    |KG:  AXL,  FORM,  ASSY
60:  3   NELEM
61:  'IE3DBEAM'  'R/C CIRCULAR MEMBER 1'  1  2  3  4  5  6  1  2  0  0  1  0  40.56  765.
62:  0  0.316  48.  0.  40.56  0.  0  0.316  48.  0.  40.56  40.56  0.  0
63:  '3D-BEAM'   'MEMBER 2'  7  2  3  0  0  1  0  0  000000  0
64:  'IE3DBEAM'  'R/C CIRCULAR MEMBER 3'  1  2  3  4  5  6  4  3  0  0  1  0  40.56  765.
65:  0  0.316  48.  0.  40.56  0.  0  0.316  48.  0.  40.56  40.56  0.  0
66:  0  0  .FALSE.  |  MASS
67:  0  0  |  DAMP
68:  'SOL01  SOLUTION'
69:  'APPLY AXIAL LOAD AT JOINT 2 AND JOINT 3'
70:  1  1   | NLOAD MAXELD
71:  1  2  0  0  'FX'  -765.    |JOINT DEAD LOAD
72:  1  3  0  0  'FX'  -765.    |JOINT DEAD LOAD
73:  0  0  0  0  'END'  0.
74:  0  0  0  0  'END'  'FZ'  0.  0.
```

```
75:    'SOLO4 SOLUTION'
76:    'INCREMEMTAL DISPLACEMENT CONTROL AT JOINT 3'
77:    1 50000 2  .TRUE.  0  |MAXELD IPRINT IWRITE UNBAL SPLIMIT
78:    'JOINT FY'  3  11  0   0
79:            'ELE  '  1  12  0   0   0
80:            'ELE  '  3  13  0   0   0
81:    'JOINT FY'  1  24  0   0
82:    'END  '  2  15  0   0
83:    1  3  0  'FY'  -35
84:    0  0  0  'END'  0        |JOINT LOAD
85:    0  0  0  'END'  'FY'  0   0  |  ELEMENT LOAD
86:    'DISP. FROM  0    TO  -35    '  1   0   0   800
87:    'END OF DISP. CONTROL           0   0   0    0
88:    'READ UNIT=11 UNIT=12 UNIT=13 UNIT=24'
89:    'STOP'
```

PLASTIC HINGE LENGTH MOMENT-ROTATION MODEL
===

MAT.	HA	VA	RATIO	MAX DUC	BETA	STIELE	PRMAX
1	0.310000E-02	39306.0	0.240000E-02	-1.00000	0.00000	0.393060E+09	0.515000E-01

TOTAL DATA POINTS FOR A0-A3 (M) = 7
TOTAL COEFS FOR POLYNOMIAL (N) = 4
DATA POINT X=AXIAL LOAD; Y=NOMINAL MOMENT

DATA POINT NO.	1:	X=	0.00000	Y=	28718.8
DATA POINT NO.	2:	X=	1068.40	Y=	42434.2
DATA POINT NO.	3:	X=	2136.81	Y=	51121.7
DATA POINT NO.	4:	X=	3205.21	Y=	55918.3
DATA POINT NO.	5:	X=	4273.61	Y=	55821.7
DATA POINT NO.	6:	X=	5342.02	Y=	51441.6
DATA POINT NO.	7:	X=	6410.42	Y=	45376.6

TOTAL DATA POINTS FOR B0-B3 (M) = 7
TOTAL COEFS FOR POLYNOMIAL (N) = 4
DATA POINT X=AXIAL LOAD; Y=PLASTIC ROTATION CAPACITY

| DATA POINT NO. | 1: | X= | 0.00000 | Y= | 0.643671E-01 |
| DATA POINT NO. | 2: | X= | 1068.40 | Y= | 0.434580E-01 |

```
DATA POINT NO.    3: X=    2136.81    Y=    0.310194E-01
DATA POINT NO.    4: X=    3205.21    Y=    0.236437E-01
DATA POINT NO.    5: X=    4273.61    Y=    0.197330E-01
DATA POINT NO.    6: X=    5342.02    Y=    0.169753E-01
DATA POINT NO.    7: X=    6410.42    Y=    0.147900E-01
```

P-MY INTERACTION EQUATION:
$$M(P)=A0 + A1*P + A2*P**2 + A3*P**3$$
```
WHERE    A0=    28622.8
         A1=    15.6001
         A2=    -0.246176E-02
         A3=    0.672645E-07
```

P-PLASTIC ROTATION CAPACITY (PRC) INTERACTION EQUATION:
$$PRC(P)=B0 + B1*P + B2*P**2 + B3*P**3$$
```
WHERE    B0=    0.642238E-01
         B1=    -0.232192E-04
         B2=    0.414012E-08
         B3=    -0.269005E-12
```

. . .

DEGREE OF FREEDOM #   15 IS READ FROM UNIT #   11   JOINT #   3, DIRECTION: FY

STEP	TIME	LOAD	DISPLACEMENT	VELOCITY	ACCELERATION
0	0.0000	-2.74601E-06	0.0000	0.0000	0.0000
2	0.0000	-4.9246	-8.75000E-02	0.0000	0.0000

. . .

100	0.0000	-246.23	-4.3750	0.0000	0.0000
2 102	0.0000	-250.16	-4.4625	0.0000	0.0000
104	0.0000	-252.55	-4.5500	0.0000	0.0000

. . .

```
       660    0.0000    -217.79    -28.875    0.0000    0.0000
*3*    662    0.0000    -217.58    -28.963    0.0000    0.0000
       664    0.0000    -217.36    -29.050    0.0000    0.0000
 . .
 . .
       798    0.0000    -203.18    -34.913    0.0000    0.0000
       800    0.0000    -202.96    -35.000    0.0000    0.0000
%5E%
```

1 STRUCTURE......: EXAMPLE: TWO-COLUMN BENT, use IE3DBEAM MAT19, plastic hinge length METHOD   TIME: 10:24:20, DATE: 31-AUG-09
  SOLUTION......: INCREMEMTAL DISPLACEMENT CONTROL AT JOINT 3   TIME: 10:24:20, DATE: 31-AUG-09

```
%5S%
ELEMENT #    1 IS READ FROM UNIT #    12
```

IE3D BEAM FORCES...

STEP	TIME	NODE	AXIAL	FY	FZ	TORSION	MY	MZ	STBFAG & FLP
0	0.0000	1	765.000	0.00000	-1.448560E-06	0.00000	2.794J90E-04	0.00000	
		2	-765.000	0.00000	1.448560E-06	0.00000	5.076500E-04	0.00000	000000
		DISP	-6.332390E-02	0.00000	0.00000	0.00000	1.171610E-11	0.00000	
				0.00000	6.326700E-09	0.00000	1.684990E-10	0.00000	0.0
2	0.0000	1	770.890	0.00000	-2.46060	0.00000	700.680	0.00000	
		2	-770.890	0.00000	2.46060	0.00000	695.340	0.00000	000000
		DISP	-6.381160E-02	0.00000	0.00000	0.00000	1.616600E-04	0.00000	
				0.00000	8.729650E-02	0.00000	1.575920E-04	0.00000	0.0
800	0.0000	1	1097.60	0.00000	-93.7560	0.00000	445-4.0	0.00000	
		2	-1097.60	0.00000	93.7560	0.00000	445-7.0	0.00000	001100
		DISP	-9.085230E-02	0.00000	0.00000	0.00000	6.479520E-02	0.00000	
				0.00000	34.9890	0.00000	6.453800E-02	0.00000	17.

```
 . . .
%5E%
```

```
1 STRUCTURE.....: EXAMPLE: TWO-COLUMN BENT,use IE3DBEAM MAT19,plastic hinge length METHOD      TIME: 10:24:20, DATE: 31-AUG-09
  SOLUTION......: INCREMEMTAL DISPLACEMENT CONTROL AT JOINT 3                                   TIME: 10:24:20, DATE: 31-AUG-09
%5S%

  ELEMENT #    3 IS READ FROM UNIT #    13

  IE3D BEAM FORCES...
```

STEP	TIME	NODE	AXIAL	FY	FZ	TORSION	MY	MZ	STBFAG & FLP
0	0.0000	4	765.000	0.00000	-1.280530E-06	0.00000	2.470730E-04	0.00000	
		3	-765.000	0.00000	1.280530E-06	0.00000	4.487110E-04	0.00000	000000
		DISP	0.00000	0.00000	0.00000	0.00000	1.040350E-11	0.00000	
			-6.332390E-02	0.00000	5.617910E-09	0.00000	1.489120E-10	0.00000	0.0
2	0.0000	4	759.110	0.00000	-2.46410	0.00000	701.130	0.00000	
		3	-759.110	0.00000	2.46410	0.00000	695.790	0.00000	000000
		DISP	0.00000	0.00000	0.00000	0.00000	1.617620E-04	0.00000	
			-6.283620E-02	0.00000	8.735120E-02	0.00000	1.580930E-04	0.00000	0.0
. . .									
800	0.0000	4	432.440	0.00000	-109.270	0.00000	37101.0	0.00000	
		3	-432.440	0.00000	109.270	0.00000	37036.0	0.00000	001100
		DISP	0.00000	0.00000	0.00000	0.00000	6.479890E-02	0.00000	
			-3.579560E-02	0.00000	34.9920	0.00000	6.459350E-02	0.00000	10.

```
%5E%
```

The output result shows that the displacement capacity of the bent is reached at Step 661 (pushover displacement$=28.9''$) at which the plastic rotation of member 1 exceeds the plastic capacity of 0.0434 rad. It can be seen that the displacement capacity of the bent reduces from 32.3'' to 28.9'' if the column axial–plastic rotation capacity interaction is included in the analysis.

2. PM Method

As shown in Figure 7.16, the post-yield slope of the idealized bilinear moment–curvature curve is close to zero (0.26%). Assume that the post-yield slope is equal to zero (i.e., SP=0).

The effect of column axial load and moment interaction is considered here. The P–M interaction curve with $A_0=28{,}989.6$, $A_1=15.31$, $A_2=-0.0024$, and $A_3=6.684E-8$ is shown in Figure 7.6 in Example 1. Since SP=0 (i.e., no isotropic hardening) is considered here, the member force point (axial load, moment) is on the P–M yield curve when a member end is in the yield stage. The output results are shown as follows:

Output (EX4_PM3.out file)

1   ECHO OF INPUT DATA

```
LINE ....|...10....|...20....|...30....|...40....|...50....|...60....|...70....|...80....|...90....|..100....|..110
  1:      'STRUCTURE DEFINITION-TEST NCHRP 12-49 EXAMPLE 8'
  2:      'EXAMPLE:TWO-COLUMN BENT,IE3DBEAM, PM method, ELAS=4 & POSTSLOPE=0 for columns'
  3:      4 1 4 0 0 1.        NNODE,NCOS , NSUPT,NCOND,NCONST    SCALE
  4:        1   0.00    0.00      .00 1 0
  5:        2 580.56    0.00      .00 1 0
  6:        3 580.56  270.0       .00 1 0
  7:        4   0.00  270.0       .00 1 0
  8:        1 0 0     0 1 0 | DIRECTION COSINE
  9:        1  1 1 1 1 1 1    0 0
 10:        2  0 0 1 1 1 0    0 0
 11:        3  0 1 1 1 1 0    0 0
 12:        4  1 1 1 1 1 1    0 0
 13:        7  |NMAT
 14: 'IA_BILN MAT#1: MYA '  4  0   3605. 109032. 39306.   -1  0   0.0515
 15: 0
 16: 28989.6  15.31   -0.0024   6.684E-8
 17: 0
 18: 'IA_BILN MAT#2: MYB '  4  0   3605. 109032. 39306.   -1  0   0.0515
 19: 0
 20: 28989.6  15.31   -0.0024   6.684E-8
 21: 0
 22: 'IA_BILN MAT#3: MZA '  0  0.001  3605. 260576. 38400.   -1  0  -1
 23: 'IA_BILN MAT#4: MZA '  0  0.001  3605. 260576. 38400.   -1  0  -1
 24: 'IA_BILN MAT#5: MXA '  0  0.001  3605. 521152. 38400.   -1  0  -1
 25: 'IA_BILN MAT#6: FXA '  0  0.001  3605. 1809.6 38400.   -1  0  -1
 26: '3D-BEAM MAT#7'  3122.  1200.  3888. 0 0 207000000. 207000000. 207000000.
 27:  2 1 1 .TRUE.    |KG: AXL, FORM, ASSY
 28:  3  NELEM
 29: 'IE3DBEAM' 'R/C CIRCULAR MEMBER 1' 1 2 3 4 5 6  1 2  0 0 1  0 40.56  765.
 30: 0 0 0 0 0 0 0   0 0 0 0 0 0 0
 31: '3D-BEAM'  'MEMBER 2' 7      2 3  0 0 1 0 0 0  000000  0
 32: 'IE3DBEAM' 'R/C CIRCULAR MEMBER 3' 1 2 3 4 5 6  4 3  0 0 1  0 40.56  765.
 33: 0 0 0 0 0 0 0   0 0 0 0 0 0 0
 34:   0 0 .FALSE. | MASS
```

```
35:      0 0   | DAMP
36:  'SOL01 SOLUTION'
37:  'APPLY AXIAL LOAD AT JOINT 2 AND JOINT 3'
38:  1 1     | NLOAD MAXELD
39:  1 2 0 0 'FX'  -765.  |JOINT DEAD LOAD
40:  1 3 0 0 'FX'  -765.  |JOINT DEAD LOAD
41:  0 0 0 0 'END' 0.
42:  0 0 0 0 'END' 'FZ'  0.   0.
43:  'SOL04 SOLUTION'
44:  'INCREMEMTAL DISPLACEMENT CONTROL AT JOINT 3'
45:  1 50000 2 .TRUE. 0 | MAXELD IPRINT IWRITE UNBAL SPLIMIT
46:  'JOINT FY' 3 11 0 0
47:            'ELE '  1 12 0 0 0
48:            'ELE '  3 13 0 0 0
49:  'JOINT FY' 1 14 0 0 0
50:  'END '     2 15 0 0 0
51:  1 3 0 0 'FY' -35
52:  0 0 0 0 'END' 0            |JOINT LOAD
53:  0 0 0 0 'END' 'FY' 0 0     | ELEMENT LOAD
54:  'DISP. FROM 0 TO -35 '   1 0 0 800
55:  'END OF LISP. CONTROL '  0 0 0 0
56:  'READ UNIT=11 UNIT=12 UNIT=13 UNIT=14'
57:  'STOP'
```

```
.
.
.

LOAD CASE:  1  JOINT:   3  DIRECTION: FY  DOF(S)  15     MAGNITUDE:  -35.0000              ...JOINT DISPLACEMENT...
1 STRUCTURE.....: EXAMPLE:TWO-COLUMN BENT, IE3DBEAM, PM method, ELAS=4 & POSTSLOPE=0 for columns    TIME: 10:34:05, DATE: 31-AUG-09
  SOLUTION......: INCREMENTAL DISPLACEMENT CONTROL AT JOINT 3                                        TIME: 10:34:05, DATE: 31-AUG-09


****  IA_BILN ELEMENT
****  THE ELEMENT FORCES DISP. FROM STEP:  103  FACTOR:  1.000    3 YIELD AT END A    ARE:                                    FLP:  3.0

3    1  DISPL  4    0.00000        0.00000       0.00000     0.00000       8.249833E-03    0.00000
        DISPL  3   -3.844846E-02   0.00000       4.45491     0.00000       8.062724E-03    0.00000
```

```
   3    1  FORCE    4    464.487        0.00000   -127.788        0.00000   35589.8         0.00000
                   3   -464.487        0.00000    127.788        0.00000   35485.0         0.00000   CHR:  001000
%8 % LIMIT STATE POINT 2 DEFINED DUE TO ELEMENT NO   3 AT STEP=  103

**** IA_BILN ELEMENT           3 YIELD AT END B
**** THE ELEMENT FORCES DISP.  FROM STEP:  104 FACTOR:  1.000    ARE:
   3    1  DISPL    4    0.00000        0.00000    0.00000        0.00000   8.330755E-03    0.00000
                   3   -3.827332E-02   0.00000    4.49861        0.00000   8.142351E-03    0.00000   FLP:   4.0
   3    1  FORCE    4    462.371        0.00000   -127.859        0.00000   35562.0         0.00000
                   3   -462.371        0.00000    127.859        0.00000   35562.0         0.00000   CHR:  001100

**** IA_BILN ELEMENT           1 YIELD AT END A
**** THE ELEMENT FORCES DISP.  FROM STEP:  124 FACTOR:  1.000    ARE:
   1    1  DISPL    1    0.00000        0.00000    0.00000        0.00000   9.943656E-03    0.00000
                   2   -9.066310E-02   0.00000    5.36958        0.00000   9.737154E-03    0.00000   FLP:   3.0
   1    1  FORCE    1    1095.28        0.00000   -147.985        0.00000   42967.0         0.00000
                   2   -1095.28        0.00000    147.985        0.00000   42826.0         0.00000   CHR:  001000

**** IA_BILN ELEMENT           1 YIELD AT END B
**** THE ELEMENT FORCES DISP.  FROM STEP:  125 FACTOR:  1.000    ARE:
   1    1  DISPL    1    0.00000        0.00000    0.00000        0.00000   1.002465E-02    0.00000
                   2   -9.070938E-02   0.00000    5.41331        0.00000   9.817779E-03    0.00000   FLP:   4.0
   1    1  FORCE    1    1095.84        0.00000   -148.162        0.00000   42967.0         0.00000
                   2   -1095.84        0.00000    148.162        0.00000   42972.8         0.00000   CHR:  001100

* ELEMENT     3 INELASTIC ROTATION IN MY  DIR. AT END A   =  0.5157E-01 > PLASTIC ROTATION
CAPACITY OF   0.5150E-01 AT STEP =  738
%9 % LIMIT STATE POINT 3 DEFINED DUE TO ELEMENT NO   3 AT STEP=  738

%5S%
```

```
DEGREE OF FREEDOM #   15 IS READ FROM UNIT #   11      JOINT #    3, DIRECTION: FY

     STEP    TIME       LOAD           DISPLACEMENT     VELOCITY      ACCELERATION
       0    0.0000    -2.74601E-06      0.0000          0.0000         0.0000
       2    0.0000    -4.9246          -8.75000E-02     0.0000         0.0000
   .
   .
   .
      102   0.0000    -251.16          -4.4625          0.0000         0.0000
*2*   104   0.0000    -253.29          -4.5500          0.0000         0.0000
      106   0.0000    -255.52          -4.6375          0.0000         0.0000
   .
   .
   .
      736   0.0000    -197.81          -32.200          0.0000         0.0000
*3*   738   0.0000    -197.56          -32.288          0.0000         0.0000
      740   0.0000    -197.31          -32.375          0.0000         0.0000
   .
   .
   .
      800   0.0000    -189.83          -35.000          0.0000         0.0000

%5E%
1 STRUCTURE.....: EXAMPLE:TWO-COLUMN BENT, IE3DBEAM, PM method, ELAS=4 & POSTSLOPE=0 for columns   TIME: 10:34:05, DATE: 31-AUG-09
  SOLUTION......: INCREMEMTAL DISPLACEMENT CONTROL AT JOINT 3                                       TIME: 10:34:05, DATE: 31-AUG-09
%5S%
ELEMENT #    1 IS READ FROM UNIT #    12

IE3D BEAM FORCES...
STEP  TIME     NODE     AXIAL          FY          FZ            TORSION       MY             MZ        STBFAG & FLP
  0  0.0000     1     765.000        0.00000   -1.448560E-06     0.00000    2.794100E-04    0.00000
                2    -765.000        0.00000    1.448560E-06     0.00000    5.076510E-04    0.00000    000000
             DISP     0.00000        0.00000    0.00000          0.00000    1.171610E-11    0.00000
                     -6.332390E-02   0.00000    6.326700E-09     0.00000    1.684990E-10    0.00000       0.0

  2  0.0000     1     770.890        0.00000   -2.46060          0.00000    700 680         0.00000
                2    -770.890        0.00000    2.46060          0.00000    695 340         0.00000    000000
             DISP     0.00000        0.00000    0.00000          0.00000    1.616600E-04    0.00000
                     -6.381160E-02   0.00000    8.729650E-02     0.00000    1.579920E-04    0.00000       0.0
```

800	0.0000	1	1082.50	0.00000	-88.5060	0.00000	42835.0	0.00000	
		2	-1082.50	0.00000	88.5060	0.00000	42835.0	0.00000	011100
	DISP		0.00000	0.00000	0.00000	0.00000	6.479570E-02	0.00000	
			-8.960670E-02	0.00000	34.9900	0.00000	6.459820E-02	0.00000	4.0

%5E%

```
1 STRUCTURE.....: EXAMPLE:TWO-COLUMN BENT, IE3DBEAM, PM method, ELAS=4 & POSTSLOPE=0 for columns    TIME: 10:34:05, DATE: 31-AUG-09
  SOLUTION.....: INCREMEMTAL DISPLACEMENT CONTROL AT JOINT 3                                          TIME: 10:34:05, DATE: 31-AUG-09
```

%5S%

```
ELEMENT #    3 IS READ FROM UNIT #    13
```

IE3D BEAM FORCES...

STEP	TIME	NODE	AXIAL	FY	FZ	TORSION	MY	MZ	STBFAG & FLP
0	0.0000	4	765.000	0.00000	-1.280530E-06	0.00000	2.470740E-04	0.00000	
		3	-765.000	0.00000	1.280530E-06	0.00000	4.487110E-04	0.00000	000000
	DISP		0.00000	0.00000	0.00000	0.00000	1.040350E-11	0.00000	
			-6.332390E-02	0.00000	5.617910E-09	0.00000	1.489120E-10	0.00000	0.0
2	0.0000	4	759.110	0.00000	-2.46410	0.00000	701.130	0.00000	
		3	-759.110	0.00000	2.46410	0.00000	695.790	0.00000	000000
	DISP		0.00000	0.00000	0.00000	0.00000	1.617620E-04	0.00000	
			-6.283620E-02	0.00000	8.735120E-02	0.00000	1.580930E-04	0.00000	0.0
800	0.0000	4	447.490	0.00000	-101.340	0.00000	35190.0	0.00000	
		3	-447.490	0.00000	101.340	0.00000	35190.0	0.00000	011100
	DISP		0.00000	0.00000	0.00000	0.00000	6.479930E-02	0.00000	
			-3.704120E-02	0.00000	34.9920	0.00000	6.460350E-02	0.00000	4.0

The output result shows that the nominal moment occurred at member 3 at Step 103 (i.e., pushover displacement = 4.5″), and the displacement capacity of the bent is reached at Step 738 (pushover displacement = 32.3″) at which the plastic rotation of member 3 exceeds the plastic capacity of 0.0515 rad. If $p-\delta$ effect is not considered, the pushover curve at joint 3 is shown in Figure 7.19b. Comparing Figure 7.19a and b, it can be seen that $p-\delta$ effect is significant.

Equilibrium check at Step = 800:
From output plot file units 11, 12, and 13
$H = 189.83$ kip; $\Delta = 35″$
Column No. 1 end forces: $M_1 = 42,835$ k-in.; $V_1 = 88.506$ kip; $A_1 = 1,082.5$ kip
Column No. 2 end forces: $M_2 - 35,190$ k-in.; $V_2 - 101.34$ kip; $A_2 = 447.49$ kip

$$\sum A = 1530 \text{ (kip)} = 2P \text{ (ok)}$$

$$\sum V = 189.85 \text{ (kip)} \approx H \text{ (ok)}$$

$$\sum M_B = 0:$$

$$\bar{M_1} + M_2 + A_1 * 270 - 580.56 * H - P\Delta - P * (270 + \Delta) = 0$$

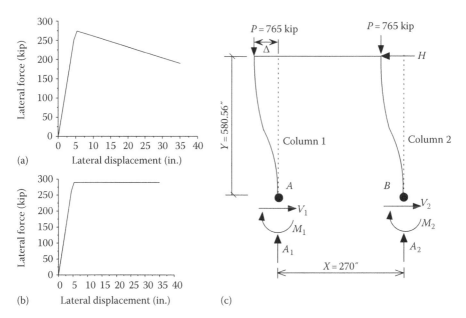

FIGURE 7.19 (a) Pushover curve at Joint 3; (b) pushover curve at Joint 3 (without $p-\delta$ effect); (c) equilibrium check.

$$\to \bar{M}_1 = P*(270+2\Delta)+580.56*(H)-270*A_1 - M_2 = 42{,}842 \approx M_1 \text{ (ok)}$$

Similarly

$$\bar{M}_2 = -P*(270-2\Delta)+580.56*(H)+270*A_2 - M_1 = 35{,}195 \approx M_2 \text{ (ok)}$$

3. FSFS Method

Each column is modeled as a "STABILITY" element with "R/CONCRETE1" material. The element is divided into eight segments. The length of each end segment is equal to the plastic hinge length of 33 in. For the FSFS method, the lateral displacement capacity can be conservatively determined when the first column confined concrete strain in the cross-sectional compression region reaches the ultimate concrete compression strain, ε_{cu}, defined in Equation 3.24, which is $\varepsilon_{cu} = 0.004 + 1.4\rho_s f_{yh}\varepsilon_{su}/f'_{cc}$, where ρ_s is the volumetric ratio of transverse steel, ε_{su} is the ultimate strain of transverse steel ($\varepsilon_{su}=0.09$), f_{yh} is yield stress of transverse steel, and f'_{cc} is the confined concrete strength.

However, a user can also input the maximum allowable plastic curvature into the program, and the lateral displacement capacity of the structure can be considered at the point when a column's plastic curvature exceeds the maximum allowable plastic curvature. For example, the moment–curvature analysis for dead load = 765 kip in Example 1 shows that $\phi_n=0.00033$ and $\phi_u=0.00166$. Therefore, the maximum plastic curvature capacity is equal to $\phi_u-\phi_n=0.00133$, which was input into the program in the FSFS pushover analysis. The output results with ε_{cu} based on Equation 3.24 are shown as follows (Figure 7.20a and b):

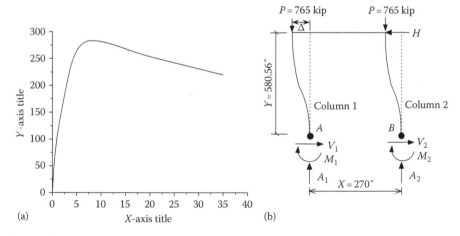

FIGURE 7.20 (a) Pushover curve at Joint 3; (b) equilibrium check.

Output (EX4_FSFS5.out file)

1   ECHO OF INPUT DATA

```
LINE  .....|.. 10.....|.. 20.....|.. 30.....|.. 40.....|.. 50.....|.. 60.....|.. 70.....|.. 80.....|.. |...100....|..110
   1: 'STRUCTURE DEFINITION-TEST NCHRP 12-49 EXAMPLE 8, use HYST17.SUB'
   2: 'EXAMPLE: TWO-COLUMN BENT'
   3:     4 1 4 0 0 1.       NNODE,NCOS , NSUPT,NCOND,NCONST     SCALE
   4:     1  0.00   0.00        .00 1 0
   5:     2  580.56 0.00        .00 1 0
   6:     3  580.56 270.0       .00 1 0
   7:     4  0.00   270.0       .00 1 0
   8:         1 0 0  0 1 0 | DIRECTION COSINE
   9:         1  1 1 1 1 1 1  0 0
  10:         2  0 0 1 1 1 0  0 0
  11:         3  0 1 1 1 0    0 0
  12:         4  1 1 1 1 1 1  0 0
  13:     3  |NMAT
  14: 'R/CONCRETE1 MAT#1 '   8 60000 29000000  2 48. 2.625 0 10 6  4000
  15:  0.625 3.25 0.95 1.27  0 33.0 0.  1809.6  0 2 0.01 0 1200780. 0.01 1
  16:  20 0 0 0 0 60000. 1 0.00133   -1 -1 0  0 -1 -1
  17: 'R/CONCRETE1 MAT#2 '   8 60000 29000000  2 48. 2.625 0 10 6  4000
  18:  0.625 3.25 0.95 1.27  0 33.0 0.  1809.6  0 2 0.01 0 1200780. 0.01 1
  19:  20 0 0 0 0 60000. 1 0.00133   -1 -1 0  0 -1 -1
  20: '3D-BEAM MAT#3' 3122020. 1200780. 3888. 0 0 207000000. 207000000. 207000000.
  21:  0 0 .FALSE.        |KG: AXL, FORM, ASSY
  22:  3  NELEM
  23: 'STABILITY'  'R/C CIRCULAR MEMBER 1' 1 1 2   0 0 1  0.  40.56  0
  24: '3D-BEAM'    'MEMBER 2' 3 2 3            0 0 1  0 0 0 000000  0
  25: 'STABILITY'  'R/C CIRCULAR MEMBER 3' 2 4 3   0 0 1  0.  40.56  0
  26:  0 0 .FALSE. |   MASS
  27:  0 0    DAMP
  28: 'SOL01  SOLUTION'
  29: 'APPLY AXIAL LOAD AT JOINT 2 AND JOINT 3'
  30:  1 1  | NLOAD MAXELD
  31:  1 2 0 0 'FX'  -765000. |JOINT DEAD LOAD
  32:  1 3 0 0 'FX'  -765000. |JOINT DEAD LOAD
  33:  0 0 0 0 'END' 0.
  34:  0 0 0 0 'END' 'FZ' 0.  0.
```

```
35:  'SOL04 SOLUTION'
36:  'INCREMENTAL DISPLACEMENT CONTROL AT JOINT 3'
37:  1 50000 2 .TRUE. 0 |MAXELD IPRINT IWRITE UNBAL SPLIMIT
38:  'JOINT FY' 3 11 0  0  0
39:          'ELE   `  1 12  0  0  0
40:          'ELE   `  3 13  0  0  0
41:  'END    `  2 14  0  0
42:  1  3  0 'FY'  -35
43:  0  0  0 'END'   0    |JOINT LOAD
44:  0  0  0 'END' 'FY'  0  0 | ELEMENT LOAD
45:  'DISP. FROM   0   TO  -35   `  1  0  0  2000
46:  'END OF DISP. CONTROL       0  0  0  0
47:  'READ UNIT=11 UNIT=12 UNIT=13'
48:  'STOP'
```
.
.
.
```
***%MEMBER  1SEG.   1ELE.    41STRAIN=  -0.159623E-01AT ISTEP  1509
     WHICH EXCEEDS ECU of  0.159607E-01
     STRESS =   -4754.05
     STRAIN =   -0.159623E-01
     FCCP   =    5502.39
     ECC    =    0.575599E-02


****  STABILITY ELEMENT      1 SEGMENT NO.     1 REACHES ULTIMATE CONCRETE COMPRESSION STRAIN.
      PLASTIC CURVATURE CURPLS =   0.110436E-02
      CUR  : TOTAL CURVATURE =    0.139570E-02
      COMPRESSION DEPTH TO N.A. IN U0 DIR.  =     0.00000

      COMPRESSION DEPTH TO N.A. IN V0 DIR.  =    14.6908

*****  STABILITY ELEMENT      1 EXTREME FIBER REACHES ULTIMATE COMPRESSION STRAIN,          PROGRAM CONTINUE
*****  THE ELEMENT FORCES DISP. FROM STEP: 1509 FACTOR:  1.000    ARE:

       1   2  DISPL    1    0.00000        0.00000       0.00000       0.00000
                DISPL    2   -0.779201       26.3730        0.00000      -3.507670E-04    FLP:   5.0
```

```
        1   2  FORCE     1   1.114250E+06   1.187199E-02   -115863.       0.00000   4.591423E+07   3.48332
               FORCE     2  -1.115236E+06  -1.187199E-02    115863.       0.00000   4.606385E+07   2.92756   SP:   0.0
%9 % LIMIT STATE POINT 3 DEFINED DUE TO ELEMENT NO   1 AT STEP= 1509

**** STABILITY ELEMENT       1 SEGMENT NO.   1 REACHES MAXIMUM PLASTIC CURVATURE OF CURPLS =   0.133049E-02
     PCMAX: MAXIMUM USER INPUT PLASTIC CURVATURE =   0.133000E-02
     CUR : TOTAL CURVATURE =   0.162183E-02   IN U0 DIR. =   0.00000
     COMPRESSION DEPTH TO N.A. IN U0 DIR. =   0.00000

     COMPRESSION DEPTH TO N.A. IN V0 DIR. =   14.8806

**** STABILITY ELEMENT       1 PLASTIC CURVATURE  REACHES MAXIMUM ALLOWABLE,        PROGRAM CONTINUE
**** THE ELEMENT FORCES DISP. FROM STEP: 1725 FACTOR:  1.000   ARE:

        1   2  DISPL     1    0.00000       0.00000        0.00000        0.00000        0.00000
               DISPL     2   -0.987333      30.1526        0.00000       -3.630802E-04   0.00000   FLP:   10.

        1   2  FORCE     1   1.115103E+06   1.325050E-02   -108527.       0.00000   4.605488E+07   3.62961
               FORCE     2  -1.116118E+06  -1.325050E-02    108527.       0.00000   4.620340E+07   3.52566   SP:   0.0
```

DEGREE OF FREEDOM #   15 IS READ FROM UNIT #   11   JOINT #   3, DIRECTION: FY

STEP	TIME	LOAD	DISPLACEMENT	VELOCITY	ACCELERATION
0	0.0000	-1.29051E-02	0.0000	0.0000	0.0000
2	0.0000	-5417.5	-3.50000E-02	0.0000	0.0000
.					
.					
.					
194	0.0000	-2.20071E+05	-3.3950	0.0000	0.0000
1 196	0.0000	-2.21506E+05	-3.4300	0.0000	0.0000
198	0.0000	-2.22921E+05	-3.4650	0.0000	0.0000
.					
.					
466	0.0000	-2.83439E+05	-8.1550	0.0000	0.0000
2 468	0.0000	-2.83433E+05	-8.1900	0.0000	0.0000
470	0.0000	-2.83425E+05	-8.2250	0.0000	0.0000

. . .

	1508	0.0000	-2.39313E+05	-26.390	0.0000	0.0000
3	1510	0.0000	-2.39235E+05	-26.425	0.0000	0.0000
	1512	0.0000	-2.39158E+05	-26.460	0.0000	0.0000

. . .

| | 2000 | 0.0000 | -2.20085E+05 | -35.000 | 0.0000 | 0.0000 |

```
%5E%
1 STRUCTURE.....: EXAMPLE: TWO-COLUMN BENT                       TIME: 10:47:00, DATE: 31-AUG-09
  SOLUTION.....: INCREMEMTAL DISPLACEMENT CONTROL AT JOINT 3     TIME: 10:47:00, DATE: 31-AUG-09
%5S%
```

ELEMENT #   1 IS READ FROM UNIT #   12

STABILITY ELEMENT FORCES...

STEP	TIME	NODE	AXIAL	FY	FZ	TORSION	MY	MZ	FLP,SP
0	0.0000	1	765000.	0.178087E-03	-0.686280E-02	0.00000	1.13830	-0.207280	
		2	-765000.	-0.178087E-03	0.686280E-02	0.00000	2.57580	0.303450	SP: 0.0
		DISP	0.00000	0.00000	0.00000	0.00000	0.00000	0.00000	
			-0.569151E-01	0.00000	0.106943E-07	0.00000	0.267420E-09	0.00000	FLP: 0.0
2	0.0000	1	771285.	-0.764127E-04	-2706.40	0.00000	751168.	-0.247190	
		2	-771285.	0.764127E-04	2706.40	0.00000	737150.	0.205930	SP: 0.0
		DISP	0.00000	0.00000	0.00000	0.00000	0.00000	0.00000	
			-0.573914E-01	0.00000	0.347948E-01	0.00000	-0.357647E-05	0.00000	FLP: 0.0
2000	0.0000	1	0.111644E+07	0.101708E-01	-99306.0	0.00000	0.462865E+08	2.85750	
		2	-0.111749E+07	-0.101708E-01	99306.0	0.00000	0.464309E+08	2.63480	SP: 0.0
		DISP	0.00000	0.00000	0.00000	0.00000	0.00000	0.00000	
			-1.29370	0.00000	34.9820	0.00000	-0.380235E-03	0.00000	FLP: 10.1

. . .

```
%5E%
1 STRUCTURE.....: EXAMPLE: TWO-COLUMN BENT                       TIME: 10:47:00, DATE: 31-AUG-09
  SOLUTION.....: INCREMEMTAL DISPLACEMENT CONTROL AT JOINT 3     TIME: 10:47:00, DATE: 31-AUG-09
```

%5S%

ELEMENT #   3 IS READ FROM UNIT #   13

STABILITY ELEMENT FORCES...

STEP	TIME	NODE	AXIAL	FY	FZ	TORSION	MY	MZ	FLP,SP
0	0.0000	4	765001.	-0.172806E-03	-0.561881E-02	0.00000	1.00-50	-0.285030	
		3	-765000.	0.172806E-03	0.561881E-02	0.00000	2.03990	0.191710	SP: 0.0
			0.00000	0.00000	0.00000	0.00000	0.00000	0.00000	
		DISP	-0.569152E-01	0.00000	0.958611E-08	0.00000	0.233444E-09	0.00000	FLP: 0.0
2	0.0000	4	758716.	-0.777721E-04	-2717.20	0.00000	7538-4.	-0.254520	
		3	-758716.	0.777721E-04	2717.20	0.00000	739930.	0.212530	SP: 0.0
			0.00000	0.00000	0.00000	0.00000	0.00000	0.00000	
		DISP	-0.564416E-01	0.00000	0.348549E-01	0.00000	-0.357701E-05	0.00000	FLP: 0.0
2000	0.0000	4	412903.	-0.240555E-01	-120787.	0.00000	0.398108E+08	-1.76810	
		3	-412512.	0.240555E-01	120787.	0.00000	0.398156E+08	-11.2220	SP: 0.0
			0.00000	0.00000	0.00000	0.00000	0.00000	0.00000	
		DISP	-1.19210	0.00000	34.9850	0.00000	-0.379035E-03	0.00000	FLP: 5.1

. . .

The output result shows that the displacement capacity of the bent base on the ultimate concrete compression strain, ε_{cu}, is reached at pushover displacement $= 26.4$ in. (at Step $= 1509$) at which the confined concrete compression strain of Column No. 1 exceeds ε_{cu} of 0.01596. If the displacement capacity of the bent is based on the maximum plastic curvature of 0.00133 (rad), it occurs at Column No. 1 with a pushover displacement of 30.15 in. (at Step $= 1725$).

Equilibrium check at Step $= 2000$:

From output units 11, 12, and 13

$H = 220$ kip; $\Delta = 35''$

Column No. 1 end forces: $M_1 = 46,286.5$ k-in.; $V_1 = 99.31$ kip; $A_1 = 1,116.4$ kip

Column No. 2 end forces: $M_2 = 39,840.8$ k-in.; $V_2 = 120.79$ kip; $A_2 = 412.9$ kip

$$\sum A = 1116.4 + 412.9 = 1529.3 \text{ (kip)} \approx 2P \text{ (ok)}$$

$$\sum V = 99.31 + 120.79 = 220.1 \text{ (kip)} \approx H \text{ (ok)}$$

$$\sum M_B = 0:$$

$$\bar{M}_1 + \bar{M}_2 + A_1 * 270 - 580.56 * H - P\Delta - P * (270 + \Delta) = 0$$

$$\rightarrow \bar{M}_1 = P * (270 + 2\Delta) + 580.56 * (H) - 270 * A_1 - \bar{M}_2 = 46,554 \approx M_1 \text{ (ok)}$$

Similarly

$$\bar{M}_2 = -P * (270 - 2\Delta) + 580.56 * (H) + 270 * A_2 - \bar{M}_1 = 39,919 \approx M_2 \text{ (ok)}$$

4. FSMC Method

Each column is modeled as a "STABILITY" element with "MOMCURVA1" material. The element is divided into eight segments. The length of each end segment is equal to the plastic hinge length of 33 in. Only the moment–curvature curve corresponding to axial load $P = 765$ kip (see Figure 7.16) is input into the program. Hence, the moment–curvature will not be adjusted due to the effect of axial load change. The lateral displacement capacity of the structure is determined at the point when a column's plastic curvature exceeds the maximum allowable plastic curvature. The moment–curvature curve for dead load $= 765$ kip shows that $\phi_n = 0.00033$ and $\phi_u = 0.00166$. Therefore, the maximum plastic curvature capacity of the column is equal to $\phi_u - \phi_n = 0.00133$, which was input into the program in the FSMC pushover analysis. The output results are shown as follows (Figure 7.21a and b):

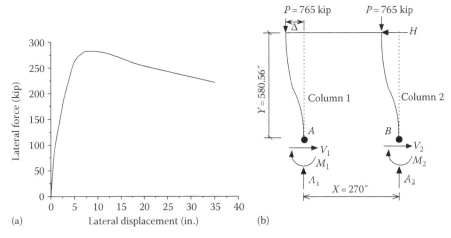

FIGURE 7.21 (a) Pushover curve at Joint 3; (b) equilibrium check.

Output (EX4_FSMC5.out file)

1   ECHO OF INPUT DATA

```
LINE ....|...10....|...20....|...30....|...40....|...50....|...60....|...70....|...80....|...90....|...100....|...110
  1: 'STRUCTURE DEFINITION-TEST NCHRP 12-49 EXAMPLE 8'
  2: 'EXAMPLE: TWO-COLUMN BENT'
  3:      4  1 4 0 0  1.              NNODE,NCOS , NSUPT,NCOND,NCONST    SCALE
  4:      1   0.00    0.00                .00 1 0
  5:      2  580.56   0.00                .00 1 0
  6:      3  580.56  270.0                .00 1 0
  7:      4   0.00   270.0                .00 1 0
  8:              1 0 0     0 1 0  | DIRECTION COSINE
  9:      1  1 1 1 1 1 1    0 0
 10:      2  0 0 1 1 1 0    0 0
 11:      3  0 1 1 1 1 0    0 0
 12:      4  1 1 1 1 1 1    0 0
 13:      3  |NMAT
 14: 'MOMCURVA1 MAT#1 '  8 1200. 3605.  1  260576. 521152.  0  0 0.00133 0.000333 0
 15:  18  1  0  0  0 33.0  0. 1809.6  0  1 0.0  0 1200. 0.011 1
 16:  765.    |AXIAL LOAD CASE
 17:  9066.  25339. 34980. 36676. 38323. 38660. 39096. 39408. 39519.
 18:  39696. 39760. 39728. 39638. 39479. 39528. 39697. 40569. 43018.  |MOMENTS
 19:  0.00001 0.00006 .00012 0.00015 0.00021 0.00024 0.0003  0.00036 0.00039
 20:  0.00045 0.00051 .0006  0.00069 0.00081 0.0009 0.00102 0.0015 0.00288 |CUR
 21: 'MOMCURVA1 MAT#2 '  8 1200. 3605.  1  260576. 521152.  0  0 0.00133 0.000333 0
 22:  18  1  0  0  0 33.0  0. 1809.6  0  1 0.0  0 1200. 0.011 1
 23:  765.    |AXIAL LOAD CASE
 24:  9066.  25339. 34980. 36676. 38323. 38660. 39096. 39408. 39519.
 25:  39696. 39760. 39728. 39638. 39479. 39528. 39697. 40569. 43018.  |MOMENTS
 26:  0.00001 0.00006 .00012 0.00015 0.00021 0.00024 0.0003  0.00036 0.00039
 27:  0.00045 0.00051 .0006  0.00069 0.00081 0.0009 0.00102 0.0015 0.00288 |CUR
 28: '3D-BEAM MAT#3'  3122.  1200.  3888.  0 0 207000000. 207000000. 207000000.
 29:  0 0 0 .FALSE.       |KG: AXL, FORM, ASSY
 30:      3  NELEM
 31: 'STABILITY'  'R/C CIRCULAR MEMBER 1'  1 1 2      0 0 1  0.  40.56    0
 32: '3D-BEAM'                'MEMBER 2'  3 2 3       0 0 1  0 0  000000   0
 33: 'STABILITY'  'R/C CIRCULAR MEMBER 3'  2 4 3      0 0 1  0.  40.56    0
 34:  0 0 .FALSE. | MASS
 35:  0 0  | DAMP
 36: 'SOL01  SOLUTION'
 37: 'APPLY AXIAL LOAD AT JOINT 2 AND JOINT 3'
```

```
38:        1  1    | NLOAD MAXELD
39:        1  2  0  'FX'  -765.   |JOINT DEAD LOAD
40:        1  3  0  'FX'  -765.   |JOINT DEAD LOAD
41:        0  0  0  0  'END'  0.
42:        0  0  0  0  'END'  'FZ'   0.   0.
43:     'SOL04 SOLUTION'
44:     'INCREMEMTAL DISPLACEMENT CONTROL AT JOINT 3'
45:     1 50000 2  .TRUE.  0  |MAXELD IPRINT IWRITE UNBAL SPLIMIT
46:     'JOINT FY'    3 10 0  0
47:     'ELE  '   1 12 0  0  0
48:     'ELE  '   3 13 0  0  0
49:     'END  '   2 14 0  0  0
50:        1  3  0  'FY'  -35
51:        0  0  0  'END'  0            |JOINT LOAD
52:        0  0  0  'END' 'FY'  0  0 | ELEMENT LOAD
53:     'DISP. FROM  0    TO  -35  '  1  0  1800
54:     'END OF DISP. CONTROL '   0  0  0  0
55:     'READ UNIT=10 UNIT=12 UNIT=13'
56:     'STOP'
```

```
**** STABILITY ELEMENT          3  SEGMENT NO.       1 REACHES MAXIMUM PLASTIC CURVATURE OF CURPLS =    0.133105E-02
     PCMAX: MAXIMUM USER INPUT PLASTIC CURVATURE =    0.133000E-02
     CUR  : TOTAL CURVATURE =    0.166405E-02
```

```
**** STABILITY ELEMENT          3  PLASTIC CURVATURE  REACHES MAXIMUM ALLOWABLE,          PROGRAM CONTINUE
**** THE ELEMENT FORCES DISP. FROM STEP: 1641 FACTOR: 1.000     ARE:

     3  1  DISPL   4       0.00000        0.00000        0.00000        0.00000     FLP:   10.
           DISPL   3      -1.01384       31.8795         0.00000        0.00000

     3  1  FORCE   4     411.887          0.00000     -135.938     -2.279244E-04    SP:    0.0
           FORCE   3    -411.903          0.00000      135.938         0.00000

                                                        43270.8
                                                        43267.1

%9 % LIMIT STATE POINT 3 DEFINED DUE TO ELEMENT NO   3 AT STEP= 1641
```

```
DEGREE OF FREEDOM #    15 IS READ FROM UNIT #    10     JOINT #      3 , DIRECTION: FY

STEP   TIME       LOAD           DISPLACEMENT      VELOCITY      ACCELERATION
  0    0.0000    -6.75075E-06     0.0000            0.0000        0.0000
  2    0.0000    -5.0726         -3.88889E-02       0.0000        0.0000
```

478	0.0000	-282.13	-9.2945	0.0000	0.0000
480	0.0000	-282.10	-9.3333	0.0000	0.0000
482	0.0000	-282.08	-9.3722	0.0000	0.0000

2

1640	0.0000	-228.46	-31.889	0.0000	0.0000
1642	0.0000	-228.38	-31.928	0.0000	0.0000
1644	0.0000	-228.30	-31.966	0.0000	0.0000

3

1798	0.0000	-221.96	-34.961	0.0000	0.0000
1800	0.0000	-221.87	-35.000	0.0000	0.0000

%5E%

```
1 STRUCTURE.....: EXAMPLE: TWO-COLUMN BENT                              TIME: 21:39:13, DATE: 07-SEP-07
  SOLUTION......: INCREMEMTAL DISPLACEMENT CONTROL AT JOINT 3           TIME: 21:39:13, DATE: 07-SEP-07

%5S%

ELEMENT #   1 IS READ FROM UNIT #   12
```

STABILITY ELEMENT FORCES...

STEP	TIME	NODE	AXIAL	FY	FZ	TORSION	MY	MZ	FLP,SP
0	0.0000	1	765.000	0.00000	-3.567900E-06	0.00000	6.855170E-04	0.00000	SP: 0.0
		2	-765.000	0.00000	3.567900E-06	0.00000	1.246270E-03	0.00000	
			0.00000	0.00000	0.00000	0.00000	0.00000	0.00000	
		DISP	-6.332390E-02	0.00000	6.693200E-09	0.00000	1.669820E-10	0.00000	FLP: 0.0
2	0.0000	1	770.910	0.00000	-2.53770	0.00000	706.080	0.00000	SP: 0.0
		2	-770.910	0.00000	2.53770	0.00000	694.090	0.00000	
			0.00000	0.00000	0.00000	0.00000	0.00000	0.00000	
		DISP	-6.381370E-02	0.00000	3.868340E-02	0.00000	-3.677250E-06	0.00000	FLP: 0.0
1800	0.0000	1	1119.30	0.00000	-87.0390	0.00000	43083.0	0.00000	SP: 0.0
		2	-1119.30	0.00000	87.0390	0.00000	43082.0	0.00000	
			0.00000	0.00000	0.00000	0.00000	0.00000	0.00000	
		DISP	-1.27500	0.00000	34.9890	0.00000	-2.304930E-04	0.00000	FLP: 11.

```
%5E%
1 STRUCTURE.....: EXAMPLE: TWO-COLUMN BENT                          TIME: 21:39:13, DATE: 07-SEP-07
  SOLUTION.......: INCREMEMTAL DISPLACEMENT CONTROL AT JOINT 3      TIME: 21:39:13, DATE: 07-SEP-07
%5S%
ELEMENT #    3 IS READ FROM UNIT #   13

STABILITY ELEMENT FORCES...
STEP   TIME     NODE      AXIAL          FY              FZ           TORSION          MY            MZ        FLP,SP

  0    0.0000    4       765.000       0.00000     -3.164220E-06     0.00000      6.083540E-06    0.00000
                 3      -765.000       0.00000      3.164220E-06     0.00000      1.104910E-05    0.00000     SP:   0.0
                DISP     0.00000       0.00000      0.00000          0.00000      0.00000         0.00000
                     -6.332390E-02                 5.997440E-09                  1.478660E-10     0.00000     FLP:  0.0

  2    0.0000    4       759.090       0.00000     -2.54240          0.00000      707.180         0.00000
                 3      -759.090       0.00000      2.54240          0.00000      695.100         0.00000     SP:   0.0
                DISP     0.00000       0.00000      0.00000          0.00000      0.00000         0.00000
                     -6.283580E-02                 3.873970E-02                 -3.677520E-06     0.00000     FLP:  0.0

1800   0.0000    4       410.650       0.00000     -134.840         0.00000      43594.0         0.00000
                 3      -410.660       0.00000      134.840          0.00000      43590.0         0.00000     SP:   0.0
                DISP     0.00000       0.00000      0.00000          0.00000      0.00000         0.00000
                       -1.21360                     34.9900                     -2.310050E-01     0.00000     FLP:  11.
```

. . .

The output result shows that the displacement capacity of the bent occurs at pushover displacement = 31.9 in. (at Step = 1641) at which Column No. 3 reaches its maximum plastic curvature of 0.00133 (rad).

Equilibrium check at Step = 1800:
    From output units 11, 12, and 13
    $H = 221.87$ kip; $\Delta = 35''$
    Column No. 1 end forces: $M_1 = 43{,}083$ k-in.; $V_1 = 87.04$ kip; $A_1 = 1{,}119.3$ kip
    Column No. 2 end forces: $M_2 = 43{,}594$ k-in.; $V_2 = 134.84$ kip; $A_2 = 410.65$ kip

$$\sum A = 1119.3 + 410.65 = 1530 \text{ (kip)} = 2P \text{ (ok)}$$

$$\sum V = 221.88 \text{ (kip)} \approx H \text{ (ok)}$$

$$\sum M_B = 0:$$

$$\bar{M}_1 + M_2 + A_1 * 270 - 580.56 * H - P\Delta - P * (270 + \Delta) = 0$$

$$\rightarrow \bar{M}_1 = P * (270 + 2\Delta) + 580.56 * (H) - 270 * A_1 - M_2 = 43{,}104 \approx M_1 \text{ (ok)}$$

Similarly

$$\bar{M}_2 = -P * (270 - 2\Delta) + 580.56 * (H) + 270 * A_2 - M_1 = 43{,}601 \approx M_2 \text{ (ok)}$$

Figure 7.22 shows the lateral force–lateral displacement relationships generated from different methods with consideration of p–δ effects. The displacement capacities of the bent determined by different methods are summarized as follows:

From Table 7.2, it can be seen that structural-displacement capacity decreases when the effect of the column axial load on the column plastic rotational capacity is considered (see Section 4.9). In the elastic dynamic analysis (such as using multiple-mode response spectrum analysis as described in Appendix H), the demand displacement Δ_d should be less than the least of the displacement capacity calculated from the above methods. For example, $1.5\Delta_d \le \Delta_c$ (ATC and MCEER, 2003).

7.3.5 EXAMPLE 5: TWO-COLUMN BENT (FORCE CONTROL)

Use force control to reanalyze the structure shown in Example 4 by the PM Method and compare the pushover curve generated by the force control with that based on displacement control in Example 4 (Figure 7.23).

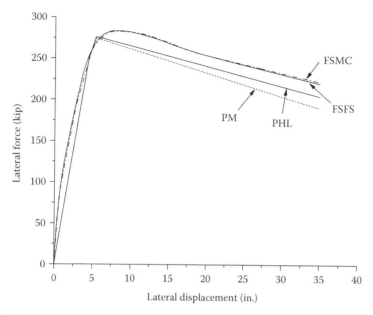

FIGURE 7.22 Lateral force–lateral displacement curves.

TABLE 7.2

Pushover Results

Method	Displacement Capacity (in.), Δ_c	Criteria to Determine Displacement Capacity
PHL (without P–PRC interaction)	32.3	$\theta_p = 0.0515$ (rad)
PHL (with P–PRC interaction)	28.9	$\theta_p = 0.0434$ (rad)
PM	32.3	$\theta_p = 0.0515$ (rad)
FSFS	26.4	$\varepsilon_{cu} = 0.004 + \dfrac{1.4\rho_s f_{yh}\varepsilon_{su}}{f'_{cc}}$
FSFS	30.2	$\phi_p = 0.00133$ (based on M–ϕ curve with axial load = 765 kip)
FSMC	31.9	$\phi_p = 0.00133$ (based on M–ϕ curve with axial load = 765 kip)

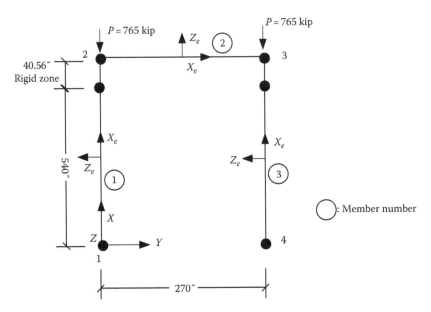

FIGURE 7.23   Structural model.

First Run Output (EX5_PM_FORCE.out file)

1   ECHO OF INPUT DATA

```
LINE  ....|..10....|..20....|..30....|..40....|..50....|..60....|..70....|..80....|..90....|..100....|..110
 1:   'STRUCTURE DEFINITION-TEST NCHRP 12-49 EXAMPLE 8'
 2:   'EXAMPLE: TWO-COLUMN BENT,IE3DBEAM, MAT10 with ELAS=4 for columns'
 3:      4 1 4 0 0 1.       NNODE,NCOS , NSUPT,NCOND,NCONST     SCALE
 4:      1   0.00   0.00    .00 1 0
 5:      2 580.56   0.00    .00 1 0
 6:      3 580.56  270.0    .00 1 0
 7:      4   0.00  270.0    .00 1 0
 8:         1 0    0    0 1 0 | DIRECTION COSINE
 9:      1  1 1 1 1 1 1   0 0
10:      2  0 0 1 1 1 0   0 0
11:      3  0 0 1 1 1 0   0 0
12:      4  1 1 1 1 1 1   0 0
13:      7 |NMAT
14:   'IA_BILN MAT#1: MYA ' 4 0. 3605. 109032. 39306.   -1 0  0.0515
15:   0
16:   28989.6 15.31 -0.0024 6.684E-8
17:   0
18:   'IA_BILN MAT#2: MYB ' 4 0. 3605. 109032. 39306.   -1 0  0.0515
19:   0
20:   28989.6 15.31 -0.0024 6.684E-8
21:   0
22:   'IA_BILN MAT#3: MZA ' 0 0.001 3605. 260576. 38400.  -1 0  -1
23:   'IA_BILN MAT#4: MZA ' 0 0.001 3605. 260576. 38400.  -1 0  -1
24:   'IA_BILN MAT#5: MXA ' 0 0.001 3605. 521152. 38400.  -1 0  -1
25:   'IA_BILN MAT#6: FXA ' 0 0.001 3605. 1809.6 38400.   -1 0  -1
26:   '3D-BEAM MAT#7' 3122. 1200. 3888. 0 0 207000000. 207000000. 207000000.
27:      2 1 1 .TRUE.  |KG: AXL, FORM, ASSY
28:      3  NELEM
29:   'IE3DBEAM'  'R/C CIRCULAR MEMBER 1' 1 2 3 4 5 6  1 2  0 0 1  0 40.56 765.
30:   0 0 0 0 0 0 0 0 0 0 0 0 0 0
31:   '3D-BEAM'  'MEMBER 2' 7  2 3  0 0 1 0 0 0 0 000000 0
32:   'IE3DBEAM'  'R/C CIRCULAR MEMBER 3' 1 2 3 4 5 6  4 3  0 0 1  0 40.56 765.
33:   0 0 0 0 0 0 0 0 0 0 0 0 0
34:      0 0 .FALSE. | MASS
```

```
35:      0 0    | DAMP
36:  'SOL01 SOLUTION'
37:  'APPLY AXIAL LOAD AT JOINT 2 AND JOINT 3'
38:     1 1    | NLOAD MAXELD
39:     1 2 0 0 'FX'  -765.   |JOINT DEAD LOAD
40:     1 3 0 0 'FX'  -765.   |JOINT DEAD LOAD
41:     0 0 0 0 'END'  0.
42:     0 0 0 0 'END'  'FZ'   0.   0.
43:  'SOL04 SOLUTION'
44:  'INCREMEMTAL FORCE CONTROL AT JOINT 3'
45:     1 50000 1 .TRUE. 0.00001  | MAXELD IPRINT IWRITE UNBAL SPLIMIT
46:  'JOINT FY'  3 11  0 0 0
47:  'ELE   '    1 12  0 0 0
48:  'ELE   '    3 13  0 0 0
49:  'END   '    2 14  0 0 0
50:     1 3 0 'FY'  -280              |JOINT LOAD
51:     0 0 0 'END'  0                |JOINT LOAD
52:     0 0 0 'END'  'FY'  0 0   | ELEMENT LOAD
53:  'DISP. FROM  0  TO  -280  ,  1 0 0  560
54:  'END OF DISP. CONTROL  ,  0   0  0  0
55:  'READ UNIT=11 UNIT=12 UNIT=13'
56:  'STOP'
```

.
.

```
LOAD CASE:  1   JOINT:    3    DIRECTION: FY    DOF(S)    5        MAGNITUDE:   -280.000
1 STRUCTURE.....: EXAMPLE: TWO-COLUMN BENT;IE3DBEAM, MAT10 with ELAS=4 for columns
  SOLUTION......: INCREMEMTAL FORCE CONTROL AT JOINT 3

@@@@ NOTE: STIFFNESS PARAMETER, SP= 0.9943534E-05< PLIMIT=  0.100000E-04AT ISTEP= 327
           UNBALANCE FORCES ARE NOT ADJUSTED AT   ISTEP =   328
```

.
.

```
**** IA_BILN ELEMENT      1 YIELD AT END B
**** THE ELEMENT FORCES DISP. FROM STEP:  553 FACTOR:  1.000        ARE:
```

TIME: 11:24:40, DATE: 31-AUG-09
TIME: 11:24:40, DATE: 31-AUG-09

```
                                                                        FLP:  4.0

                                                                        CHR:  001100

1  1  DISPL  1    0.00000        0.00000    0.00000    1.011303E-02   0.00000
      DISPL  2   -9.090001E-02   5.46107    0.00000    9.904815E-03   0.00000

1  1  FORCE  1    1098.14       -148.108    0.00000    42979.0        0.00000
      FORCE  2   -1098.14        148.108    0.00000    42996.4        0.00000
```

@@@@ STABILITY PARAMETER IS NEGATIVE OF -0.856483E-08 AT ISTEP = 553
PROGRAM WILL STOP. CHANGE SIGN OF LOAD INCREMENT AT ISTEP = 553 AND RERUN THE PROGRAM

. . .

From the above output, it shows that the structure's SP is less than the user-defined limit (i.e., SPLIMIT=0.00001) at Step 327. Therefore, the unbalanced forces are not adjusted after Step 327. At Step 553, the SP becomes negative and the program stops. In order to obtain the descending curve, a second run is performed by changing the sign of load increment (in this example, change from negative to positive) at Step 553 and following steps after Step 553 in the input data file.

Second Run Output (EX5_PM_FORCE2.out file)

```
1   ECHO OF INPUT DATA

LINE   ....|.. 10....|.. 20....|.. 30....|.. 40....|.. 50....|.. 60....|.. 70....|.. 80....|.. 90....|..100....|..110
   1:      'STRUCTURE DEFINITION-TEST NCHRP 12-49 EXAMPLE 8'
   2:      'EXAMPLE: TWO-COLUMN BENT, IE3DBEAM, PM method, ELAS=4 for columns'
   3:      4 1 4 0 0 1.      NNODE,NCOS , NSUPT,NCOND,NCONST    SCALE
   4:      1   0.00    0.00     .00 1 0
   5:      2   580.56  0.00     .00 1 0
   6:      3   580.56  270.0    .00 1 0
   7:      4   0.00    270.0    .00 1 0
   8:      1 0 0    0 1 0 | DIRECTION COSINE
   9:      1  1 1 1 1 1 1   0 0
  10:      2  0 0 1 1 1 0   0 0
  11:      3  0 0 1 1 1 0   0 0
  12:      4  1 1 1 1 1 1   0 0
  13:      7   |NMAT
  14:      'IA_BILN MAT#1: MYA `  4  0.   3605. 109032. 39306.   -1   0   0.0515
  15:      0
  16:      28989.6   15.31   -0.0024   6.684E-8
  17:      0
  18:      'IA_BILN MAT#2: MYB `  4  0.   3605. 109032. 39306.   -1   0   0.0515
  19:      0
  20:      28989.6   15.31   -0.0024   6.684E-8
  21:      0
  22:      'IA_BILN MAT#3: MZA `  0  0.001  3605. 260576. 38400.   -1   0
  23:      'IA_BILN MAT#4: MZA `  0  0.001  3605. 260576. 38400.   -1   0
  24:      'IA_BILN MAT#5: MXA `  0  0.001  3605. 521152. 38400.   -1   0
  25:      'IA_BILN MAT#6: FXA `  0  0.001  3605. 1809.6  38400.   -1   0
  26:      '3D-BEAM MAT#7'  3122. 1200. 3888. 0 207000000. 207000000. 207000000.
  27:      2 1 1 .TRUE.     |KG: AXL, FORM, ASSY
  28:      3   NELEM
  29:      'IE3DBEAM'   'R/C CIRCULAR MEMBER 1' 1 2 3 4 5 6   1 2   0 0 1   0 40.56   765.
  30:      0 0 0 0 0 0 0   0 0 0 0 0 0   0
  31:      '3D-BEAM'   'MEMBER 2' 7   2 3   0 0 1 0 0   0 000000   0
  32:      'IE3DBEAM'   'R/C CIRCULAR MEMBER 3' 1 2 3 4 5 6   4 3   0 0 1   0 40.56   765.
  33:      0 0 0 0 0 0 0   0 0 0 0 0 0   0
  34:      0 0 .FALSE. | MASS
```

```
35:      0  0  | DAMP
36:   'SOL01  SOLUTION'
37:   'APPLY AXIAL LOAD AT JOINT 2 AND JOINT 3'
38:    1  1    | NLOAD MAXELD
39:    1  2  0  0  'FX'   -765.   |JOINT DEAD LOAD
40:    1  3  0  0  'FX'   -765.   |JOINT DEAD LOAD
41:    0  0  0  0  'END'  0.
42:    0  0  0  0  'END'  'FZ'  0.  0.
43:   'SOL04 SOLUTION'
44:   'INCREMEMTAL FORCE CONTROL AT JOINT 3'
45:    1 50000 1  .TRUE.  0.00001  | MAXELD IPRINT IWRITE UNBAL SPLIMIT
46:   'JOINT FY'  3  11  0  0
47:   'ELE   `   1  12  0  0  0
48:   'ELE   `   3  13  0  0  0
49:   'END   `   2  14  0  0  0
50:    1  3  0  0  'FY'  -276.0
51:    0  0  0  0  'END'  0        |JOINT LOAD
52:    0  0  0  0  'END'  'FY'  0  0  | ELEMENT LOAD
53:   'DISP. FROM   0      TO  -276.0  `  1  0  0  552
54:   'DISP. FROM  -276.0  TO  -190.0  `  0.688406  0  0  172
55:   'END OF DISP. CONTROL   `  0  0  0  0
56:   'READ UNIT=11 UNIT=12 UNIT=13'
57:   'STOP'
```

.
.

LOAD CASE: 1 JOINT: 3 DIRECTION: FY   DOF(S)   5   MAGNITUDE: -276.000
1 STRUCTURE.....: EXAMPLE: TWO-COLUMN BENT, IE3DBEAM, PM method, ELAS=4 for columns   TIME: 11:29:10, DATE: 31-AUG-09
SOLUTION......: INCREMEMTAL FORCE CONTROL AT JOINT 3   TIME: 11:29:10, DATE: 31-AUG-09

@@@@ NOTE: STIFFNESS PARAMETER, SP= 0.9943470E-05< PLIMIT= 0.100000E-04AT ISTEP= 327
           UNBALANCE FORCES ARE NOT ADJUSTED AT   ISTEP = 328

.
.

**** IA_BILN ELEMENT   1 YIELD AT END B
**** THE ELEMENT FORCES DISP. FROM STEP: 553 FACTOR: 0.6884   ARE:

     1 1 DISPL 1   0.00000   0.00000   0.00000   1.011307E-02   0.00000
           DISPL 2 -9.090000E-02 0.00000   5.46107   9.904793E-03   0.00000   FLP: 4.0

```
   1   1   FORCE   1    1098.14      0.00000    -148.108      0.00000    42979.0      0.00000    CHR: 001100
           FORCE   2   -1098.14      0.00000     148.108      0.00000    42996.4      0.00000

* ELEMENT   3 INELASTIC ROTATION IN MY  DIR. AT END A  =  0.5158E-01 > PLASTIC ROTATION
CAPACITY OF  0.5150E-01AT STEP =  704
%9 % LIMIT STATE POINT 3 DEFINED DUE TO ELEMENT NO   3 AT STEP=  704
.
.
.
DEGREE OF FREEDOM #   5 IS READ FROM UNIT #   11        JOINT #   3, DIRECTION: FY

STEP    TIME       LOAD       DISPLACEMENT     VELOCITY    ACCELERATION
  0    0.0000     0.0000       4.60371E-08      0.0000       0.0000
  1    0.0000    -0.50000     -8.88375E-03      0.0000       0.0000
.
.
     501    0.0000    -250.50      -4.4508         0.0000       0.0000
*2*  502    0.0000    -251.00      -4.4655         0.0000       0.0000
     503    0.0000    -251.50      -4.4801         0.0000       0.0000
.
.
     703    0.0000    -200.50     -32.118          0.0000       0.0000
*3*  704    0.0000    -200.00     -32.295          0.0000       0.0000
     705    0.0000    -199.50     -32.471          0.0000       0.0000
.
.
.
```

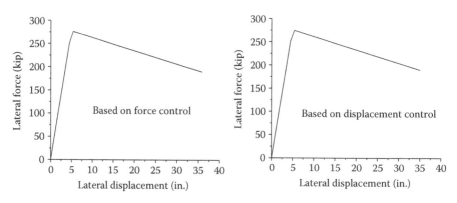

FIGURE 7.24 Pushover curve comparison.

The pushover curves based on force control and displacement control are shown in Figure 7.24. They are in agreement with each other.

7.3.6 EXAMPLE 6: COLUMN WITH RECTANGULAR SECTION

This example compares the numerical results with test results for a column specimen (Unit 7, ZAHN86U7.WK1) from NISTIR 5984 report (Taylor et al., 1997). The test was performed at the University of Canterbury, New Zealand (Zahn et al., 1986). The test setup and structural model are shown in Figure 7.25. The height of the

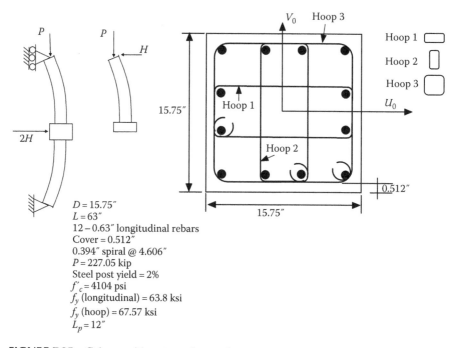

$D = 15.75''$
$L = 63''$
12 – 0.63'' longitudinal rebars
Cover = 0.512''
0.394'' spiral @ 4.606''
$P = 227.05$ kip
Steel post yield = 2%
$f'_c = 4104$ psi
f_y (longitudinal) = 63.8 ksi
f_y (hoop) = 67.57 ksi
$L_p = 12''$

FIGURE 7.25 Column with rectangular section.

column was 63″, and width of the column was 15.75″. The lateral load was applied to a central stub at the column midheight. The height of the central stub was 15.75″. For simplicity, the numerical solutions were based on the simple cantilever model. The longitudinal bar diameter and cross-sectional area are 0.63 (in.) and 0.312 (in.^{2}). The transverse rebar diameter and area (A_{sp}) are 0.394 (in.) and 0.122 (in.^{2}). Other material properties are shown in Figure 7.25.

From Chapter 3, the volumetric ratios of transverse hoops, ρ_X and ρ_Y, can be expressed as $\rho_X = N_X A_{sp}/sh_Y''$; $\rho_Y = N_Y A_{sp}/sh_X''$, in which h_X'' and h_Y'' are the confined core dimensions in the X and Y directions, respectively. $h_X'' = h_Y'' = 14.33''$. N_X and N_Y are the total number of transverse hoop legs in the X and Y directions, which are 4.67 and 4.67, respectively. Transverse loop spacing, s, is 4.606 (in.). Therefore, $\rho_X = \rho_Y = 0.00863$.

The input data are shown as follows:

Input Data (EX6_pries7_5.dat file)

```
1   ECHO OF INPUT DATA

LINE ....|.. 10....|.. 20....|.. 30....|.. 40....|.. 50....|.. 60....|.. 70....|.. 80....|.. 90....|.. 100....|...110
1:     'STRUCTURE DEFINITION- SINGLE COLUMN TEST'
2:     'TEST SMALL-SCALE RECTANGULAR COLUMN SPECIMEN UNIT 7 W/ CENTER STUB'
3:     3  1 2 0 0 1.        NNODE,NCOS , NSUPT,NCOND,NCONST    SCALE
4:     1  0.00   0.00    .00 1 0
5:     2  7.875  0.00    .00 1 0
6:     3  70.875 0.00    .00 1 0
7:     1 0 0   0 1 0 | DIRECTION COSINE
8:     1     1 1 1 1 1 1   0 0
9:     3     0 1 1 1 0   0 0
10:    2  |NMAT
11:    'R/CONCRETE1 MAT#1'  1 63800.  29000000  3  15.75  15.75 0  0  10   4104.
12:    0.0087  0.75  0.512  0  7.875 0. 0.  248.06 0 3  0.02  0  1200000. 0.01 1
13:    4 0.63  4 0.63 2 0.63 67570.  0  -1  -1   0.394  4.606  0  0
14:    'R/CONCRETE1 MAT#1'  6 63800.  29000000  3  15.75  15.75 0  0  10   4104.
15:    0.0087  0.75  0.512  0  12.0 0. 0.  248.06 0 3  0.02  0  1200000. 0.01 1
16:    4 0.63  4 0.63 2 0.63 67570.  1  -1  -1   0.394  4.606  0  0
17:    0 0 0 .FALSE.       |KG: AXL, FORM, ASSY
18:    2  NELEM
19:    'STABILITY'  'R/C CIRCULAR MEMBER 1'  1 1 2   0 0 1  0 0 0
20:    'STABILITY'  'R/C CIRCULAR MEMBER 1'  2 2 3   0 0 1  0 0 0
21:    0 0 .FALSE. |  MASS
22:    0 0 | DAMP
23:    'SOLO1  SOLUTION'
24:    'APPLY AXIAL LOAD AT JOINT 2'
25:    1 1     | NLOAD MAXELD
26:    1 3  0  0 'FX'  -227050.  |JOINT DEAD LOAD
27:    0 0 0 0 'END' 0.
28:    0 0 0 0 'END' 'FZ'  0.  0.
29:    'SOL04 SOLUTION'
30:    'INCREMEMTAL DISPLACEMENT CONTROL AT JOINT 3'
31:    1 50000 2 .TRUE. 0 |MAXELD IPRINT IWRITE UNBAL SPLIMIT
32:    'JOINT FY'  3 11 0  0 0
33:    'ELE  '  2 12 0  0 0
34:    'END  '  2 14 0  0 0
```

```
35:   1   3   0   0   'FY'   5           |JOINT LOAD
36:   0   0   0   0   'END'  0           |JOINT LOAD
37:   0   0   0   0   'END'  'FY'  0  0  | ELEMENT LOAD
38:   'DISP. FROM  0  TO  5   '   1   0   0   1000
39:   'END OF DISP. CONTROL     '   0   0   0   0
40:   'READ UNIT=11 UNIT=12'
41:   'STOP'
```

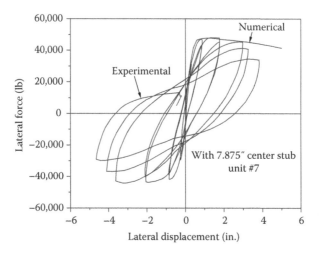

FIGURE 7.26 Pushover curve comparison.

The output results are shown in Figure 7.26.

It can be seen that the lateral force–lateral displacement curve generated by the FSFS method is in agreement with the test results when lateral displacement is between 0 and 3 in.. The numerical calculation shows that the concrete ultimate compression strain, ε_{cu}, is not developed when pushover displacement reaches 5 in.

7.3.7 EXAMPLE 7: THREE-COLUMN BENT (WITH 3D-BEAM, IE3DBEAM, SPRING, PLATE, AND POINT ELEMENTS)

A three-column intermediate bent as shown in Figure 7.27 contains circular R/C columns, collision walls, and bracing members. The bent has 63.44° skew from the longitudinal direction of the beam to the bridge transverse direction. The bracing members are treated as "SPRING" elements with bilinear material model, and the collision walls are treated as plate elements. The cross-sectional properties of the columns are as follows: diameter=32″, cross-sectional area=804.25 (in.²), $f_c' = 4$ ksi, $E=3,605,000$ psi, $G=1,442,000$ psi, $I=51,472$ (in.⁴), and $J=102,943$ (in.⁴), $M_n=23,500$ in-kip, $\theta_n=0.000127$ (rad). The height of the collision wall is 6 ft. The thickness of the wall is 30 in. The elastic modulus of the wall is $E=3320.6$ ksi. The spring element is an L5 × 3.5 × 0.375 angle member with $\sigma_y=36$ ksi. The post-yield axial stiffness of the angle member is 5% of its elastic axial stiffness. The POINT elements are located at the ground level, which represents the stiffnesses of the foundation–soil interaction. The translational stiffnesses of a point element corresponding to element coordinate system (ECS) (X_e, Y_e, Z_e) are 5415, 6937.5, and 6937.5 (k/in.), respectively. The rotational stiffnesses of a point element corresponding to ECS (X_e, Y_e, Z_e) are $1.42E+8$, $3.57E+7$, and $3.57E+7$ (k-in./rad), respectively. Find the displacement capacity of the bent by applying pushover displacement at joint 6 in the joint coordinate system (JCS) X_j direction. Assume column plastic rotation capacity, θ_p, is 0.035 rad.

Input Data (EX7_Test3b.dat)

1   ECHO OF INPUT DATA

```
LINE    ....|.. 10....|.. 20....|.. 30....|.. 40....|.. 50....|.. 60....|.. 70....|.. 80....|.. 90....|..100....|..110
  1:  'STRUCTURAL MODEL:Three-Column Bent'
  2:  'Elements: 3D-beam, IB3D-beam, Plate, spring, and point elements'
  3:  9   2   2   0   1   1
  4:  1  -240  -120    0   1   2
  5:  3   240   120    0
  6:  2  -240  -120   72   1   2
  7:  3   240   120    0
  8:  3  -240  -120  192   2   2
  9:  3   240   120    0
 10:  1   0   0   1   0    | Direction Cosine
 11:  0.8944   0.4472   0   -0.4472   0.8944   0    | Direction Cosine
 12:  6   1   1   0   1   1   0   0
 13:  1   0   0   0   0   1   3   2
 14:  0   6   3   1   6
 15:  10    | Number of Material
 16:  '3D-BEAM  MAT#1'   3605000   1442000   1764    0     0   518616   259308   259308
 17:  'IA_BILN  MAT#2'   1   0   3605   51472   23500   -1
 18:  0
 19:  'IA_BILN  MAT#3'   1   0   3605   51472   23500   -1   0   0.035
 20:  0
 21:  'IA_BILN  MAT#4'   0   0.001   3605   51472   23500   23500   -1   0   -1
 22:  'IA_BILN  MAT#5'   0   0.001   3605   51472   23500   23500   -1   0   -1
 23:  'IA_BILN  MAT#6'   0   0.001   1442   102943   23500   -1   0   -1
 24:  'IA_BILN  MAT#7'   0   0.001   3605   804.25   3271   -1   0   -1
 25:  'PLATE   MAT#8'   3320.6   0.3   30
 26:  'BILINEAR  MAT#9'   176610   219   0   10   0
 27:  'POINT   MAT#10'
 28:  5145   6940   6940
 29:  1.42E+8   3.57E+7   3.57E+7
 30:  0   0   0
 31:  0   0   0
 32:  0   0   0
 33:  0   0   0
 34:  0   0   0
 35:  0   0   0   .FALSE.    | Geometric stiffness
```

```
36:   15       | Number of Element
37:  '3D-BEAM'   'Member 1'   1  1  2   -0.4472  0.8944  0  0  0  0  000000  0
38:  '3D-BEAM'   'Member 2'   1  4  5   -0.4472  0.8944  0  0  0  0  000000  0
39:  '3D-BEAM'   'Member 3'   1  7  8   -0.4472  0.8944  0  0  0  0  000000  0
40:  'IE3DBEAM'  'Member 4'   2  3  4  5  6  7   2  3   -0.4472  0.8944  0
41:   0  0  0  0  0  0  0  0
42:  'IE3DBEAM'  'Member 5'   2  3  4  5  6  7   5  6   -0.4472  0.8944  0
43:   0  0  0  0  0  0  0  0
44:  'IE3DBEAM'  'Member 6'   2  3  4  5  6  7   8  9   -0.4472  0.8944  0
45:   0  0  0  0  0  0  0  0
46:  'SPRING'    'Member 7'   9  2  6  1   294   0.4472  -0.8944  0  0  0
47:  'SPRING'    'Member 8'   9  5  9  1   294   0.4472  -0.8944  0  0  0
48:  'SPRING'    'Member 9'   9  3  5  1   294   0.4472  -0.8944  0  0  0
49:  'SPRING'    'Member 10'  9  6  8  1   294   0.4472  -0.8944  0  0  0
50:  'PLATE'     'Member 11'  8  5  2  1  4  0
51:  'PLATE'     'Member 12'  8  8  5  4  7  0
52:  'POINT'     'Member 13'  10  1  0.8944  0.4472  0  0  1  0
53:  'POINT'     'Member 14'  10  4  0.8944  0.4472  0  0  1  0
54:  'POINT'     'Member 15'  10  7  0.8944  0.4472  0  0  1  0
55:   0  0  .FALSE.   | Mass
56:   0  0            | Damp

1   ECHO OF INPUT DATA

LINE  ....|...10....|...20....|...30....|...40....|...50....|...60....|...70....|...80....|..90....|..100....|..110
57:  'SOL04 Inelastic Incremental Pushover Analysis'
58:  'Incremental displacement control at Joint 6'
59:   1 50000 3 .TRUE. 0  |MAXELD IPRINT IWRITE UNBAL SPLIMIT
60:  'JOINT FX'  6  11  0  0  0
61:  'JOINT FY'  6  12  0  0  0
62:  'ELE'       1  13  4  3  1
63:  'ELE'      11  18  0  0  0
64:  'END'       0   0  0  0
65:   1  6  0  0  'FX'   5           |JOINT LOAD
66:   0  0  0  0  'END'  0
67:   0  0  0  0  'END'  'FY'  0  0  | ELEMENT LOAD
68:  'DISP. FROM  0  TO  5 '  1  0  0  100
69:  'END OF DISP. CONTROL '  0  0  0  0
70:  'READ UNIT=11 UNIT=12 UNIT=13 UNIT=14'
71:  'READ UNIT=15 UNIT=16 UNIT=17 UNIT=18'
72:  'STOP'
```

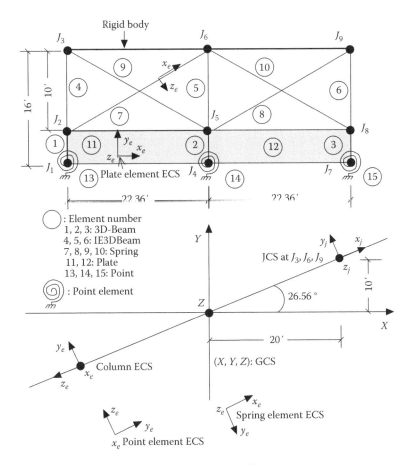

FIGURE 7.27 Three-column bent with skew of 63.44°.

The pushover curve at joint 6 is shown in Figure 7.28. The output result shows that the nominal moment occurred at member 4 at Step 10 (i.e., pushover displacement=0.5″), and the displacement capacity of the bent is reached at Step 93 (pushover displacement=4.65″) at which the plastic rotation of member 4 exceeds the plastic capacity of 0.035 rad.

7.3.8 EXAMPLE 8: FOUR-COLUMN BENT

A four-column intermediate bent with circular R/C columns is shown in Figure 7.29. Five AASHTO Type IV P/S I-girders are placed on the cap beam. The dead load reaction of each girder line is 150 kip. The cross-sectional properties of the columns are diameter=32″, cross-sectional area=804.25 (in.²), $f_c' = 4000$ psi, $E_c = 3,605,000$ psi, $G = 1,442,000$ psi, longitudinal bars=18 – #11, spiral bar=#5 @3″ pitch, $f_y = 60$ ksi, $E_s = 29,000$ ksi, and plastic hinge length=25″. The superstructural center of the mass is 78″ above the centerline of the cap beam. The cap beam is 42″ deep and 42″ wide and is assumed to be elastic. The properties of the cap beam are cross-sectional

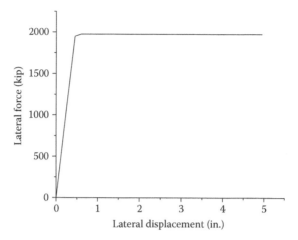

FIGURE 7.28   Pushover curve at Joint 6.

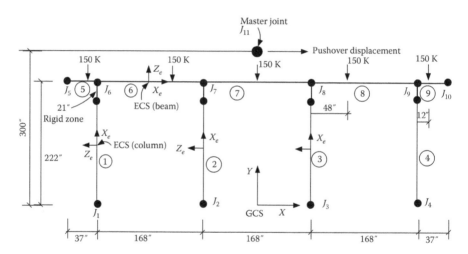

FIGURE 7.29   Four-column bent.

area = 1764 (in.²), $f_c' = 4$ ksi, $E = 3,605,000$ psi, $G = 1,442,000$ psi, $I = 259,308$ (in.⁴), and $J = 518,616$ (in.⁴). Use the FSFS method to find the displacement capacity of the bent by applying pushover displacement at the "master" joint 11 (i.e., at the super-structural mass center) in the GCS' X direction. Joints 6, 7, 8, and 9 are "slave" joints and constrained by the "master" joint. The lateral displacement capacity is determined when the first column confined concrete strain in the cross-sectional compression region reaches the ultimate concrete compression strain, ε_{cu}, defined in Equation 3.24, which is $\varepsilon_{cu} = 0.004 + (1.4\rho_s f_{yh} \varepsilon_{su}/f_{cc}')$, where ρ_s is the volumetric ratio of transverse steel, ε_{su} is the ultimate strain of transverse steel ($\varepsilon_{su} = 0.09$), f_{yh} is yield stress of transverse steel, and f_{cc}' is the confined concrete strength.

Input Data (EX8_Four_Column_bent.dat)

1 ECHO OF INPUT DATA

```
LINE   ....|..10....|..20....|..30....|..40....|..50....|..60....|..70....|..80....|..90....|..100....|..110
   1: 'STRUCTURE DEFINITION: Four-column bent with AASHTO girders.'
   2: 'PROJECT INFORMATION: Project No, 888888'
   3: 11  1  3  0  1  1    |NNODE,NCOS , NSUPT,NCOND,NCONST    SCALE
   4: 1  -252  0  0
   5: 1   168  0  0
   6: 5  -289  0  222  1  1
   7: 5   578  0
   8: 6  -252  0  222  1  3
   9: 1   168  0  0
  10: 11  0  0  300  1  0
  11: 1  0  0  0  1  0    |Direction Cosine
  12: 1  1  1  1  1  1  3  1
  13: 5  0  1  0  1  5  1
  14: 11  1  1  0  1  0  1  0
  15: 0  11  6  3  1
  16: 5    | Number of Material
  17: 'R/CONCRETE1 MAT#1 '  10  60000  29000000  2  32.  2.  0  20  10  4000
  18: 0.625  3.  0.95  1.375   0  25.  0.  804.25  1  2  0.01  1  1442000.  0.01  1
  19: 18  0  0  0  0  60000.  1   -1  -1  0  0  -1  -1
  20: 'R/CONCRETE1 MAT#2 '  10  60000  29000000  2  32.  2.  0  20  10  4000
  21: 0.625  3.  0.95  1.375   0  25.  0.  804.25  1  2  0.01  1  1442000.  0.01  1
  22: 18  0  0  0  0  60000.  1   -1  -1  0  0  -1  -1
  23: 'R/CONCRETE1 MAT#3 '  10  60000  29000000  2  32.  2.  0  20  10  4000
  24: 0.625  3.  0.95  1.375   0  25.  0.  804.25  1  2  0.01  1  1442000.  0.01  1
  25: 18  0  0  0  0  60000.  1   -1  -1  0  0  -1  -1
  26: 'R/CONCRETE1 MAT#4 '  10  60000  29000000  2  32.  2.  0  20  10  4000
  27: 0.625  3.  0.95  1.375   0  25.  0.  804.25  1  2  0.01  1  1442000.  0.01  1
  28: 18  0  0  0  0  60000.  1   -1  -1  0  0  -1  -1
  29: '3D-BEAM MAT#5 '  3605000.  1442000.  1764.  0  0  518616.  259308.  259308.
  30: 0  0  .FALSE.    | Geometric stiffness
  31: 9    | Number of Element
  32: 'STABILITY'  'R/C CIRCULAR MEMBER 1'  1  6  0  1  0  0.  21.0  0
  33: 'STABILITY'  'R/C CIRCULAR MEMBER 2'  2  7  0  1  0  0.  21.0  0
  34: 'STABILITY'  'R/C CIRCULAR MEMBER 3'  3  8  0  1  0  0.  21.0  0
  35: 'STABILITY'  'R/C CIRCULAR MEMBER 4'  4  9  0  1  0  0.  21.0  0
```

```
36:        '3D-BEAM'    'MEMBER  5'  5  5  6    0 1 0   0 0 0 0   000000   0
37:        '3D-BEAM'    'MEMBER  6'  5  6  7    0 1 0   0 0 0 0   000000   0
38:        '3D-BEAM'    'MEMBER  7'  5  7  8    0 1 0   0 0 0 0   000000   0
39:        '3D-BEAM'    'MEMBER  8'  5  8  9    0 1 0   0 0 0 0   000000   0
40:        '3D-BEAM'    'MEMBER  9'  5  9  10   0 1 0   0 0 0 0   000000   0
41:   0   0   .FALSE.        | Mass
42:   0   0                  | Damp
43:  'SOL01 SOLUTION'
44:  'APPLY DEAD LOADS AT BEAM'
45:   1 5              | NLOAD MAXELD
46:   0   0   0 'END' 0.     |JOINT DEAD LOAD
47:   1  5  0  0 'CONC'  'FZ'    -150000.  0.6756
48:   1  6  0  0 'CONC'  'FZ'    -150000.  0.7143
49:   1  7  0  0 'CONC'  'FZ'    -150000.  0.5
50:   1  8  0  0 'CONC'  'FZ'    -150000.  0.2857
51:   1  9  0  0 'CONC'  'FZ'    -150000.  0.3244
52:  'SOL04 SOLUTION'
53:  'INCREMEMTAL DISPLACEMENT CONTROL AT MASTER JOINT 11'
54:   1 50000 3 .TRUE. 0  |MAXELD IPRINT IWRITE UNBAL SPLIMIT
55:  'JOINT FX'  11  11 0 0 0
56:  'ELE '   1  12 0 0 0
  1  ECHO OF INPUT DATA

LINE  ....|.. 10....|.. 20....|.. 30....|.. 40....|.. 50....|.. 60....|.. 70....|.. 80....|.. 90....|..100....|..110
57:  'ELE '   2  13 0 0 0
58:  'ELE '   3  14 0 0 0
59:  'ELE '   4  15 0 0 0
60:  'ELE '   5  16 0 0 0
61:  'ELE '   6  17 0 0 0
62:  'END '   2  18 0 0 0
63:   1 11  0  0 'FX'  15
64:   0  0  0 'END'  0        |JOINT LOAD
65:   0  0  0 'END'  'FY'  0  0   | ELEMENT LOAD
66:  'DISP. FROM  0  TO  15  '  1  0  0  500
67:  'END OF DISP. CONTROL '   0  0  0  0
68:  'READ UNIT=11 UNIT=12 UNIT=13 UNIT=14 UNIT=15 UNIT=16 UNIT=17'
69:  'STOP'
```

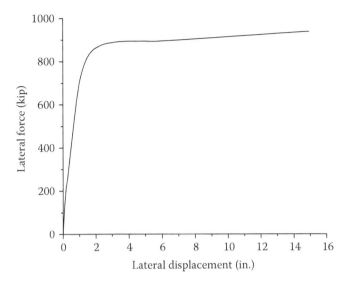

FIGURE 7.30 Pushover curve at master Joint 11.

The pushover curve at master joint 11 is shown in Figure 7.30. The output result shows that the nominal moment occurred at member 4 at Step 73 (i.e., pushover displacement = 2.18″), and the displacement capacity of the bent is reached at Step 397 (pushover displacement = 11.88″) at which the confined concrete compression strain of member 4 reaches its ε_{cu} of 0.021.

7.3.9   EXAMPLE 9: PILE CAP BENT

A pile-cap intermediate bent shown in Figure 7.31 contains four steel HP 12×53 piles, diagonal angle braces $\left(L5 \times 3\frac{1}{2} \times (3/8) \right)$, and horizontal braces $\left(L5 \times 3\frac{1}{2} \times (3/8) \right)$. The unsupported length of the pile is 10 ft above ground, and the spacing between the two piles is 8 ft. The axial load–moment interaction of the steel pile is considered in the pushover analysis. The horizontal bracing members are 12″ above ground line. The brace members have yield stress of 36 ksi. The postbuckling of brace members is considered in the analysis. The stiffnesses of pile–soil interaction are modeled by using POINT element for each pile. The structural model is shown in Figure 7.31. The properties of HP pile are $A = 15.5$ (in.²), $f_y = 36$ ksi, $E = 29,000$ ksi, $G = 11,300$ ksi, I_x (strong axis) = 393 (in.⁴), I_y (weak axis) = 127 (in.⁴), $J = 1.12$ (in.⁴), $M_{px} = 2,664$ in-kip, $M_{py} = 1,159$ in-kip, and $F_y = 558$ kip. The stiffnesses of point element are $S(1,1) = 1498.33$ kip/in., $S(2,2) = 159$ kip/in., $S(3,3) = 142.9$ kip/in., $S(4,4) = 41.44$ kip-in./rad, $S(5,5) = 240,000$ kip-in./rad, $S(6,6) = 538,460$ kip-in./rad, $S(2,6) = -10,460$ kip/rad, and $S(3,5) = 5,200$ kip/rad. The superstructure has five girder lines, and the dead load reaction of each girder line is 100 kip. The performance-based criteria for this

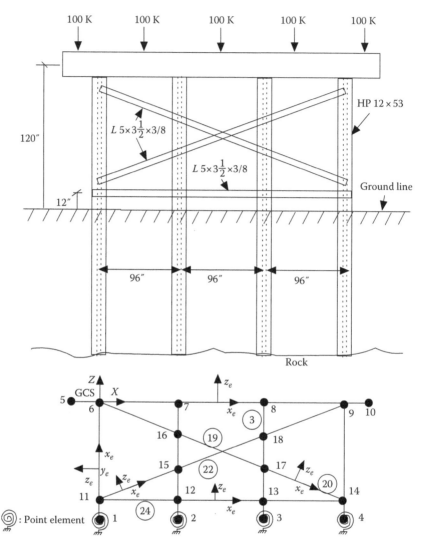

FIGURE 7.31 Pile cap bent.

example are as follows: pile plastic rotation capacity, θ_p, is 0.05 rad. The maximum allowable brace tensile elongation is 10 times that of the brace yield elongation. Find the displacement capacity of the bent by applying incremental pushover displacement at joint 5 in the GCS's X direction.

The output results are shown as follows:

Output (Ex9_Pilecap_4c_flex_cap.out file)

1 ECHO OF INPUT DATA

```
LINE  ....|....10....|....20....|....30....|....40....|....50....|....60....|....70....|....80....|....90....|...100....|...110
  1:  'STRUCTURE DEFINITION - Four-Pile bent w/flexible foundation'
  2:  'Pushover at cap beam'
  3:  18   1   4   0   1
  4:   1     0   0  -120   1   0
  5:   2    96   0  -120   1   0
  6:   3   192   0  -120   1   0
  7:   4   288   0  -120   1   0
  8:   5   -30   0     0   1   0
  9:   6     0   0     0   1   0
 10:   7    96   0     0   1   0
 11:   8   192   0     0   1   0
 12:   9   288   0     0   1   0
 13:  10   318   0     0   1   0
 14:  11     0   0  -108   1   0
 15:  12    96   0  -108   1   0
 16:  13   192   0  -108   1   0
 17:  14   288   0  -108   1   0
 18:  15    96   0   -72   1   0
 19:  16    96   0   -36   1   0
 20:  17   192   0   -72   1   0
 21:  18   192   0   -36   1   0
 22:   1     0   0     0   1   0   | Direction Cosine
 23:   1     0   0     0   1   3   1
 24:   5     1   1     0   1   0   0
 25:   6     0   1     0   1   4   1
 26:  11     0   1     0   1   3   1
 27:  10   | NMAT
 28:  '3D-BEAM   MAT#1'   3605   1442   1764   0     0   198533   198533   397066
 29:  'IA_BILN   MAT#2'   4   0.005   29000.   127.   1159.   -1   0   0.05
 30:   0
 31:  1146.79   1.10979   -0.554461E-02   -0.272039E-06   |A0 - A3
 32:   0
```

```
33:  'IA_BILN   MAT#3'    4    0.005   29000.   127.   1159.        -1    0   0.05
34:  0
35:  1146.79   1.10979   -0.554461E-02   -0.272039E-06   |A0 - A3
36:  0
37:  'IA_BILN   MAT#4'    0    0    1    3683000    0    0    0    -1
38:  'IA_BILN   MAT#5'    0    0    1    3683000    0    0    0    -1
39:  'IA_BILN   MAT#6'    0    0    1    7366000    0    0    0    -1
40:  'IA_BILN   MAT#7'    0    0    1    449500     0    0    0    -1
41:  'BRACE   MAT#8(DIA)'    29000.   3.05   0.762   36.   1   10   -1
42:  'BRACE   MAT#9(HOR)'    29000.   3.05   0.762   36.   1   10   -1
43:  'POINT   MAT#10'
44:  1498.33   159   142.9
45:  41.44   240000   538460
46:  0    0    0
47:  0    0
48:  0    0    -8313
49:  0    5200    0
50:  0    0    0
51:  2    1    1    .TRUE.     | Geometric stiffness
52:  30    | Number of Element
53:  'IE3DBEAM'  'Member 1'   2   3   4   5   6   7   11   6   0   1   0   0   18.0   0
54:  0    0.   0.   0.   0.   0.   0.   0.   0.   0.   0.   0
55:  'IE3DBEAM'  'Member 2'   2   3   4   5   6   7   16   7   0   1   0   0   18.0   0
56:  0    0.   0.   0.   0.   0.   0.   0.   0.   0.   0.   0

ECHO OF INPUT DATA

LINE  ....|..10....|..20....|..30....|..40....|..50....|..60....|..70....|..80....|..90....|..100....|..110
57:  'IE3DBEAM'  'Member 3'   2   3   4   5   6   7   18   8   0   1   0   0   18.0   0
58:  0    0.   0.   0.   0.   0.   0.   0.   0.   0.   0.   0
59:  'IE3DBEAM'  'Member 4'   2   3   4   5   6   7   14   9   0   1   0   0   18.0   0
60:  0    0.   0.   0.   0.   0.   0.   0.   0.   0.   0.   0
61:  '3D-BEAM'   'Member 5'   1   5   6   0   1   0   0   0   000000   0   0
62:  '3D-BEAM'   'Member 6'   1   6   7   0   1   0   0   0   000000   0   0
63:  '3D-BEAM'   'Member 7'   1   7   8   0   1   0   0   0   000000   0   0
64:  '3D-BEAM'   'Member 8'   1   8   9   0   1   0   0   0   000000   0   0
65:  '3D-BEAM'   'Member 9'   1   9   10   0   1   0   0   0   000000   0   0
66:  'IE3DBEAM'  'Member 10'   2   3   4   5   6   7   1   11   0   1   0   0   0   0

1
```

```
 67:              0.  0.  0.  0.  0.  0.
 68: 'IE3DBEAM'  'Member 11'   2  3  4  5  6  7  0    2  12   0  1  0  0  0
 69:              0.  0.  0.  0.  0.  0.
 70: 'IE3DBEAM'  'Member 12'   2  3  4  5  6  7  0    3  13   0  1  0  0  0
 71:              0.  0.  0.  0.  0.  0.
 72: 'IE3DBEAM'  'Member 13'   2  3  4  5  6  7  0    4  14   0  1  0  0  0
 73:              0.  0.  0.  0.  0.  0.
 74: 'IE3DBEAM'  'Member 14'   2  3  4  5  6  7  0   12  15   0  1  0  0  0
 75:              0.  0.  0.  0.  0.  0.
 76: 'IE3DBEAM'  'Member 15'   2  3  4  5  6  7  0   15  16   0  1  0  0  0
 77:              0.  0.  0.  0.  0.  0.
 78: 'IE3DBEAM'  'Member 16'   2  3  4  5  6  7  0   13  17   0  1  0  0  0
 79:              0.  0.  0.  0.  0.  0.
 80: 'IE3DBEAM'  'Member 17'   2  3  4  5  6  7  0   17  18   0  1  0  0  0
 81:              0.  0.  0.  0.  0.  0.
 82: 'BRACE'  'Member 18'   8    6  16    1  1  0  1  0  0
 83: 'BRACE'  'Member 19'   8   16  17    1  1  0  1  0  0
 84: 'BRACE'  'Member 20'   8   17  14    1  1  0  1  0  0
 85: 'BRACE'  'Member 21'   8   11  15    1  1  0  1  0  0
 86: 'BRACE'  'Member 22'   8   15  18    1  1  0  1  0  0
 87: 'BRACE'  'Member 23'   8   18   9    1  1  0  1  0  0
 88: 'BRACE'  'Member 24'   9   11  12    1  1  0  1  0  0
 89: 'BRACE'  'Member 25'   9   12  13    1  1  0  1  0  0
 90: 'BRACE'  'Member 26'   9   13  14    1  1  0  1  0  0
 91: 'POINT'  'Member 27'  10    1   0   -1  0  0  1  0
 92: 'POINT'  'Member 28'  10    2   0   -1  0  0  1  0
 93: 'POINT'  'Member 29'  10    3   0   -1  0  0  1  0
 94: 'POINT'  'Member 30'  10    4   0   -1  0  0  1  0
 95: 0  0    .FALSE.       | Mass
 96: 0  0                  | Damp
 97: 'SOL01 Elastic Static Analysis'
 98: 'APPLY DEAD LOADS AT CAPBEAM'
 99: 1  10
100: 1  5  0  0  'FZ'  0    | Joint Load
101: 1  6  0  0  'FZ'  0    | Joint Load
102: 1  7  0  0  'FZ'  0    | Joint Load
103: 1  8  0  0  'FZ'  0    | Joint Load
```

```
104: 1    9   0   0   'FZ'    0            | Joint Load
105: 1   10   0   0   'FZ'    0            | Joint Load
106: 0    0   0   0   'END'   0            | Joint Load
107: 1    5   0   0   'CONC'  'FZ'  -100  0.4000  | Element Load
108: 1    6   0   0   'CONC'  'FZ'  -100  0.6000  | Element Load
109: 1    7   0   0   'CONC'  'FZ'  -100  0.5000  | Element Load
110: 1    8   0   0   'CONC'  'FZ'  -100  0.4000  | Element Load
111: 1    9   0   0   'CONC'  'FZ'  -100  0.6000  | Element Load
112: 1    5   0   0   'UNIF'  'FZ'  0            | Element Load

ECHO OF INPUT DATA

LINE ....|.. 10....|.. 20....|.. 30....|.. 40....|.. 50....|.. 60....|.. 70....|.. 80....|.. 90....|..100....|..110
113: 1    6   0   0   'UNIF'  'FZ'  0    | Element Load
114: 1    7   0   0   'UNIF'  'FZ'  0    | Element Load
115: 1    8   0   0   'UNIF'  'FZ'  0    | Element Load
116: 1    9   0   0   'UNIF'  'FZ'  0    | Element Load
117: 'SOL04 Inelastic Incremental Pushover Analysis'
118: 'INCREMENTAL DISPLACEMENT CONTROL AT CAPBEAM'
119: 0   50000   2   .TRUE.   0
120: 'JOINT FX'   5   11   0   0
121: 'ELE  '  1   12   0   0
122: 'ELE  '  2   13   0   0
123: 'ELE  '  3   14   0   0
124: 'ELE  '  4   15   0   0
125: 'ELE  '  5   16   0   0
126: 'ELE  '  6   17   0   0
127: 'ELE  '  7   18   0   0
128: 'ELE  '  8   19   0   0
129: 'ELE  '  9   20   0   0
130: 'ELE  '  10  21   0   0
131: 'ELE  '  11  22   0   0
132: 'ELE  '  12  23   0   0
133: 'ELE  '  13  24   0   0
134: 'ELE  '  14  25   0   0
135: 'ELE  '  15  26   0   0
136: 'ELE  '  16  27   0   0
137: 'ELE  '  17  28   0   0
138: 'ELE  '  18  29   0   0
139: 'ELE  '  19  30   0   0
140: 'ELE  '  20  31   0   0
```

1

```
141: 'ELE  `  21  32   0   0   0   0
142: 'ELE  `  22  33   0   0   0   0
143: 'ELE  `  23  34   0   0   0   0
144: 'ELE  `  24  35   0   0   0   0
145: 'ELE  `  25  36   0   0   0   0
146: 'ELE  `  26  37   0   0   0   0
147: 'ELE  `  27  38   0   0   0   0
148: 'ELE  `  28  39   0   0   0   0
149: 'ELE  `  29  40   0   0   0   0
150: 'ELE  `  30  41   0   0   0   0
151: 'END'
152: 1   5   0   0   0   0  'FX'   8   0  | Joint Load
153: 0   0   0   0   0   0  'END'   0      | Joint Load
154: 'DISP. FROM  0  TO  8'   1   0   0  400
155: 'END'   0   0   0   0
156: 'READ UNIT=11 UNIT=31'
157: 'STOP'
```

TIME: 12:48:28, DATE: 28-AUG-09

1 STRUCTURE.....: Pushover at cap beam
SOLUTION......: INCREMEMTAL DISPLACEMENT CONTROL AT CAPBEAM

TIME: 12:48:28, DATE: 28-AUG-09

**** BRACE ELEMENT   20 BUCKLING LOAD = -48.2194

**** BRACE ELEMENT   20 BUCKLED AT STEP =   26. DISP. = -0.578410E-01

**** BRACE ELEMENT   19 BUCKLING LOAD = -48.2194

**** BRACE ELEMENT   19 BUCKLED AT STEP =   36. DISP. = -0.568199E-01

**** BRACE ELEMENT   24 BUCKLING LOAD = -55.0001

**** BRACE ELEMENT   24 BUCKLED AT STEP =   47. DISP. = -0.604826E-01

```
**** IA_BILN ELEMENT      3 YIELD AT END B
**** THE ELEMENT FORCES DISP. FROM STEP:  65 FACTOR:  1.000    ARE:

    3   1  DISPL   18  -0.114368    0.00000    -1.22615    0.00000    1.176101E-03    0.00000     FLP:  4.0
           DISPL    8  -0.118619    0.00000    -1.26996    0.00000   -2.056564E-03    0.00000

    3   1  FORCE   18   106.156     0.00000     59.7867    0.00000    120.981         0.00000     CHR: 000100
           FORCE    8  -106.156     0.00000    -59.7867    0.00000   -1201.79         0.00000

%8 % LIMIT STATE POINT 2 DEFINED DUE TO ELEMENT NO   3 AT STEP=  65
    .
    .
**** BRACE ELEMENT       22 ELONGATION AT STEP =   329 IS    1.27689   WHICH IS GREATER THAN MAXIMUM ALLOWABLE ELONGATION OF
1.27276

%9 % LIMIT STATE POINT 3 DEFINED DUE TO ELEMENT NO  22 AT STEP=  329
```

The pushover curve of the bent is shown in Figure 7.32. The output results indicate that the brace element no. 20 (with start joint 17 and end joint 14) buckled first at Step 26 (corresponding pushover displacement=0.52 in.). Consequently, brace elements 19 and 24 buckled. The buckling behavior of brace element no. 20 is plotted in Figure 7.33, which shows the buckling load is about 48.2 kip. The first plastic hinge developed at the top of pile element no. 3 at Step 65 (pushover displacement = 1.32 in.). The displacement capacity of the bent is reached at Step 329 (pushover displacement=6.6 in.) at which the elongation of brace element no. 22 exceeds the allowable elongation of 1.27 in. (i.e., 10 times of yield deformation).

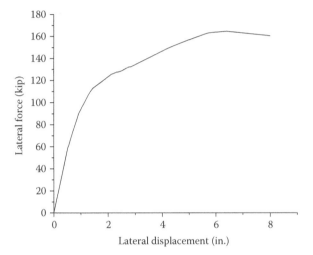

FIGURE 7.32   Pushover curve at Joint 5.

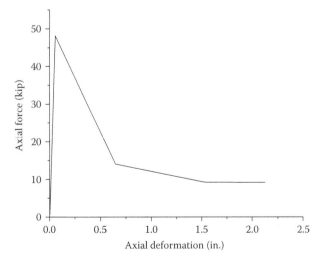

FIGURE 7.33   Postbuckling of brace element 20.

7.3.10 EXAMPLE 10: CROSS FRAME ANALYSIS

Figure 7.34 shows cross frame details. The components of the cross frame include top and bottom chords and diagonal angle members. The top and bottom chords are L5×5×5/16 angles, diagonal angle members are L3×3×5/16, and the bearing stiffener size is 8.5″ × 0.75″. The sizes of top flange, bottom flange, and girder web are 12″ × 1.125″, 14″ × 2″, and 0.5″, respectively. The yield stress of steel is 36 ksi, and the girder spacing is 117″. The bearing stiffeners provide the main stiffness for girders, and only 20 in. of the girder length is considered in the structural model (Zahrai and Bruneau, 1999) (see Figure 7.34b). The structural model is shown in Figure 7.35.

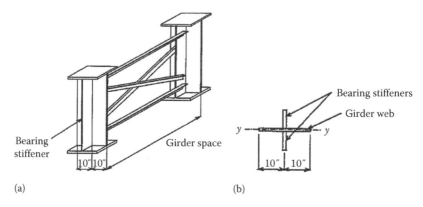

FIGURE 7.34 Cross frame details: (a) cross flame; (b) stub girder.

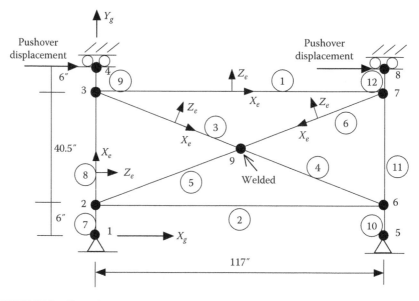

FIGURE 7.35 Cross frame model.

Two identical pushover displacements are applied at joints 4 and 8, simultaneously. Find the displacement capacity at which the first angle member buckles. The stub girders are assumed to be elastic. Use "STABILITY" elements for the diagonal members, top chord, and bottom chord in the pushover analysis. The initial imperfection ratio of diagonal members is assumed to be 0.01. After the analysis, replace "STABILITY" elements with "BRACE" elements and perform the same analysis. Compare both analytical results.

The pushover curves at joints 4 and 8 are shown in Figure 7.36. When pushover displacement reached 0.18 in., the diagonal angle member 4 buckled (see Figure 7.37). As shown in Figure 7.36, once diagonal angle member buckled, the

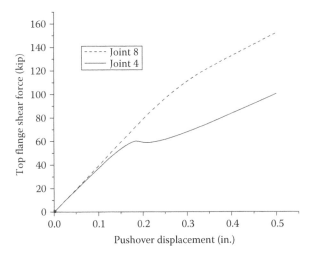

FIGURE 7.36   Pushover curves at Joints 4 and 8 (use STABILITY elements).

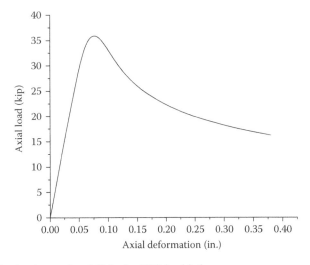

FIGURE 7.37   Angle member 4 (L3×3×5/16) buckled.

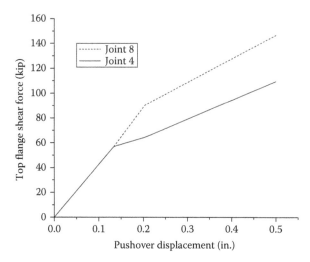

FIGURE 7.38 Pushover curves at Joints 4 and 8 (use BRACE elements).

earthquake inertial shear forces transferred from deck to girders are not equally distributed. Therefore, bearings resist different shear forces after the diagonal angle member buckled. The output shows that the maximum shear forces at elements 7 and 10 (bearing locations) are 145.6 and 107.3 kip, respectively. The nonuniformed shear distribution due to buckling of cross frame members may damage the bearing, gusset plate connection, or cause the transverse stability of the girder. Note that if stability elements in the structural model are replaced by the BRACE elements, the pushover curve based on the structural model with BRACE elements is shown in Figure 7.38, which is in agreement with that in Figure 7.36.

Input Data (EXAMPLE_10.dat file)

```
1   ECHO OF INPUT DATA

LINE   ....|..10....|..20....|..30....|..40....|..50....|..60....|..70....|..80....|..90....|..100....|..110
 1:  'STRUCTURE DEFINITION CROSS FRAME OVER INTERMEDIATE BENT'
 2:  'LATERAL STRENGTH OF CROSS FRAME'
 3:  9  15  0  0  1.          NNODE,NCOS  , NSUPT,NCOND,NCONST   SCALE
 4:  1  0.00   0.00    .00 1 0
 5:  2  0.0    6.0     0.0 1 0
 6:  3  0.00   46.5    .00 1 0
 7:  4  0.0    52.5    0.0 1 0
 8:  5  117.0  0.00    .00 1 0
 9:  6  117.0  6.0     0.0 1 0
10:  7  117.0  46.5    .00 1 0
11:  8  117.0  52.5    0.0 1 0
12:  9  58.5   26.25   .00 1 0
13:  1  0  0  1  0  | DIRECTION COSINE
14:  1  1 1 1 1 1 0   1 4
15:  4  1 1 1 1 1 1   0 0
16:  8  1 1 1 1 1 1   0 0
17:  2  0 0 1 1 1 0   1 1
18:  6  0 0 1 1 1 0   1 1
19:  7  |NMAT
20:  'STABILITY1 L5x5x.3125 MAT1'  16  36  29000  7  5.0  0  -1.37   -1.13   0  0  0.3125
21:  41  4  40  4  1  4. 0.01  0.01  3.125  1  0  0.01  0  11300 0.01 1 1 0.05
22:  'STABILITY1 L5x5x.3125 MAT2'  16  36  29000  7  5.0  0  -1.37   -1.13   0  0  0.3125
23:  41  4  40  4  1  4. 0.01  0.01  3.125  1  0  0.01  0  11300 0.01 1 1 0.05
24:  'STABILITY1 L3x3x.3125 MAT3'  16  36  29000  7  3.0  0  -0.865  -0.635  0  0  0.3125
25:  41  4  40  4  1  2. 0.01  0.01  1.875  1  0  0.01  0  11300 0.01 1 1 0.05
26:  'STABILITY1 L3x3x.3125 MAT4'  16  36  29000  7  3.0  0  -0.865  -0.635  0  0  0.3125
27:  41  4  40  4  1  2. 0.01  0.01  1.875  1  0  0.01  0  11300 0.01 1 1 0.05
28:  'STABILITY1 L3x3x.3125 MAT5'  16  36  29000  7  3.0  0  0.865   -0.635  0  0  0.3125
29:  41  4  40  4  1  2. 0.01  0.01  1.875  1  0  0.01  0  11300 0.01 1 1 0.05
30:  'STABILITY1 L3x3x.3125 MAT6'  16  36  29000  7  3.0  0  0.865   -0.635  0  0  0.3125
31:  41  4  40  4  1  2. 0.01  0.01  1.875  1  0  0.01  0  11300 0.01 1 1 0.05
32:  '3D-BEAM MAT#7'  29000. 11300.  22.75  0 0  614.  307.  307.
33:  0 0 0 .FALSE.  KG: AXL, FORM, ASSY
34:  12 NELEM   ELE MAT SJ EJ VYI,VYJ,VYK XS XE
35:  'STABILITY'  'ELE#1'  1  3  7  0  0  -1  0  0  0
```

```
36:   'STABILITY'   'ELE#2'    2   2   6   0   0  -1   0   0   0
37:   'STABILITY'   'ELE#3'    3   3   9   0   0  -1   0   0   0
38:   'STABILITY'   'ELE#4'    4   9   6   0   0  -1   0   0   0
39:   'STABILITY'   'ELE#5'    5   9   2   0   0   1   0   0   0
40:   'STABILITY'   'ELE#6'    6   7   9   0   0   1   0   0   0
41:   '3D-BEAM'     'ELE#7'    7   1   2   0   0  -1   0   0   0   000000   0
42:   '3D-BEAM'     'ELE#8'    7   2   3   0   0  -1   0   0   0   000000   0
43:   '3D-BEAM'     'ELE#9'    7   3   4   0   0  -1   0   0   0   000000   0
44:   '3D-BEAM'     'ELE#10'   7   5   6   0   0  -1   0   0   0   000000   0
45:   '3D-BEAM'     'ELE#11'   7   6   7   0   0  -1   0   0   0   000000   0
46:   '3D-BEAM'     'ELE#12'   7   7   8   0   0  -1   0   0   0   000000   0
47:   0  0 .FALSE. | MASS
48:   0  0         | DAMP
49:   'SOL04 SOLUTION'
50:   'INCREMENTAL DISPLACEMENT CONTROL'
51:   1 50000 2 .FALSE. 0.00001 |MAXELD IPRINT IWRITE UNBAL SPLIMIT
52:   'JOINT FX'   4  11   0   0
53:   'JOINT FX'   8  12   0   0
54:   'ELE     `   1  13   0   0
55:   'ELE     `   2  14   0   0
56:   'ELE     `   3  15   0   0
    ECHO OF INPUT DATA

LINE  ....|..10....|..20....|..30....|..40....|..50....|..60....|..70....|..80....|..90....|..100....|..110
57:   'ELE     `   4  16   0   0
58:   'ELE     `   5  17   0   0
59:   'ELE     `   6  18   0   0
60:   'ELE     `   7  19   0   0
61:   'ELE     `  10  20   0   0
62:   'END     `  11  20   0   0   0
63:   1  4  0  0 'FX'  0.5
64:   1  8  0  0 'FX'  0.5
65:   0  0  0  0 'END'  0.5              |JOINT LOAD
66:   0  0  0  0 'END' 'FY'   0   0  | ELEMENT LOAD
67:   'DISP. FROM 0 TO 0.5 `  1  0  0  400
68:   'END OF DISP. CONTROL     1  0  0  60
69:   'READ UNIT=11 UNIT=12 UNIT=13 UNIT=14 UNIT=15'
70:   'READ UNIT=16 UNIT=17 UNIT=18 UNIT=19 UNIT=20'
71:   'STOP'
```

1

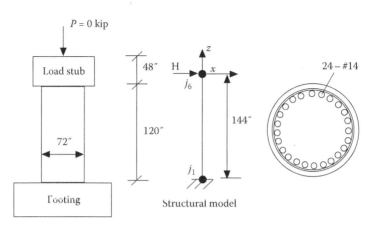

FIGURE 7.39   Shear column specimen.

7.3.11   EXAMPLE 11: COLUMN WITH SHEAR FAILURE

Figure 7.39 shows a short column tested at the University of California, San Diego, CA (Ohtaki et al., 1996). The column diameter is 6 ft, and the height is 12 ft. The longitudinal reinforcement consists of 24 – #14 rebars with $f_y = 73.8$ ksi. The concrete cover is 2.5 in. The transverse reinforcement is #4 @ 12" with $f_y = 43.3$ ksi. Concrete strength is $f'_c = 4.29$ ksi. The footing and the load stub were designed to be strong enough to resist flexural or shear failure prior to column failure. There is no axial load applied to the column. Perform pushover analysis and identify the failure mode.

1. Output 1 (shear1_MC.out file)

   The moment–curvature analysis is performed first in order to determine the input parameters for the HINGE material to be used for the column pushover analysis. The bilinear moment–curvature parameters, φ_n, M_n, φ_u, and M_u, were obtained from the moment–curvature analysis. The M–φ curve is shown in Figure 7.40. The analysis also calculates the initial concrete shear strength, $V_{ci} = 746.7$ kip, the concrete shear strength, $V_{cf} = 128$ kip, when the rotational ductility (μ) of the member end is greater than or equal to 15 and the transverse steel shear strength, $V_s = 130.1$ kip, which are shown in the output.

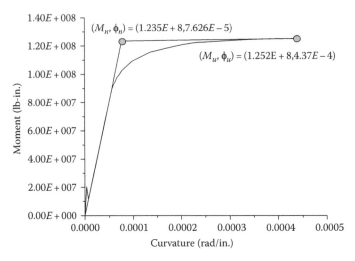

FIGURE 7.40   Moment–curvature curve.

1   ECHO OF INPUT DATA

```
LINE  ....|..10....|..20....|..30....|..40....|..50....|..60....|..70....|..80....|..90....|..100....|..110
   1: 'STRUCTURE DEFINITION - Moment-curvature analysis of R/C columns'
   2: Moment-curvature analysis under dead load'
   3: 2 1 2 0 0 1
   4: 1 0.0 0.0 0.0 1 0
   5: 2 2.0 0.0 0.0 1 0
   6: 1 0 0 1 0    | Direction Cosine
   7: 1 0 1 1 1 1 0 0
   8: 2 1 1 1 1 1 0 0
   9: 1 |NMAT
  10: 'R/CONCRETE1    MAT#1'  1   73800   29000000   2   72   2.5   0   0   20   10   4290
  11: 0.5   12   0.95   1.75   0   -1   0.   0.   4071.50   1   2   0.01   0   1500000
  12: 0.01   1   24   0   0   0   43300   1   -1   -1   -1   -1   -1
  13: 0   0   .FALSE.    | Geometric stiffness
  14: | Number of Element
  15: 'STABILITY'   'R/C CIRCULAR SECTION ELE.'   1   1   2   0   1   0   0   0   0
  16: 0   0   .FALSE.    | Mass
  17: 0   0    | Damp
  18: 'SOL01 Elastic Static Analysis'
  19: 'APPLY AXIAL LOAD AT JOINT 1'
  20: 1   0    |SOL01_VALUE
  21: 1   1   0   0   'FX'   0    | Joint Load
  22: 0   0   0   'END'   0    | Joint Load
  23: 'SOL04 Inelastic Incremental Pushover Analysis'
  24: 'incremental disp. control'
  25: 0   50000   1   .TRUE.   0
  26: 'JOINT MZ'   1   13   0   0   0
  27: 'ELE  '   1   11   0   0   0
  28: 'END'   0   0   0   0
  29: 1   0   0   'MZ'   0.007   0    | Joint Load
  30: 1   2   0   0   'MZ'   -0.007   0    | Joint Load
  31: 0   0   0   0   'END'   0    | Joint Load
  32: 'disp. from 0 to 0.007'   1   0   0   2000
  33: 'END'   0   0   0   0
  34: 'READ   UNIT=13'
  35: 'STOP'
```

```
MAT. NO.  =  1
INITIAL CONCRETE SHEAR CAPACITY (VCI in lb)   = 7.467E+05
FINAL CONCRETE SHEAR CAPACITY (VCF in lb)     = 1.280E+05
TRANS.STEEL SHEAR CAPACITY,U0 DIR (VSu in lb) = 1.301E+05
TRANS.STEEL SHEAR CAPACITY,VO DIR (VSv in lb) = 1.301E+05
```

DEGREE OF FREEDOM # 6 IS READ FROM UNIT # 13   JOINT # 1, DIRECTION: MZ

	STEP	TIME	LOAD	DISPLACEMENT	VELOCITY	ACCELERATION
	0	0.0000	0.0000	0.0000	0.0000	0.0000
	1	0.0000	2.02760E+07	3.50000E-06	0.0000	0.0000
	2	0.0000	1.15081E+07	7.00000E-06	0.0000	0.0000
1	14	0.0000	7.95350E+07	4.90000E-05	0.0000	0.0000
	15	0.0000	8.50287E+07	5.25000E-05	0.0000	0.0000
	16	0.0000	8.99098E+07	5.60000E-05	0.0000	0.0000
2	80	0.0000	1.23436E+08	2.80000E-04	0.0000	0.0000
	81	0.0000	1.23506E+08	2.83500E-04	0.0000	0.0000
	82	0.0000	1.23575E+08	2.87000E-04	0.0000	0.0000
3	125	0.0000	1.25213E+08	4.37500E-04	0.0000	0.0000
	126	0.0000	1.25216E+08	4.41000E-04	0.0000	0.0000
	127	0.0000	1.25219E+08	4.44500E-04	0.0000	0.0000

From the above output, the bilinear parameters φ_n, M_n, φ_u, and M_u are calculated per Section 4.2 and equal to $7.626E-5$ rad., $1.235E+5$ k-in., $4.37E-4$ rad., and $1.252E+5$ k-in., respectively. The plastic hinge length L_p is 38.74″ per Equation 4.5. Therefore, the yield rotation, θ_n, and the plastic rotational capacity, θ_p, at the hinge top, are $2.56E-3$ rad and 0.0139 rad based on Equations 4.16 and 4.17, respectively.

2. Output 2 (shear1.out file)
   Once the parameters of θ_n, M_n, θ_p, and M_u for HINGE material were obtained from Output1 above, they were used as the input data for the column pushover analysis. The pushover output results are shown as follows:

```
1     ECHO OF INPUT DATA

LINE  ....|...10....|...20....|...30....|...40....|...50....|...60....|...70....|...80....|...90....|...100....|..110
   1: 'STRUCTURE DEFINITION - Test shear failure'
   2: 'Compare test results from UC-San Diego.'
   3: 4   1   3   0   0   1
   4: 1   0   0   -144   1   0
   5: 5   -30   0   0   1   0
   6: 6   0   0   0   1   0
   7: 7   30   0   0   1   0        | Direction Cosine
   8: 1   0   0   1   0   1   1   1   0   0
   9: 1   1   1   1   1   1   1   0   0
  10: 5   1   1   0   1   1   0
  11: 6   0   1   0   1   1   1
  12: 7   | NMAT
  13: '3D-BEAM   MAT#1'   3122   1200   7000   0   0   20700000   41400000
  14: 'HINGE   MAT#2'   0.002558531   123504   0.002526851   0   1.618667E+09   0.01399   0
  15: 1
  16: 746.7   128.0   130.1   22.6
  17: 'HINGE   MAT#3'   0.002558531   123504   0.002526851   0   1.618667E+09   0.01399   0
  18: 1
  19: 746.7   128.0   130.1   22.6
  20: 'IA_BILN   MAT#4'   0   1   8.458694E+08   0   0   -1
  21: 'IA_BILN   MAT#5'   0   1   8.458694E+08   0   0   -1
  22: 'IA_BILN   MAT#6'   0   1   3.166002E+09   0   0   -1
  23: 'IA_BILN   MAT#7'   0   1   1.585248E+07   0   0   -1
  24: 2   1   1   .TRUE.   | Geometric stiffness
  25: 3   | Number of Element
  26: 'IE3DBEAM'   'Member 1'   2   3   4   5   6   7   1   6   0   1   0   0   24   0
  27: 0   0.316   48.   0.   40.56   40.56   0.   0   0.316   48.   0.   40.56   40.56   0.
  28: '3D-BEAM'   'Member 2'   1   5   6   0   1   0   0   0   000000   0
  29: '3D-BEAM'   'Member 3'   1   6   7   0   1   0   0   0   000000   0
  30: 0   0   .FALSE.   | Mass
  31: 0   0   | Damp
  32: 'SOL01 Elastic Static Analysis'
  33: 'APPLY DEAD LOADS AT CAPBEAM'
  34: 1   2
  35: 1   5   0   0   'FZ'   0   0   | Joint Load
  36: 1   6   0   0   'FZ'   0   0   | Joint Load
  37: 1   7   0   0   'FZ'   0   0   | Joint Load
```

```
38: 0   0   0   'END'    0  | Joint Load
39: 1   2   0   'UNIF',  'FZ',  0  | Element Load
40: 1   3   0   'UNIF',  'FZ',  0  | Element Load
41: 'SOL04 Inelastic Incremental Pushover Analysis'
42: 'INCREMEMTAL DISPLACEMENT CONTROL AT CAPBEAM'
43: 0   50000  1  .TRUE.  0
44: 'JOINT FX'  5  11  0  0  0
45: 'ELE  '  1  12  0  0  0
46: 'ELE  '  2  13  0  0  0
47: 'ELE  '  3  14  0  0  0
48: 'END'  0  0  0  0
49: 1   5   0   'FX',   2  | Joint Load
50: 0   0   0   'END',  0  | Joint Load
51: 'DISP. FROM  0  TO  2'  1  0  0  1000
52: 'END'  0  0  0
53: 'READ UNIT=11 UNIT=12'
54: 'STOP'
```

PLASTIC HINGE LENGTH MOMENT-ROTATION MODEL
===

MAT.	HA	VA	RATIO	MAX DUC	BETA	STIELE	PRMAX
2	0.255853E-02	123504.	0.252685E-02	0.00000	0.00000	0.161867E+10	0.00000

PARAMETERS FOR MEMBER SHEAR CAPACITIES:
    VCI= 746.700
    VCF= 128.000
    VS=   130.100
    ALFA (degree) =   22.6000

MAT.	HA	VA	RATIO	MAX DUC	BETA	STIELE	PRMAX
3	0.255853E-02	123504.	0.252685E-02	0.00000	0.00000	0.161867E+10	0.00000

PARAMETERS FOR MEMBER SHEAR CAPACITIES:
    VCI= 746.700
    VCF= 128.000
    VS=   130.100
    ALFA (degree) =   22.6000

```
                                                       TIME: 09:29:56, DATE: 30-MAR-09
                                                       TIME: 09:29:56, DATE: 30-MAR-09

 1 STRUCTURE.....: Compare test results from UC-San Diego.
   SOLUTION......: INCREMEMTAL DISPLACEMENT CONTROL AT CAPBEAM

**** IA_BILN ELEMENT       1 YIELD AT END A
**** THE ELEMENT FORCES DISP. FROM STEP:  265 FACTOR:  1.000     ARE:

    1   2  DISPL  1    0.00000        0.00000     0.00000   -3.317976E-03   0.00000    FLP:  3.0
             DISPL  6  2.586920E-07  -0.398156    0.00000    2.042972E-03   0.00000

    1   2  FORCE  1   -3.417425E-02   856.557     0.00000   -123505.        0.00000    CHR: 001000
             FORCE  6  3.417425E-02  -856.557     0.00000    20718.2        0.00000

%8 % LIMIT STATE POINT 2 DEFINED DUE TO ELEMENT NO   1 AT STEP=  265

**** IA_BILN ELEMENT       1 FAILED DUE TO SHEAR

                                                             AT END A
**** THE ELEMENT FORCES DISP. FROM STEP:  775 FACTOR:  1.000     ARE:

    1   1  DISPL  1    0.00000        0.00000     0.00000   -1.041259E-02   0.00000    FLP:  12.
             DISPL  6  5.863909E-07  -1.24949     0.00000    2.059160E-03   0.00000

    1   1  FORCE  1   -7.746466E-02   861.924     0.00000   -124371.        0.00000    CHR: 001000
             FORCE  6  7.746466E-02  -861.924     0.00000    20940.6        0.00000

* ELEMENT   1 SHEAR IN Z DIR. AT END A  =    861.9   > SHEAR
CAPACITY OF   861.5   AT STEP =  775
%9 % LIMIT STATE POINT 3 DEFINED DUE TO ELEMENT NO   1 AT STEP=  775

DEGREE OF FREEDOM #   15 IS READ FROM UNIT #   11    JOINT #   5, DIRECTION: FX

STEP   TIME      LOAD    DISPLACEMENT   VELOCITY   ACCELERATION
  0    0.0000   0.0000   0.0000         0.0000     0.0000
  1    0.0000   3.2597   2.00000E-03    0.0000     0.0000
  2    0.0000   6.5195   4.00000E-03    0.0000     0.0000
```

264	0.0000	0.52800	0.0000	859.91	0.0000
2 265	0.0000	0.53000	0.0000	857.12	0.0000
266	0.0000	0.53200	0.0000	857.13	0.0000

774	0.0000	1.5480	0.0000	861.87	0.0000
3 775	0.0000	1.5500	0.0000	861.88	0.0000
776	0.0000	1.5520	0.0000	861.89	0.0000

ELEMENT #   1 IS READ FROM UNIT #   12

IE3D BEAM FORCES...

STEP	TIME	NODE	AXIAL	FY	FZ	TORSION	MY	MZ	STBFAG & FLP
264	0.0000	1	-3.417420E-02	0.00000	856.560	0.00000	-123505.	0.00000	
		6	3.417420E-02	0.00000	-856.560	0.00000	20718.0	0.00000	000000
		DISP	0.00000	0.00000	0.00000	0.00000	-3.317980E-03	0.00000	
			2.586920E-07	0.00000	-0.399160	0.00000	2.042970E-03	0.00000	0.0
265	0.0000	1	-3.425030E-02	0.00000	857.130	0.00000	-123507.	0.00000	
		6	3.425030E-02	0.00000	-857.130	0.00000	20652.0	0.00000	001000
		DISP	0.00000	0.00000	0.00000	0.00000	-3.321600E-03	0.00000	
			2.592680E-07	0.00000	-0.399860	0.00000	2.041350E-03	0.00000	3.0
774	0.0000	1	-7.738040E-02	0.00000	861.920	0.00000	-124370.	0.00000	
		6	7.738040E-02	0.00000	-861.920	0.00000	20540.0	0.00000	001000

775   0.0000

```
DISP    0.00000    0.00000    0.00000    0.00000    -1.039870E-02    0.00000    3.0
        5.857530E-07  0.00000  -1.24780   0.00000   2.059130E-03    0.00000

     1  -7.746470E-02  0.00000   861.920   0.00000   -124371.        0.00000    001000
     6   7.746470E-02  0.00000  -861.920   0.00000   20941.0         0.00000
DISP    0.00000       0.00000   0.00000    0.00000   -1.041260E-02   0.00000    12.
        5.863910E-07  0.00000  -1.24950   0.00000   2.059160E-03    0.00000
```

. .

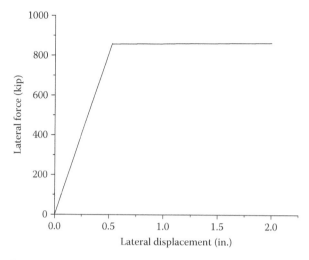

FIGURE 7.41 Pushover curve.

The pushover curve at the center of the load stub is shown in Figure 7.41. The above output shows that the shear failure occurs at incremental Step 775, at which the lateral shear force is 861.9 kip, and the lateral displacement is 1.55 in. The column shear strength capacity at Step 775 can be checked from the above element output. It shows that the column yields at Step 265 with the yield rotation $\theta_y = 0.0033$ rad, and the rotation at Step 775 is $\theta = 0.0104$ rad. Therefore, the rotational ductility at Step 775 is $\mu = 0.0104/0.0033 = 3.12$. From Figure 4.17 and $V_{ci} = 746.7$ kip, $V_{cf} = 128$ kip, and $V_s = 130.1$ kip obtained from Output1, V_{c12} corresponding to $k = 1.2$ is $V_{c12} = V_{ci}(1.2/3.5) = 746.7 * 0.342 = 256$ kip. V_c corresponding to the ductility of $\mu = 3.12$ is $V_c = V_{c12} + ((V_{c12} - V_{ci})/(7-3))(\mu-7) = 731.5$ kip. Therefore, the total shear strength at Step 775 is $V = V_c + V_s = 731.5 + 130.1 = 861.6$ kip. Since 861.6 kip is less than the shear demand of 861.9 kip, shear failure occurs at Step 775. The calculated shear strength of 861.6 kip from this analysis is in agreement with that of 856.1 kip based on UCSD predictive model (Ohtaki et al., 1996).

7.3.12   EXAMPLE 12: BEAM–COLUMN JOINT FAILURE

Figure 7.42a and b shows the full-scale inversed cap beam–column test specimen #1 and #2, respectively. The specimens were tested at the University of California, San Diego, CA (Seible et al., 1994). The geometries and material properties of the specimens are described as follows.

7.3.12.1   For Test Specimen #1

The column diameter (h_c or D) is 5 ft and the height is 25ft. The cap beam width (b_b) is 5 ft-6 in., and the cap beam depth (h_b) is 72 in. The test specimen #1 was post tensioned with the prestressed force (P_b) of 1726 kip. The column axial load (P) is 600 kip applied at the load stub. The longitudinal reinforcement consists of 20 – #18 rebars with $f_y = 77.5$ ksi. The concrete cover is 2 in. The transverse reinforcement is #6 @ 3.5" with $f_y = 62.3$ ksi. Concrete strength is $f_c' = 6$ ksi. The load stub was designed to be strong

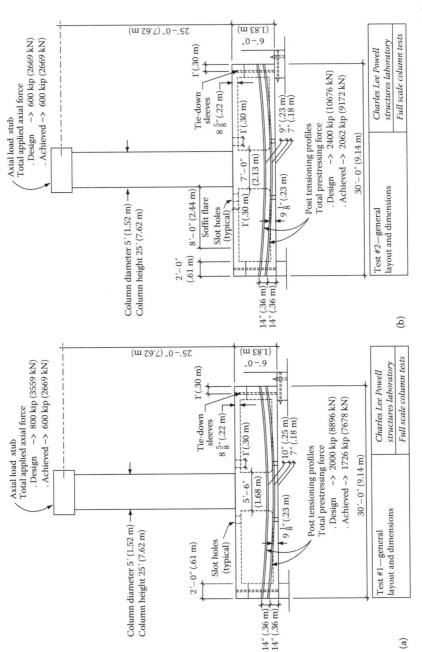

FIGURE 7.42 (a) Test specimen #1. (b) Test Specimen #2. (Copied from Seible, F. et al., Full-scale bridge column/superstructure connection tests under simulated longitudinal seismic loads, Report No. SSRP-94/14, University of California, San Diego, 1994. With permission)

enough to resist flexural or shear failure prior to column failure. The plastic hinge length of the column can be calculated using Equation 4.5, which is $L_p = 52.47''$. Based on L_p and moment–curvature analysis of the column, the bilinear moment and rotational parameters, θ_n, M_n, θ_p, and M_u, are 0.00495 rad, 149,760 k-in., 0.0495 rad, and 160,162 k-in., respectively. These parameters are used to define the HINGE material for the column. Perform pushover analysis and identify the failure mode for test specimen #1.

7.3.12.2 For Test Specimen #2

The geometry and reinforcement for test specimen #2 are the same as those for test specimen #1, with the exception of (1) an increased cap beam width (h_b) from 5′–6″ to 7′–0″, (2) column $f_c' = 4.67$ ksi, and (3) $f_y = 70$ ksi for longitudinal reinforcement and $f_y = 66.8$ ksi for the transverse reinforcement. The plastic hinge length of column per Equation 4.5 is $L_p = 47.9''$. Based on L_p and moment–curvature analysis of the column, the bilinear moment and rotational parameters, θ_n, M_n, θ_p, and M_u, are 0.00422 rad, 135,480 k-in., 0.0538 rad, and 148,776 k-in., respectively. These parameters are used to define the HINGE material for the column. Perform pushover analysis and identify the failure mode for test specimen #2.

1. Output 1 (Joint_shear1.out file)

1   ECHO OF INPUT DATA

```
LINE   ....|..10....|..20....|..30....|..40....|..50....|..60....|..70....|..80....|..90....|..100....|..110
 1: 'STRUCTURE DEFINITION - Test Joint shear failure, Sample #1'
 2: 'Compare test results from UC-San Diego (SSRP-94/14).'
 3: 4   1   3   0   0   1
 4: 1   0   0   -300   1   0
 5: 5   -30   0   0   1   0
 6: 6   0   0   0   1   0
 7: 7   30   0   0   1   0
 8: 1   0   0   0   1   0        | Direction Cosine
 9: 1   1   1   1   1   0
10: 5   1   1   0   1   1
11: 6   0   1   1   0   1   1
12: 7        | NMAT
13: '3D-BEAM  MAT#1'   3122   1200   7000   0   0   41400000   20700000   20700000   0
14: 'HINGE   MAT#2'   0.004954254   149760   0.006946992   0   0   1.44752E+09   0.04949   0
15: 0
16: 'HINGE   MAT#3'   0.004954254   149760   0.006946992   0   0   1.44752E+09   0.04949   0
17: 0
18: 'IA_BILN  MAT#4'   0   0   1   1.44752E+09   0   0   -1
19: 'IA_BILN  MAT#5'   0   0   1   1.44752E+09   0   0   -1
20: 'IA_BILN  MAT#6'   0   0   1   1.526814E+09   0   0   -1
21: 'IA_BILN  MAT#7'   0   0   1   1.445089E+07   0   0   -1
22: 2   1   1   .TRUE.        | Geometric stiffness
23: 3        | Number of Element
24: 'IE3DBEAM'  'Member 1'   2   3   4   5   6   7   1   6   0   1   0   24   0
25: 1   0.39   60   0   72   66   1726.   0   0.39   60.   0   72   66   1726.   0
26: '3D-BEAM'  'Member 2'   1   5   6   0   1   0   000000   0
27: '3D-BEAM'  'Member 3'   1   6   7   0   1   0   000000   0
28: 0   0   .FALSE.   | Mass
29: 0   0        | Damp
30: 'SOL01 Elastic Static Analysis'
31: 'APPLY DEAD LOADS AT CAPBEAM'
32: 1   2
```

```
33:  1   5   0   0   'FZ'    0   | Joint Load
34:  1   6   0   0   'FZ'  -600  | Joint Load
35:  1   7   0   0   'FZ'    0   | Joint Load
36:  0   0   0   0   'END'   0   | Joint Load
37:  1   2   0   0   'UNIF'  'FZ'   0   | Element Load
38:  1   3   0   0   'UNIF'  'FZ'   0   | Element Load
39:  'SOL04 Inelastic Incremental Pushover Analysis'
40:  'INCREMEMTAL DISPLACEMENT CONTROL AT CAPBEAM'
41:  0   50000   2   .TRUE.   0
42:  'JOINT FX'   5   11   0   0   0
43:  'ELE  `      1   12   0   0   0
44:  'ELE  `      2   13   0   0   0
45:  'ELE  `      3   14   0   0   0
46:  'END'   0   0   0   0
47:  1   5   0   0   'FX'   20   | Joint Load
48:  0   0   0   0   'END'   0   | Joint Load
49:  'DISP. FROM 0 TO 20'   1   0   1000
50:  'END'   0   0   0   0
51:  'READ UNIT=11 UNIT=12'
52:  'STOP'
```

```
.
.
.
 1  STRUCTURE.....:  Compare test results from UC-San Diego (SSRP-94/14).
    SOLUTION......:  INCREMEMTAL DISPLACEMENT CONTROL AT CAPBEAM

                                              TIME: 09:26:45,  DATE: 02-APR-09
                                              TIME: 09:26:45,  DATE: 02-APR-09

*****  IA_BILN ELEMENT          1 YIELD AT END A
*****  THE ELEMENT FORCES DISP. FROM STEP:  157  FACTOR:  1.000     ARE:

    1  2  DISPL  1    0.00000        0.00000    0.00000    0.00000    -9.953722E-03
          DISPL  6   -1.146027E-02  -2.74722    0.00000    0.00000     5.550232E-03      FLP:  3.0

    1  2  FORCE  1    600.041        0.00000    493.075    0.00000    -149766.
          FORCE  6   -600.041        0.00000   -493.075    0.00000     12028.5           CHR: 001000

%8 % LIMIT STATE POINT 2 DEFINED DUE TO ELEMENT NO      1 AT STEP=  157
```

```
**** IA_BILN ELEMENT          1 FAILED DUE TO POSSIBLE JOINT SHEAR CRACK      AT END A

**** THE ELEMENT FORCES DISP. FROM STEP:  650 FACTOR:  1.000        ARE:

           1    1  DISPL   1        0.00000        0.00000        0.00000    -4.286712E-02    0.00000    FLP:  13.
                    DISPL   6   -1.145942E-02      0.00000      -11.8313      5.802284E-03      0.00000

           1    1  FORCE   1      599.995          0.00000       496.255      -156628.          0.00000    CHR: 001000
                    FORCE   6     -599.995         0.00000      -496.255      12563.3           0.00000

* ELEMENT     1 JOINT SHEAR STRESS IN Z  DIR. AT END A   =   0.5493      > JOINT SHEAR
STRESS CAPACITY OF   0.5493    AT STEP =   650
%9 % LIMIT STATE POINT 3 DEFINED DUE TO ELEMENT NO       1 AT STEP=   650

DEGREE OF FREEDOM #     15 IS READ FROM UNIT #    11        JOINT #      5, DIRECTION: FX
STEP     TIME          LOAD        DISPLACEMENT       VELOCITY        ACCELERATION
   0    0.0000       9.26422E-05      0.0000           0.0000           0.0000
   2    0.0000       6.3689        4.00000E-02         0.0000           0.0000

     .
     .
 156    0.0000      496.16          3.1200            0.0000            0.0000
*2* 158  0.0000     493.41          3.1600            0.0000            0.0000
 160    0.0000      493.42          3.2000            0.0000            0.0000

     .
     .
 648    0.0000      497.29         12.960             0.0000            0.0000
*3* 650  0.0000     497.31         13.000             0.0000            0.0000
 652    0.0000      497.33         13.040             0.0000            0.0000

     .
     .
ELEMENT #     1 IS READ FROM UNIT #    12

IE3D BEAM FORCES...
STEP     TIME       NODE      AXIAL          FY           FZ         TORSION           MY             MZ        STBFAG & FLP
 156    0.0000        1     600.040        0.00000     493.080      0.00000       -149766.        0.00000       000000
                      6    -600.040        0.00000    -493.080      0.00000        12029.0        0.00000
                    DISP    0.00000        0.00000      0.00000     0.00000     -9.953720E-03     0.00000
     .
     .
```

```
158   0.0000
      DISP   -1.146030E-02   0.00000   -2.74720   0.00000    5.550230E-03    0.00000    0.0
       1      600.040        0.00000    493.330   0.00000   -149794.         0.00000    001000
       6     -600.040        0.00000   -493.330   0.00000    11965.0         0.00000
      DISP   -1.146030E-02   0.00000   -2.78410   0.00000   -1.008730E-02    0.00000    3.0
                                                             5.547060E-03

648   0.0000
       1      599.990        0.00000    496.240   0.00000   -156601.         0.00000    001000
       6     -599.990        0.00000   -496.240   0.00000    12561.0         0.00000
      DISP   -1.145940E-02   0.00000   -11.7950   0.00000   -4.273390E-02    0.00000    3.0
                                                             5.801250E-03

650   0.0000
       1      599.990        0.00000    496.250   0.00000   -156628.         0.00000    001000
       6     -599.990        0.00000   -496.250   0.00000    12563.0         0.00000
      DISP   -1.145940E-02   0.00000   -11.8310   0.00000   -4.286710E-02    0.00000    13.
                                                             5.802280E-03
```

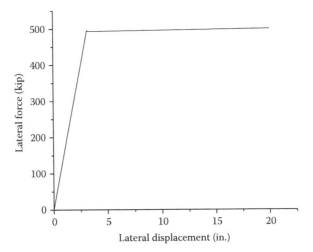

FIGURE 7.43   Pushover curve for test specimen #1.

The pushover curve for test specimen #1 is shown in Figure 7.43. The above output shows that the joint shear crack occurs at Step 650 (i.e., pushover displacement = 13 in.) at which joint shear at cap beam–column joint reaches the shear stress capacity of 0.549 ksi. The predicted joint shear failure mode is consistent with the UCSD full-scale test result. After joint shear failure occurs, INSTRUCT does not check whether or not the shear reinforcement is sufficient to prevent rapid shear strength degradation. The user should check the adequacy of shear reinforcement at the joint. INSTRUCT calculated the joint shear stress capacity per Equation 4.47, which is

$$v_j(\mu) = \sqrt{p_t^2 - p_t(f_v + f_h) + f_v f_h} = 0.549 \text{ ksi}$$

at which

$$f_v = \frac{P}{(h_c + h_b)b_{je}} = \frac{600}{(60 + 72)66} = -0.0689 \text{ ksi (compression)}$$

from Equation 4.48

and

$$f_h = \frac{P_b}{b_b h_b} = \frac{1726}{66 * 72} = -0.363 \text{ ksi (compression) from Equation 4.49}$$

At Step = 650, the ductility μ is equal to 4.3, calculated as $\mu = \theta$ (Step = 650)/ θ_y(Step = 157) = (−4.286E−2)/(−9.953E−3) = 4.3 in which θ_y (Step = 157) is the column yield rotation occurred at Step=157, and θ (Step=650) is the column rotation at Step=650. From Equations 4.52 and 4.53, the principal tensile stresses for $\mu \le 3$ and $\mu \ge 7$ are equal to 0.39 and 0.273 ksi, respectively. Therefore, from Figure 4.19, p_t corresponding to $\mu = 4.3$ is equal to 0.353 ksi. The joint shear stress at Step=650 can be obtained from Equation 4.54, which is $v_{jh} = (M/h_b)/(h_c b_{je}) = (156, 628/72)/(60 * 66) = 0.5493$ ksi > 0.549 ksi. Therefore, joint shear failure occurs at Step 650.

2. Output 2 (Joint_shear2.out file)

Perform pushover analysis for test specimen #2. The output results are shown as follows:

```
1   ECHO OF INPUT DATA

LINE   ....|...10....|...20....|...30....|...40....|...50....|...60....|...70....|...80....|...90....|..100....|..110
   1:  'STRUCTURE DEFINITION - Test Joint shear failure, Sample #2'
   2:  'Compare test results from UC-San Diego (SSRP-94/14).'
   3:  4  1  3  0  0  1
   4:  1  0  0  -300  1  0
   5:  5  -30  0  0  1  0
   6:  6  0  0  0  1  0
   7:  7  30  0  0  1  0   | Direction Cosine
   8:  1  0  0  1  1  0  1  1  0
   9:  1  1  1  1  1  0  0
  10:  5  1  1  0  1  1  1
  11:  6  0  1  1  1  1  1
  12:  7   | NMAT
  13:  '3D-BEAM'  MAT#1'  3122  1200  7000    0    0    41400000    20700000   20700000
  14:  'HINGE'    MAT#2'  0.004223643  135480  0.007698497  0  0  1.408373E+09  0.05384  0
  15:  0
  16:  'HINGE'    MAT#3'  0.004223643  135480  0.007698497  0  0  1.408373E+09  0.05384  0
  17:  0
  18:  'IA_BILN'  MAT#4'  0  0  1  1.44752E+09    0    0   -1
  19:  'IA_BILN'  MAT#5'  0  0  1  1.44752E+09    0    0   -1
  20:  'IA_BILN'  MAT#6'  0  0  1  1.52814E+09    0    0   -1
  21:  'IA_BILN'  MAT#7'  0  0  1  1.445089E+07   0    0   -1
  22:  2  1  1  .TRUE.   | Geometric stiffness
  23:  3   | Number of Element
  24:  'IE3DBEAM'  'Member 1'  2  3  4  5  6  7  1  6  0  1  0  0  24  0
  25:  1  0.342  60  0  72  84  2062.  0  0.342  0  72  60  0  1  6  84  2062.  0
  26:  '3D-BEAM'  'Member 2'  1  5  6  1  0  1  0  0  000000  0
  27:  '3D-BEAM'  'Member 3'  1  6  7  0  1  0  0  0  000000  0
  28:  0  0  .FALSE.   | Mass
  29:  0  0   | Damp
  30:  'SOL01 Elastic Static Analysis'
  31:  'APPLY DEAD LOADS AT CAPBEAM'
  32:  1  2
  33:  1  5  0  0  'FZ'     0   | Joint Load
  34:  1  6  0  0  'FZ'  -600   | Joint Load
  35:  1  7  0  0  'FZ'     0   | Joint Load
  36:  0  0  0  0  'END'    0   | Joint Load
  37:  1  2  0  0  'UNIF'  'FZ'  0   | Element Load
```

```
38: 1  3  0  0  'UNIF'    'FZ'    0    | Element Load
39: 'SOL04 Inelastic Incremental Pushover Analysis'
40: 'INCREMEMTAL DISPLACEMENT CONTROL AT CAPBEAM'
41: 0   50000  2   .TRUE.   0
42: 'JOINT FX'   5   11   0   0   0
43: 'ELE'    '   1   12   0   0   0
44: 'ELE'    '   2   13   0   0   0
45: 'ELE'    '   3   14   0   0   0
46: 'END'    0    0    0   0
47: 1    5    0    'FX'    20         | Joint Load
48: 0    0    0    'END'   0          | Joint Load
49: 'DISP. FROM 0 TO 20'   1   0   1000
50: 'END'    0    0    0   0
51: 'READ UNIT=11 UNIT=12'
52: 'STOP'
```

1 STRUCTURE.....: Compare test results from UC-San Diego (SSRP-94/14).
SOLUTION.....: INCREMEMTAL DISPLACEMENT CONTROL AT CAPBEAM

TIME: 16:01:15, DATE: 21-JUN-09
TIME: 16:01:15, DATE: 21-JUN-09

FLP: 3.0

CHR: 001000

***** IA_BILN ELEMENT   1 YIELD AT END A
**** THE ELEMENT FORCES DISP. FROM STEP: 146 FACTOR: 1.000   ARE:

```
1   2   DISPL   1    0.00000        0.00000    0.00000    -9.2519-5E-03    0.00000
        DISPL   6   -1.146033E-02  -2.55353    0.00000     5.158725E-03    0.00000

1   2   FORCE   1    600.043        445.938    0.00000    -135486.         0.00000
        FORCE   6   -600.043       -445.938    0.00000     10874 4         0.00000
```

%8 % LIMIT STATE POINT 2 DEFINED DUE TO ELEMENT NO   1 AT STEP= 146

* ELEMENT   1 INELASTIC ROTATION IN MY DIR. AT END A   =   0.5389E-01 > PLASTIC ROTATION
CAPACITY OF   0.5384E-01 AT STEP = 961
%9 % LIMIT STATE POINT 3 DEFINED DUE TO ELEMENT NO   1 AT STEP= 961

DEGREE OF FREEDOM #   15 IS READ FROM UNIT #   11   JOINT #   5, DIRECTION: FX

STEP	TIME	LOAD	DISPLACEMENT	VELOCITY	ACCELERATION
0	0.0000	9.40331E-05	0.0000	0.0000	0.0000
2	0.0000	6.1892	4.00000E-02	0.0000	0.0000
. . .					
2					
144	0.0000	445.22	2.8800	0.0000	0.0000
146	0.0000	445.97	2.9200	0.0000	0.0000
148	0.0000	446.00	2.9600	0.0000	0.0000
. . .					
3					
960	0.0000	458.36	19.200	0.0000	0.0000
962	0.0000	458.39	19.240	0.0000	0.0000
964	0.0000	458.42	19.280	0.0000	0.0000
. . .					

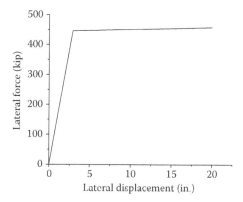

FIGURE 7.44 Pushover curve for test specimen #2.

The pushover curve for test specimen #2 is shown in Figure 7.44. The above output shows that the joint shear failure mode does not occur during the pushover analysis. By increasing the cap beam width from 5′–6″ to 7′–0″, the failure mode changes from the joint shear failure to the column compression failure of confined concrete at Step 961 (i.e., pushover displacement = 19.22″) with plastic rotation greater than the plastic rotation capacity of 0.05384 rad. The predicted failure mode of compression failure of confined concrete is consistent with the UCSD test result.

7.3.13   EXAMPLE 13: CYCLIC RESPONSE OF A CANTILEVER BEAM

A cantilever beam is shown in Figure 7.45. The concrete properties of the beam are $f_c'=3$ ksi, $f_{cr}=0.3$ ksi, $\varepsilon_{cu}=0.03$, and $E_c=3320$ ksi. The material properties of steel

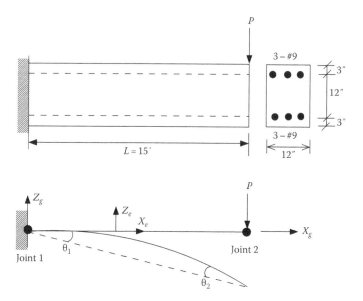

FIGURE 7.45 Reinforced concrete cantilever beam.

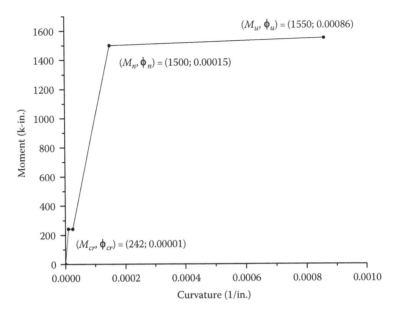

FIGURE 7.46   Moment–curvature curve.

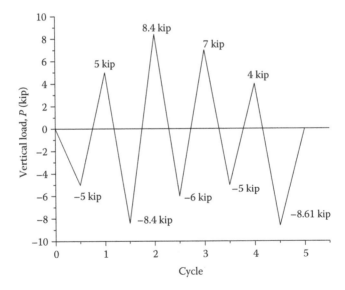

FIGURE 7.47   Vertical loading diagram.

reinforcement are $f_y = 40\,\text{ksi}$ and $E_s = 29,000\,\text{ksi}$. The steel rebar size is #9 with cross section area of $A_s = 1\,\text{in.}^2$ per bar. The moment–curvature curve of the cross section is shown in Figure 7.46, in which M_{cr}, M_n, and M_u are the crack moment, nominal moment, and ultimate moment, respectively. The beam is subjected to cyclic static vertical load, P, applying at joint 2 as shown in Figure 7.47. Using the Takeda Model, find the vertical load–vertical displacement of the beam.

In order to use the Takeda model for the nonlinear analysis of the beam, the moment–total rotation curves at two ends of the member must be determined first. The moment–total rotation relationship at each member end can be obtained by the conjugate beam method as described in Section 4.3. The moment and curvature diagrams of the beam corresponding to M_{cr}, M_n, and M_u are shown in Figure 7.48. The reactions at two ends of the member shown in the curvature diagrams, based on conjugate beam method, represent the total end rotations, θ_1 and θ_2, of the member.

From Figure 7.48, the moment–total rotation relationship at each end of member is shown in Figure 7.49. The data corresponding to each control points in the figure were input into INSTRUCT for the cyclic pushover analysis. The output results are shown in Figure 7.50.

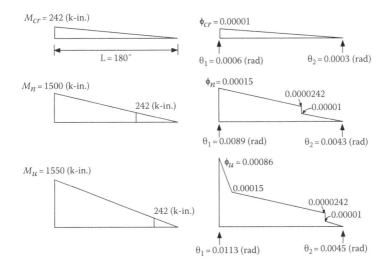

FIGURE 7.48   Moment and curvature diagrams.

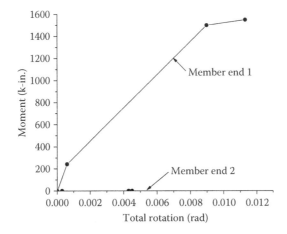

FIGURE 7.49   Moment–total rotations at member ends.

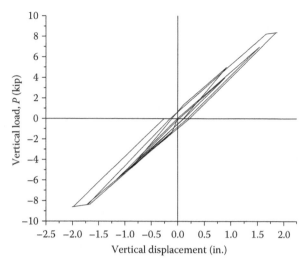

FIGURE 7.50   Response of cantilever beam subjected to cyclic loading shown in Figure 7.47.

Input (EX13_TAKEDA2.dat file)

```
1  ECHO OF INPUT DATA

LINE  ....|....10....|....20....|....30....|....40....|....50....|....60....|....70....|....80....|....90....|....100....|...110
   1: 'STRUCTURE DEFINITION - R/C CONCRETE BEAM'
   2: 'INELASTIC CYCLIC BEHAVIOR OF BEAM BASED ON TAKEDA MODEL'
   3: 2   1   2   0   0   12
   4: 1   0.00   0.00   .00   1   0
   5: 2   15.0   0.00   .00   1   0
   6: 1   0   0   1   0   | Direction Cosine
   7: 1   1   1   1   1   1   0   0
   8: 2   1   1   0   1   0   1   0
   9: 6   | Number of Material
  10: 'TAKEDA  MAT#1'   24200000   242   3.333E-6   1500   5.12167E-5   1550   6.28E-5   -1
  11: 'TAKEDA  MAT#2'   24200000   0   1.666E-6   0   2.43889E-5   0   2.497E-5   -1
  12: 'IA_BILN MAT#3: MZA `   0   0.001   3320.   3012   1125.   -1   0   -1
  13: 'IA_BILN MAT#4: MZA `   0   0.001   3320.   3012   1125.   -1   0   -1
  14: 'IA_BILN MAT#5: MXA `   0   0.001   3320.   6024   1125.   -1   0   -1
  15: 'IA_BILN MAT#6: FXA `   0   0.001   3320.   216   1125.   -1   0   -1
  16: 0   0   .FALSE.   | Geometric stiffness
  17: 1   | Number of Element
  18: 'IE3DBEAM'   'R/C CIRCULAR MEMBER 1'   1   2   3   4   5   6   1   2   0   1   0   0   0   0.
  19: 0   0   0   0   0   0   0   0   0   0   0   0   0   0
  20: 0   0   .FALSE.   | Mass
  21: 0   0   | Damp
  22: 'SOL04 Inelastic Incremental Pushover Analysis'
  23: 'INCREMEMTAL FORCE CONTROL AT JOINT 2'
  24: 0   50000   1   .TRUE.   0.0001
  25: 'JOINT FZ'   2   11   0   0   0
  26: 'ELE  `   1   12   0   0   0
  27: 'END'   0   0   0   0   0
  28: 1   2   0   'FZ'   -8.61   0   | Joint Load
  29: 0   0   0   'END'   0   | Joint Load
  30: 'DISP. FROM   0   TO   -5   `   0.580   0   290
  31: 'DISP. FROM   -5   TO   5   `   -0.580   0   580
  32: 'DISP. FROM   5   TO   -8.4  `   -0.975   0   578
  33: 'DISP. FROM   -8.4 TO   8.4  `   -0.975   0   975
  34: 'DISP. FROM   8.4  TO   -6   `   0.696   0   836
  35: 'DISP. FROM   -6   TO   7    `   -0.812   0   755
```

```
36: 'DISP. FROM   7    TO  -5    '   0.580   0   0   697
37: 'DISP. FROM  -5    TO   4    '  -0.464   0   0   522
38: 'DISP. FROM   4    TO  -8.61'    1       0   0   734
39: 'DISP. FROM  -8.61TO   0     '   0       0   0   500
40: 'END'      0      0      0
41: 'READ   UNIT=11   UNIT=12'
42: 'STOP'
```

Appendix A: Stiffness Matrix Formulation for Bilinear PM Method

Based on the moment–curvature model in Figure 4.1, assume that an inelastic member has two components, linear and elastoplastic, as shown in Figure A.1. θ_i and θ_j are member-end total rotations; α_i and α_j are plastic rotations at each end of elastoplastic component (Cheng, 2000). From Figure A.1, the end rotations of the elastoplastic component are

$$\theta_i' = \theta_i - \alpha_i; \quad \theta_j' = \theta_j - \alpha_j \tag{A.1}$$

or in incremental form

$$\Delta\theta_i' = \Delta\theta_i - \Delta\alpha_i; \quad \Delta\theta_j' = \Delta\theta_j - \Delta\alpha_j \tag{A.2}$$

The incremental forces and deformations at both ends of these two components in Figure A.1 may be expressed in terms of stiffness coefficients as follows:

1. *Linear component*

$$\begin{Bmatrix} \Delta M_{pi} \\ \Delta M_{pj} \\ \Delta V_{pi} \\ \Delta V_{pj} \end{Bmatrix} = p \begin{bmatrix} a & b & -c & c \\ & a & -c & c \\ & & d & -d \\ Symm & & & d \end{bmatrix} \begin{Bmatrix} \Delta\theta_i \\ \Delta\theta_j \\ \Delta Y_i \\ \Delta Y_j \end{Bmatrix} \tag{A.3}$$

    in which $a = 4EI/L$, $b = 2EI/L$, $c = 6EI/L^2$, and $d = 12EI/L^3$.
2. *Elastoplastic component*

$$\begin{Bmatrix} \Delta M_{qi} \\ \Delta M_{qj} \\ \Delta V_{qi} \\ \Delta V_{qj} \end{Bmatrix} = q \begin{bmatrix} a & b & -c & c \\ & a & -c & c \\ & & d & -d \\ Symm & & & d \end{bmatrix} \begin{Bmatrix} \Delta\theta_i' \\ \Delta\theta_j' \\ \Delta Y_i \\ \Delta Y_j \end{Bmatrix} \tag{A.4}$$

Moments and shears of the nonlinear member are the combination of the component end forces according to the state of the yield. The state of yield may be one of the following four conditions: (a) both ends linear, (b) i-end nonlinear and j-end linear, (c) i-end linear and j-end nonlinear, and (d) both ends nonlinear.

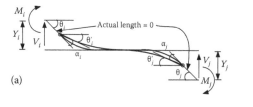

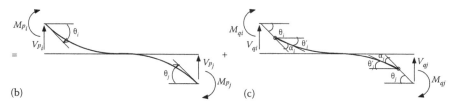

FIGURE A.1 Bilinear beam: (a) nonlinear beam, (b) linear component, and (c) elastoplastic component.

(a) *Both ends linear*: Member is still elastic and plastic hinges (plastic rotations) do not exist. From Equations A.1 and A.2,

$$\alpha_i = \alpha_j = 0; \quad \theta_i' = \theta_i; \quad \text{and} \quad \theta_j' = \theta_j \tag{A.5}$$

or in incremental form

$$\Delta\alpha_i = \Delta\alpha_j = 0; \quad \Delta\theta_i' = \Delta\theta_i; \quad \text{and} \quad \Delta\theta_j' = \Delta\theta_j \tag{A.6}$$

Combining Equations A.3 and A.4 with $p + q = 1$ leads to

$$\begin{bmatrix} \Delta M_i \\ \Delta M_j \\ \Delta V_i \\ \Delta V_j \end{bmatrix} = \left\{ \begin{bmatrix} \Delta M_{pi} \\ \Delta M_{pj} \\ \Delta V_{pi} \\ \Delta V_{pj} \end{bmatrix} + \begin{bmatrix} \Delta M_{qi} \\ \Delta M_{qj} \\ \Delta V_{qi} \\ \Delta V_{qj} \end{bmatrix} \right\} = \begin{bmatrix} a & b & -c & c \\ b & a & -c & c \\ -c & -c & d & -d \\ c & c & -d & d \end{bmatrix} \begin{bmatrix} \Delta\theta_i \\ \Delta\theta_j \\ \Delta Y_i \\ \Delta Y_j \end{bmatrix} \tag{A.7}$$

(b) *i-End nonlinear and j-end linear*: From Equations A.1 and A.2,

$$\alpha_i \neq 0; \quad \alpha_j = 0; \quad \theta_i' = \theta_i - \alpha_i; \quad \text{and} \quad \theta_j' = \theta_j \tag{A.8}$$

or in incremental form

$$\Delta\alpha_i \neq 0; \quad \Delta\alpha_j = 0; \quad \Delta\theta_i' = \Delta\theta_i - \Delta\alpha_i; \quad \text{and} \quad \Delta\theta_j' = \Delta\theta_j \tag{A.9}$$

Combining Equations A.3 and A.4 gives

$$
\begin{bmatrix} \Delta M_i \\ \Delta M_j \\ \Delta V_i \\ \Delta V_j \end{bmatrix} = \begin{bmatrix} a & b & -c & c \\ b & a & -c & c \\ -c & -c & d & -d \\ c & c & -d & d \end{bmatrix} \begin{bmatrix} \Delta \theta_i \\ \Delta \theta_j \\ \Delta Y_i \\ \Delta Y_j \end{bmatrix} + \begin{Bmatrix} -a \\ -b \\ c \\ -c \end{Bmatrix} q \Delta \alpha_i \qquad (A.10)
$$

Since the moment at the *i*-end of the elastoplastic component is constant, the increase in ΔM_i at the end of the member must be due to ΔM_{pi} of the linear component, that is,

$$
\Delta M_i = \Delta M_{pi} \qquad (A.11)
$$

which means that ΔM_{pi} of Equation A.3 is equal to ΔM_i of Equation A.10. This equality yields

$$
\Delta \alpha_i = \Delta \theta_i + \frac{b}{a} \Delta \theta_j - \frac{c}{a} \Delta Y_i + \frac{c}{a} \Delta Y_j = \Delta \theta_i + \frac{1}{2} \Delta \theta_j - \frac{3}{2L} (\Delta Y_i - \Delta Y_j)
$$

$$
(A.12)
$$

Substituting Equation A.12 into Equation A.10 gives the following matrix:

$$
\begin{bmatrix} \Delta M_i \\ \Delta M_j \\ \Delta V_i \\ \Delta V_j \end{bmatrix} = \begin{bmatrix} pa & pb & -pc & pc \\ pb & pa+qe & -pc-qf & pc+qf \\ -pc & -pc-qf & pd+qg & -pd-qg \\ pc & pc+qf & -pd-qg & pd+qg \end{bmatrix} \begin{bmatrix} \Delta \theta_i \\ \Delta \theta_j \\ \Delta Y_i \\ \Delta Y_j \end{bmatrix} \qquad (A.13)
$$

in which $e = 3EI/L$, $f = 3EI/L^2$, and $g = 3EI/L^3$.

(c)   *i-End linear and j-end nonlinear:* Similar to item (b), the stiffness matrix for condition (c) is

$$
\begin{bmatrix} \Delta M_i \\ \Delta M_j \\ \Delta V_i \\ \Delta V_j \end{bmatrix} = \begin{bmatrix} pa+qe & pb & -pc-qf & pc+qf \\ pb & pa & -pc & pc \\ -pc-qf & -pc & pd+qg & -pd-qg \\ pc+qf & pc & -pd-qg & pd+qg \end{bmatrix} \begin{bmatrix} \Delta \theta_i \\ \Delta \theta_j \\ \Delta Y_i \\ \Delta Y_j \end{bmatrix} \qquad (A.14)
$$

(d) *Both ends nonlinear:* Since both ends have plastic hinges, substitute
Equation A.2 into Equation A.4 and then combine Equations A.3 and
A.4 as follows:

$$
\begin{bmatrix} \Delta M_i \\ \Delta M_j \\ \Delta V_i \\ \Delta V_j \end{bmatrix} = \begin{bmatrix} a & b & -c & c \\ b & a & -c & c \\ -c & -c & d & -d \\ c & c & -d & d \end{bmatrix} \begin{bmatrix} \Delta\theta_i \\ \Delta\theta_j \\ \Delta Y_i \\ \Delta Y_j \end{bmatrix} + q \begin{bmatrix} -a & -b \\ -b & -a \\ c & c \\ -c & -c \end{bmatrix} \begin{Bmatrix} \Delta\alpha_i \\ \Delta\alpha_j \end{Bmatrix} \quad (A.15)
$$

Equating $\Delta M_i = \Delta M_{pi}$ and $\Delta M_j = \Delta M_{pj}$ in Equations A.3 and A.15 yields

$$
\Delta\alpha_i = \Delta\theta_i - \frac{1}{L}(\Delta Y_i - \Delta Y_j) \quad (A.16)
$$

$$
\Delta\alpha_j = \Delta\theta_j - \frac{1}{L}(\Delta Y_i - \Delta Y_j) \quad (A.17)
$$

Thus, Equation A.15 can be expressed as follows:

$$
\begin{bmatrix} \Delta M_i \\ \Delta M_j \\ \Delta V_i \\ \Delta V_j \end{bmatrix} = p \begin{bmatrix} a & b & -c & c \\ b & a & -c & c \\ -c & -c & d & -d \\ c & c & -d & d \end{bmatrix} \begin{bmatrix} \Delta\theta_i \\ \Delta\theta_j \\ \Delta Y_i \\ \Delta Y_j \end{bmatrix} \quad (A.18)
$$

which are actually the incremental forces of the linear component as
shown in Equation A.3.

Appendix B: Stiffness Matrix Formulation for Finite Segment

B.1 SECTION PROPERTIES OF FINITE SEGMENT

As shown in Figures 4.13 and B.4, section properties of the cross section corresponding to segment reference axes U_0 and V_0 can be expressed as follows:

$$EA_0 = \sum_{i=1}^{N} E_{ti} A_i \tag{B.1}$$

$$ES_{U0} = \sum_{i=1}^{N} E_{ti} U_{0i} A_i \tag{B.2}$$

$$EI_{U0} = \sum_{i=1}^{N} E_{ti} V_{0i}^2 A_i \tag{B.3}$$

$$ES_{V0} = \sum_{i=1}^{N} E_{ti} V_{0i} A_i \tag{B.4}$$

$$EI_{V0} = \sum_{i=1}^{N} E_{ti} U_{0i}^2 A_i \tag{B.5}$$

$$EI_{U0V0} = \sum_{i=1}^{N} E_{ti} U_{0i} V_{0i} A_i \tag{B.6}$$

where
   N represents the total number of cross-sectional elements
   U_{0i} and V_{0i} represent the location of the ith cross-sectional element in the segment reference coordinate U_0 and V_0 directions, respectively
   A_i and E_{ti} are the area and tangent modulus of the ith cross-sectional element, respectively

For each load step during the pushover analysis, the value of E_{ti} is determined by the instantaneous strain of the ith cross-sectional element in accordance with the material stress–strain relationship, such as bilinear, Ramberg–Osgood, or confined concrete material models described in Chapter 3. The instantaneous centroid location, $C'(U_{c0},V_{c0})$, and the rotation angle, β, the angle between reference axis, U_0, and instantaneous principal axis, U, are calculated by

$$U_{c0} = \frac{ES_{U0}}{EA_0} \tag{B.7}$$

$$V_{c0} = \frac{ES_{V0}}{EA_0} \tag{B.8}$$

$$\beta = \frac{1}{2}\tan^{-1}\left[\frac{2EI_{U'V'}}{(EI_{V'} - EI_{U'})}\right] \tag{B.9}$$

in which

$$EI_{U'} = EI_{U0} - V_{c0}^2 EA_0 \tag{B.10}$$

$$EI_{V'} = EI_{V0} - U_{c0}^2 EA_0 \tag{B.11}$$

$$EI_{U'V'} = EI_{U0V0} - U_{c0}V_{c0}EA_0 \tag{B.12}$$

The segment reference coordinates directions, (U_0,V_0,W_0), for different cross-sectional shapes are defined in Figure 3.19. The origin of the U_0 and V_0 axes (i.e., point C in Figure 4.13 or B.4) is the geometric centroid of the elastic cross section. The sectional properties about instantaneous principal axes U and V can be obtained as follows:

$$EA = EA_0 \tag{B.13}$$

$$EI_U = \frac{EI_{U'} + EI_{V'}}{2} + \frac{EI_{U'} - EI_{V'}}{2}\cos(2\beta) - EI_{U'V'}\sin(2\beta) \tag{B.14}$$

$$EI_V = \frac{EI_{U'} + EI_{V'}}{2} - \frac{EI_{U'} - EI_{V'}}{2}\cos(2\beta) + EI_{U'V'}\sin(2\beta) \tag{B.15}$$

Theoretically, EI_{UV} corresponding to principal axes U and V is equal to zero. Since finite cross-sectional elements are used here, the EI_{UV} value may not be equal to zero, but should be close to zero. Once EA, EI_U, and EI_V are obtained, the segment material stiffness matrix, $[sk_m]$, can be formulated as shown in Equation B.32.

Example B.1

Find the principal axes (U,V), cross-sectional properties EA, EI_U, EI_V, and GJ, and centroid location, $C'(U_{c0},V_{c0})$, of the $L2 \times 2 \times (1/4)$ equal-leg angle section shown in Figure B.1 for the (a) elastic condition and (b) inelastic condition. For simplicity, the cross section is only divided into seven elements. The elastic modulus, $E = 29{,}000$ ksi, and shear modulus, $G = 11{,}300$ ksi.

Solution

The segment reference coordinate system (U_0, V_0, W_0) is defined in Figure B.1. The coordinates and areas of the cross-sectional elements are

(a) *Elastic condition:* From Table B.1,

$$EA_0 = \sum_{i=1}^{7} E_{ti}A_i = (29{,}000)(4 \times 0.125 + 3 \times 0.146) = 27{,}187.5$$

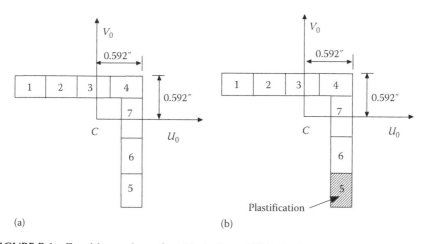

(a)   (b)

FIGURE B.1 Equal-leg angle section: (a) elastic and (b) inelastic.

TABLE B.1
Element Data

Element i	Coordinate (U_{0i}, V_{0i})	Area A_i (in.²)
1	(−1.16, 0.47)	0.125
2	(−0.66, 0.47)	0.125
3	(−0.16, 0.47)	0.125
4	(0.34, 0.47)	0.125
5	(0.47, −1.12)	0.146
6	(0.47, −0.53)	0.146
7	(0.47, 0.05)	0.146

$$ES_{U0} = \sum_{i=1}^{7} E_{ti}U_{0i}A_i = (29,000)[0.125(-1.16 - 0.66 - 0.16 + 0.34)$$

$$+ 3 \times 0.146 \times 0.47] = 1.98$$

$$EI_{U0} = \sum_{i=1}^{7} E_{ti}V_{0i}^2 A_i = (29,000)[0.125 \times 4(0.47)^2 + 0.146[(-1.12)^2$$

$$+ (-0.53)^2 + (0.05)^2]] = 9,648$$

$$ES_{V0} = \sum_{i=1}^{7} E_{ti}V_{0i}A_i = -1.248$$

$$EI_{V0} = \sum_{i=1}^{7} E_{ti}U_{0i}^2 A_i = 9709.5$$

$$EI_{U0V0} = \sum_{i=1}^{7} E_{ti}U_{0i}V_{0i}A_i = -5921$$

The centroid location, $C'(U_{c0}, V_{c0})$, is calculated as follows:

$$U_{c0} = \frac{ES_{U0}}{EA_0} = \frac{1.98}{27,187.5} \cong 0$$

$$V_{c0} = \frac{ES_{V0}}{EA_0} = \frac{-1.248}{27,187.5} \cong 0$$

$$EI_{U'} = EI_{U0} - V_{c0}^2 EA_0 = 9648$$

$$EI_{V'} = EI_{V0} - U_{c0}^2 EA_0 = 9709$$

$$EI_{U'V'} = EI_{U0V0} - U_{c0}V_{c0}EA_0 = -5921$$

$\beta = (1/2)\tan^{-1}[2EI_{U'V'}/(EI_{V'} - EI_{U'})] = 0.5 \tan^{-1}(2(-5921)/(9709-9648)) = -44.8°$ (clockwise), which is close to actual, $-45°$, considering only seven cross-sectional elements are used here. Therefore, the principal axis U is $44.8°$ from the U_0-axis, as shown in Figure B.2a. The section properties, EA, EI_U, and EI_V, are calculated from Equations B.13 through B.15 as follows:

$$EA = EA_0 = 27,187.5 \text{ (kip)}$$

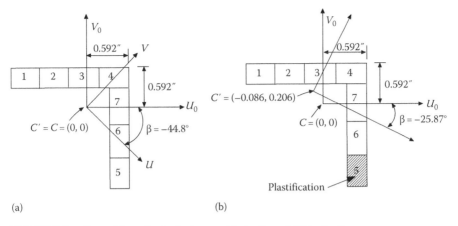

FIGURE B.2 Centroids and principal axes: (a) elastic and (b) inelastic.

$$EI_U = \frac{EI_{U'} + EI_{V'}}{2} + \frac{EI_{U'} - EI_{V'}}{2}\cos(2\beta) - EI_{U'V'}\sin(2\beta)$$

$$= 9678.5 + 30.5 \times 0 - 5927 \times 1 = 3752 \text{ (k-in.}^2)$$

$$EI_V = \frac{EI_{U'} + EI_{V'}}{2} - \frac{EI_{U'} - EI_{V'}}{2}\cos(2\beta) + EI_{U'V'}\sin(2\beta)$$

$$= 9,678.5 - 0 + 5,927 = 15,605 \text{ (k-in.}^2)$$

$$GJ = \sum_{i=1}^{7}\frac{1}{3}Gb_i t_i^3 = \left(\frac{11,300}{3}\right)\left[4(0.5)(0.25)^3 + 3(0.583)(0.25)^3\right]$$

$$= 220.7 \text{ (k-in.}^2),$$

where b_i is the length of cross-sectional element i.

(b) *Inelastic condition:* From Table B.1,

$$EA_0 = \sum_{i=1}^{7}E_{ti}A_i = (29,000)(4 \times 0.125 + 2 \times 0.146)$$

$$+ 0.29 \times 0.146 = 22,958$$

$$ES_{U0} = \sum_{i=1}^{7}E_{ti}U_{0i}A_i = (29,000)[0.125(-1.16 - 0.66 - 0.16 + 0.34)$$

$$+ 2 \times 0.146 \times 0.47] + 0.29(0.146)(0.47) = -1,974$$

$$EI_{U0} = \sum_{i=1}^{7}E_{ti}V_{0i}^2A_i = (29,000)[0.125 \times 4(0.47)^2 + 0.146[(-0.53)^2 + (0.05)^2]]$$

$$+ 0.29(-1.12)^2(0.146) = 4,371$$

$$ES_{V0} = \sum_{i=1}^{7} E_{ti} V_{0i} A_i = 4,723$$

$$EI_{V0} = \sum_{i=1}^{7} E_{ti} U_{0i}^2 A_i = 8,791$$

$$EI_{U0V0} = \sum_{i=1}^{7} E_{ti} U_{0i} V_{0i} A_i = -3,717$$

The centroid location, $C'(U_{c0}, V_{c0})$, is calculated as follows:

$$U_{c0} = \frac{ES_{U0}}{EA_0} = \frac{-1,974}{22,958} = -0.086 \text{ (in.)}$$

$$V_{c0} = \frac{ES_{V0}}{EA_0} = \frac{4,723}{22,958} = 0.206 \text{ (in.)}$$

$$EI_{U'} = EI_{U0} - V_{c0}^2 EA_0 = 3400$$

$$EI_{V'} = EI_{V0} - U_{c0}^2 EA_0 = 8621$$

$$EI_{U'V'} = EI_{U0V0} - U_{c0} V_{c0} EA_0 = -3311$$

$\beta = 1/2 \tan^{-1}[2EI_{U'V'}/(EI_{V'} - EI_{U'})] = 0.5 \tan^{-1}(-6622/5221) = -25.87°$ (clockwise). Therefore, the principal axis U is $25.87°$ from the U_0-axis, as shown in Figure B.2b. The section properties, EA, EI_U, and EI_V, are calculated from Equations B.13 through B.15 as follows:

$$EA = EA_0 = 22,958 \text{ (kip)}$$

$$EI_U = \frac{EI_{U'} + EI_{V'}}{2} + \frac{EI_{U'} - EI_{V'}}{2} \cos(2\beta) - EI_{U'V'} \sin(2\beta) = 1794 \text{ (k-in.}^2)$$

$$EI_V = \frac{EI_{U'} + EI_{V'}}{2} - \frac{EI_{U'} - EI_{V'}}{2} \cos(2\beta) + EI_{U'V'} \sin(2\beta) = 10,227 \text{ (k-in.}^2)$$

$$GJ = \sum_{i=1}^{7} \frac{1}{3} Gb_i t_i^3 = \frac{11,300}{3} [4(0.5)(0.25)^3 + 3(0.583)(0.25)^3] = 220.7 \text{ (k-in.}^2)$$

B.2   SEGMENT'S ROTATION MATRIX, $[\bar{R}]_{12\times12}$, AND STIFFNESS MATRIX

As shown in Figures 4.13 and B.3, the member stiffness matrix is established by stacking up the segment stiffness matrices for which a rotation matrix $[R]_{3\times3}$ (Chen and Atsuta, 1977) is required by transforming a vector from the segment global coordinate system (X_R, Y_R, Z_R) to the segment local coordinate system (U, V, W):

$$\begin{bmatrix} \vec{e}_u \\ \vec{e}_v \\ \vec{e}_w \end{bmatrix} = \begin{bmatrix} (UX) & (UY) & (UZ) \\ (VX) & (VY) & (VZ) \\ (WX) & (WY) & (WZ) \end{bmatrix} \begin{bmatrix} \vec{e}_x \\ \vec{e}_y \\ \vec{a}_z \end{bmatrix} = [R]_{3\times3} \begin{bmatrix} \vec{e}_x \\ \vec{e}_y \\ \vec{e}_z \end{bmatrix} \tag{B.16}$$

where

(UX) is the direction cosine between the local U-axis and the segment global X_R-axis

$\vec{e}_u$ represents the unit vector in the U-axis direction

$\vec{e}_x$ represents the unit vector in the X_R-axis direction

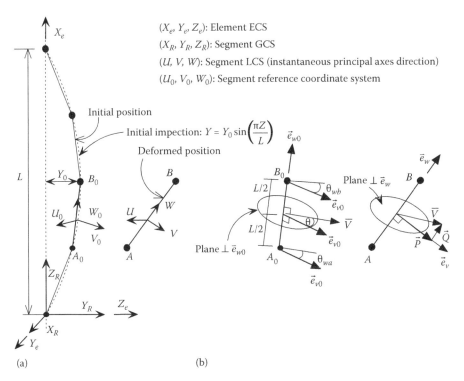

(X_e, Y_e, Z_e): Element ECS
(X_R, Y_R, Z_R): Segment GCS
(U, V, W): Segment LCS (instantaneous principal axes direction)
(U_0, V_0, W_0): Segment reference coordinate system

Initial position

Initial impection: $Y = Y_0 \sin\left(\dfrac{\pi Z}{L}\right)$

Deformed position

(a)   (b)

FIGURE B.3   Relationship between original and deformed position: (a) deformation of segment and (b) direction of V-axis.

Segment local axes U, V, and W represent the instantaneous principal axes of the cross section with consideration of sectional plasticity. Consider a segment AB deformed from the initial point A_0B_0 in (X_R, Y_R, Z_R) system as shown in Figure B.3a, the initial rotation matrix can be expressed as follows:

$$\begin{bmatrix} \vec{e}_{u0} \\ \vec{e}_{v0} \\ \vec{e}_{w0} \end{bmatrix} = \begin{bmatrix} (U_0X) & (U_0Y) & (U_0Z) \\ (V_0X) & (V_0Y) & (V_0Z) \\ (W_0X) & (W_0Y) & (W_0Z) \end{bmatrix} \begin{bmatrix} \vec{e}_x \\ \vec{e}_y \\ \vec{e}_z \end{bmatrix} \tag{B.17}$$

where $\vec{e}_{u0}$ represents the unit vector in the initial U_0-axis direction when the segment is in the undeformed position, as shown in Figure B.3a. The length AB is

$$L = \sqrt{(X_B - X_A)^2 + (Y_B - Y_A)^2 + (Z_B - Z_A)^2} \tag{B.18}$$

and the new W-axis is determined by its unit vector

$$\vec{e}_w = \begin{bmatrix} (WX) \\ (WY) \\ (WZ) \end{bmatrix} = \begin{bmatrix} (X_B - X_A)/L \\ (Y_B - Y_A)/L \\ (Z_B - Z_A)/L \end{bmatrix} \tag{B.19}$$

To determine the orientation of the new V-axis, first consider a $\bar{V}$-axis rotating about the W_0-axis by the average angle of rotation at ends A_0 and B_0, θ_{wa} and θ_{wb}, plus the rotation of the principal axes, β; thus

$$\theta = \frac{1}{2}(\theta_{w0a} + \theta_{w0b}) + \beta \tag{B.20}$$

in which β is the angle from reference axis U_0 to principal axis U (see Figure B.4 for example).

Then, the $\bar{V}$-axis shown in Figure B.3b is obtained as follows:

$$\bar{V} = (\vec{e}_{v0})\cos\theta - (\vec{e}_{u0})\sin\theta = \begin{bmatrix} (\bar{V}_0X) \\ (\bar{V}_0Y) \\ (\bar{V}_0Z) \end{bmatrix} \tag{B.21}$$

As shown in Figure B.3, $\bar{V}$ is the vector located at θ angle from the V_0-axis. θ_{wa} and θ_{wb} are the total rotations at ends A and B, respectively, at each load step. From Figure B.3,

$$\vec{Q} = (\bar{V} \cdot \vec{e}_w)\vec{e}_w \tag{B.22}$$

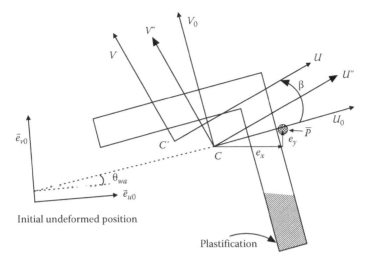

FIGURE B.4   Deformed angle segment at *A* end (*B* end is similar).

or

$$\vec{Q} = [(\bar{V}_0X)(WX) + (\bar{V}_0Y)(WY) + (\bar{V}_0Z)(WZ)]\vec{e}_w \qquad (B.23)$$

Let

$$M = (\bar{V}_0X)(WX) + (\bar{V}_0Y)(WY) + (\bar{V}_0Z)(WZ) \qquad (B.24)$$

then

$$\vec{P} = \vec{V} - \vec{Q} = \begin{bmatrix} (\bar{V}_0X) \\ (\bar{V}_0Y) \\ (\bar{V}_0Z) \end{bmatrix} - M \begin{bmatrix} (WX) \\ (WY) \\ (WZ) \end{bmatrix} \qquad (B.25)$$

The unit vector in the new *V*-axis can be expressed as follows:

$$\vec{e}_v = \frac{\vec{P}}{|\vec{P}|} = \frac{1}{L_p} \begin{bmatrix} (\bar{V}_0X) - M(WX) \\ (\bar{V}_0Y) - M(WY) \\ (\bar{V}_0Z) - M(WZ) \end{bmatrix} = \begin{bmatrix} (VX) \\ (VY) \\ (VZ) \end{bmatrix} \qquad (B.26)$$

where

$$L_p = \sqrt{[(\bar{V}_0X) - M(WX)]^2 + [(\bar{V}_0Y) - M(WY)]^2 + [(\bar{V}_0Z) - M(WZ)]^2} \qquad (B.27)$$

The unit vector $\vec{e}_u$ is obtained as the cross product of $\vec{e}_v$ and $\vec{e}_w$:

$$\vec{e}_u = \begin{bmatrix} (UX) \\ (UY) \\ (UZ) \end{bmatrix} = \begin{bmatrix} (VY)(WZ) - (VZ)(WY) \\ (VZ)(WX) - (VX)(WZ) \\ (VX)(WY) - (VY)(WX) \end{bmatrix} \qquad (B.28)$$

Therefore, the stiffness matrix of segment j, $[\overline{sk}]_j$, corresponding to the segment global coordinate system (X_R, Y_R, Z_R) can be expressed as follows:

$$[\overline{sk}]_j = [\overline{R}]_j^T [sk]_j [\overline{R}]_j \tag{B.29}$$

where

$$[\overline{R}]_j = \begin{bmatrix} [R] & & & \\ & [R] & & \\ & & [R] & \\ & & & [R] \end{bmatrix}_{12 \times 12} \tag{B.30}$$

in which $[sk]_j$ is the jth segment stiffness matrix corresponding to the (U,V,W) local coordinate system. Matrix $[sk]_j$ is expressed as follows:

$$[sk]_j = [sk_m]_j + [sk_g]_j \tag{B.31}$$

in which $[sk_m]_j$ and $[sk_g]_j$ are the segment material stiffness matrix and geometric stiffness matrix, respectively, given by

$$[sk_m]_j = \begin{bmatrix}
\frac{12EI_V}{L^3} & 0 & 0 & 0 & \frac{6EI_V}{L^2} & 0 & \frac{-12EI_V}{L^3} & 0 & 0 & 0 & \frac{6EI_V}{L^2} & 0 \\
& \frac{12EI_U}{L^3} & 0 & \frac{-6EI_U}{L^2} & 0 & 0 & 0 & \frac{-12EI_U}{L^3} & 0 & \frac{-6EI_U}{L^2} & 0 & 0 \\
& & \frac{EA}{L} & 0 & 0 & 0 & 0 & 0 & \frac{-EA}{L} & 0 & 0 & 0 \\
& & & \frac{4EI_U}{L} & 0 & 0 & 0 & \frac{6EI_U}{L^2} & 0 & \frac{2EI_U}{L} & 0 & 0 \\
& & & & \frac{4EI_V}{L} & 0 & \frac{-6EI_V}{L^2} & 0 & 0 & 0 & \frac{2EI_V}{L} & 0 \\
& & & & & \frac{GJ}{L} & 0 & 0 & 0 & 0 & 0 & \frac{-GJ}{L} \\
& & & & & & \frac{12EI_V}{L^3} & 0 & 0 & 0 & \frac{-6EI_V}{L^2} & 0 \\
& & \text{Symm.} & & & & & \frac{12EI_U}{L^3} & 0 & \frac{6EI_U}{L^2} & 0 & 0 \\
& & & & & & & & \frac{EA}{L} & 0 & 0 & 0 \\
& & & & & & & & & \frac{4EI_U}{L} & 0 & 0 \\
& & & & & & & & & & \frac{4EI_V}{L} & 0 \\
& & & & & & & & & & & \frac{GJ}{L}
\end{bmatrix} \tag{B.32}$$

and

$$[sk_g]_j = \begin{bmatrix}
\dfrac{-6P}{5L} & 0 & 0 & 0 & \dfrac{-P}{10} & 0 & \dfrac{6P}{5L} & 0 & 0 & 0 & \dfrac{-P}{10} & 0 \\
 & \dfrac{-6P}{5L} & 0 & \dfrac{P}{10} & 0 & 0 & 0 & \dfrac{6P}{5L} & 0 & \dfrac{P}{10} & 0 & 0 \\
 & & 0 & 0 & 0 & 0 & 0 & 0 & 0 & 0 & 0 & 0 \\
 & & & \dfrac{-2PL}{15} & 0 & 0 & 0 & \dfrac{-P}{10} & 0 & \dfrac{PL}{30} & 0 & 0 \\
 & & & & \dfrac{-2PL}{15} & 0 & \dfrac{P}{10} & 0 & 0 & 0 & \dfrac{PL}{30} & 0 \\
 & & & & & 0 & 0 & 0 & 0 & 0 & 0 & 0 \\
 & & & & & & \dfrac{-6P}{5L} & 0 & 0 & 0 & \dfrac{P}{10} & 0 \\
 & \text{Symm.} & & & & & & \dfrac{-6P}{5L} & 0 & \dfrac{-P}{10} & 0 & 0 \\
 & & & & & & & & 0 & 0 & 0 & 0 \\
 & & & & & & & & & \dfrac{-2PL}{15} & 0 & 0 \\
 & & & & & & & & & & \dfrac{-2PL}{15} & 0 \\
 & & & & & & & & & & & 0
\end{bmatrix}$$

$$(B.33)$$

The geometric stiffness matrix in Equation B.33 considers the effect of a compression force, P, to the member lateral deflection and rotation. Equations B.32 and B.33 are approximate, but provide reasonable accuracy for structural engineering practice. For a more accurate stiffness matrix that accounts for the change in member bending stiffness due to the presence of an axial force see Chen and Atsuta (1977) and Chen and Lui (1991). At each incremental load step during the pushover analysis, the section properties of each segment, EA, EI_U, and EI_V, need to be checked and recalculated if the plastification of segment is changed. EA, EI_U, and EI_V can be obtained from Equations B.13 through B.15, respectively. Adding all segment stiffness matrices in (X_R, Y_R, Z_R) together, the member stiffness matrix is obtained as follows:

$$[\bar{K}] = \sum_{j=1}^{K} [\overline{sk}]_j \tag{B.34}$$

where
    j is the jth segment
    k is the total number of segments

In order to provide computational efficiency, the internal degrees of freedom are condensed out by Gaussian elimination and only the degrees of freedom at both ends are maintained. Thus, the condensed member stiffness matrix, $[\bar{k}]$, in (X_R, Y_R, Z_R) has a dimension of 12×12. As shown in Figure B.3, the condensed matrix $[\bar{k}]$ also needs to be transferred from (X_R, Y_R, Z_R) to the element coordinate system (X_e, Y_e, Z_e), denoted as $[k_e]$ in Equation 5.55. The structural global stiffness matrix is then formulated per Section 5.6.3 and is illustrated in Example 5.1. If a member is subject to eccentric axial load, $\bar{P}$ (see Figure B.4), $[\bar{k}]$ will need to be transferred to the location of $\bar{P}$ first, before being transformed to $[k_e]$ in (X_e, Y_e, Z_e). The stiffness matrix transformation from the end segment's geometric centroid "C" to the location of $\bar{P}$ is given as follows:

$$[\bar{k}] = [T][\bar{k}][T]^T \tag{B.35}$$

where

$$[T] = \begin{bmatrix} [T_A] & \\ & [T_B] \end{bmatrix} \tag{B.36}$$

and

$$[T_A] = \begin{bmatrix} 1 & 0 & 0 & 0 & 0 & 0 \\ 0 & 1 & 0 & 0 & 0 & 0 \\ 0 & 0 & 1 & 0 & 0 & 0 \\ 0 & 0 & -e_y & 1 & 0 & 0 \\ 0 & 0 & e_x & 0 & 1 & 0 \\ e_y & -e_x & 0 & 0 & 0 & 1 \end{bmatrix} \tag{B.37}$$

in which e_x and e_y are the eccentricities, positive in the X_R and Y_R directions, respectively, shown in Figure B.4. The matrix $[T_A]$ transfers the force vector from the segment's geometric centroid C to the $\bar{P}$ location. Equation B.37 is similar to the rigid body transformation described in Section 5.1.3.

At load step t during the pushover analysis, a finite segment element's displacement increment vector, $\{\Delta\delta_e^t\}_{12\times1}$, corresponding to the member's two ends in the element coordinate system (X_e, Y_e, Z_e) is transferred back to the segment's global coordinate system (X_R, Y_R, Z_R) as $\{\Delta\delta_R^t\}_{12\times1}$. Once $\{\Delta\delta_R^t\}_{12\times1}$ is known, the individual segment displacement increment vector, $\{\Delta\delta_{R,SEG\,j}^t\}_{12\times1}$, corresponding to (X_R, Y_R, Z_R) can be obtained by a Gaussian back-substitution process. $\{\Delta\delta_{R,SEG\,j}^t\}_{12\times1}$ is then transformed to the segment local coordinate system (U, V, W) as $\{\Delta\delta_{LOC,SEG\,j}^t\}_{12\times1}$ through rotation matrix $[\bar{R}]_j$ transformation. The segment force increment vector, $\{\Delta F_{LOC,SEG\,j}^t\}_{12\times1}$, in the (U, V, W) coordinate system is calculated by multiplying $\{\Delta\delta_{LOC,SEG\,j}^t\}_{12\times1}$ with the jth segment stiffness matrix, $[sk]_j$. The jth segment total displacement and force vectors at step t are

$$\{\delta_{LOC,SEG\,j}^{t}\}_{12\times1} = \{\delta_{LOC,SEG\,j}^{t-1}\}_{12\times1} + \{\Delta\delta_{LOC,SEG\,j}^{t}\}_{12\times1} \qquad (\text{B.38})$$

$$\{F_{LOC,SEG\,j}^{t}\}_{12\times1} = \{F_{LOC,SEG\,j}^{t-1}\}_{12\times1} + \{\Delta F_{LOC,SEG\,j}^{t}\}_{12\times1} \qquad (\text{B.39})$$

The segment local displacement increment vector, $\{\Delta\delta_{LOC,SEG\,j}^{t}\}_{12\times1}$, at step t is expressed as follows:

$$\{\Delta\delta_{LOC,SEG\,j}^{t}\}_{12\times1} = \{\Delta U_a, \Delta V_a, \Delta W_a, \Delta\theta_{ua}, \Delta\theta_{va}, \Delta\theta_{wa}, \Delta U_b, \Delta V_b, \Delta W_b, \Delta\theta_{ub}, \Delta\theta_{vb}, \Delta\theta_{wb}\}^{T}$$

$$(\text{B.40})$$

In Equation B.40, ΔW_a, ΔW_b, $\Delta\theta_{ua}$, $\Delta\theta_{ub}$, $\Delta\theta_{va}$, and $\Delta\theta_{vb}$ are used to calculate the segment cross-sectional strain increment based on Equations 4.23 through 4.26, in order to obtain the strain of each cross-sectional element, ε^{ij}, in Equation 4.27. Once ε^{ij} is known, the stress σ^{ij} and tangent modulus E_{ti} of cross-sectional element i are determined in accordance with the material stress–strain relationship of the member. Subsequently, the new section properties and instantaneous principal axes are calculated from Equations B.1 through B.15. They will be used to calculate each segment's new stiffness matrix $[sk]_j$, rotation matrix $[\bar{R}]_j$, etc. for the next load step $t+1$.

The calculated total segment force vector, $\{F_{LOC,SEG\,j}^{t}\}_{12\times1}$, will be used to calculate the unbalanced force vector as described in Appendix C.

Example B.2

An angle member $L2\times2\times\frac{1}{4}$ shown in Figure B.5 has initial imperfection of 0.1% of its total length. The member length is 34.9 in. Assume the member is divided into two segments, and the cross section of each segment is divided into seven cross-sectional elements as shown in Example B.1a. Find the rotation matrix, $[R]$, for each segment.

Solution

    1. Find the segment rotation matrices corresponding to reference coordinate (U_0, V_0, W_0): As shown in Figure B.5, the rotation matrix with

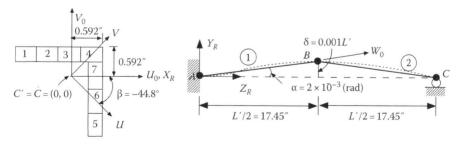

FIGURE B.5 Angle member with two segments.

consideration of initial imperfection for segment 1 can be calculated as follows:

$$(W_0Z) = \cos(-\alpha) = \cos(-2 \times 10^{-3}) \cong 1 = (V_0Y)$$

$$(V_0Z) = \cos\left(-\alpha - \frac{\pi}{2}\right) = -2 \times 10^{-3}$$

$$(W_0Y) = \cos\left(\frac{\pi}{2} - \alpha\right) = 2 \times 10^{-3}$$

$$[R]_{1,initial} = \begin{bmatrix} (U_0X) & (U_0Y) & (U_0Z) \\ (V_0X) & (V_0Y) & (V_0Z) \\ (W_0X) & (W_0Y) & (W_0Z) \end{bmatrix} = \begin{bmatrix} 1 & 0 & 0 \\ 0 & 1 & -2 \times 10^{-3} \\ 0 & 2 \times 10^{-3} & 1 \end{bmatrix}$$
(B.41)

Similarly, the rotation matrix for segment 2 is

$$[R]_{2,initial} = \begin{bmatrix} (U_0X) & (U_0Y) & (U_0Z) \\ (V_0X) & (V_0Y) & (V_0Z) \\ (W_0X) & (W_0Y) & (W_0Z) \end{bmatrix} = \begin{bmatrix} 1 & 0 & 0 \\ 0 & 1 & 2 \times 10^{-3} \\ 0 & -2 \times 10^{-3} & 1 \end{bmatrix}$$
(B.42)

2. Find the segment rotation matrices corresponding to (U,V,W):
Based on Equations B.18 through B.28, the rotation matrix for segment 1 is calculated as follows:

$$L = \sqrt{(X_B - X_A)^2 + (Y_B - Y_A)^2 + (Z_B - Z_A)^2}$$
$$= \sqrt{0 + (0.0349)^2 + (17.45)^2} \cong 17.45''$$

$$\vec{e}_w = \begin{bmatrix} (WX) \\ (WY) \\ (WZ) \end{bmatrix} = \begin{bmatrix} (X_B - X_A)/L \\ (Y_B - Y_A)/L \\ (Z_B - Z_A)/L \end{bmatrix} = \begin{bmatrix} 0 \\ 2 \times 10^{-3} \\ 1 \end{bmatrix}$$
(B.43)

Since there is no deformation for the member at the initial position, the rotations at ends A and B, θ_{wa} and θ_{wb}, are zero. Equation B.20 becomes

$$\theta = \frac{1}{2}(\theta_{wa} + \theta_{wb}) + \beta = 0 + (-44.8°) = -0.782 \ \text{(rad)}$$

From Equation B.21

$$\vec{V} = (\vec{e}_{v0})\cos\theta - (\vec{e}_{u0})\sin\theta = \begin{bmatrix} (\vec{V}_0 X) \\ (\vec{V}_0 Y) \\ (\vec{V}_0 Z) \end{bmatrix} = \begin{bmatrix} (V_0 X) \\ (V_0 Y) \\ (V_0 Z) \end{bmatrix}\cos\theta - \begin{bmatrix} (U_0 X) \\ (U_0 Y) \\ (U_0 Z) \end{bmatrix}\sin\theta$$

$$= \begin{bmatrix} 0 \\ 1 \\ -2\times10^{-3} \end{bmatrix}(0.7091) - \begin{bmatrix} 1 \\ 0 \\ 0 \end{bmatrix}(-0.7051) = \begin{bmatrix} 0.7051 \\ 0.7091 \\ -1.41\times10^{-3} \end{bmatrix}$$

From Equation B.24

$$M = (\vec{V}_0 X)(WX) + (\vec{V}_0 Y)(WY) + (\vec{V}_0 Z)(WZ)$$

$$= 0 + (0.7091)(2\times10^{-3}) - (1.41\times10^{-3})(1) \cong 0$$

From Equations B.26 and B.27

$$L_p = \sqrt{[(\vec{V}_0 X) - M(WX)]^2 + [(\vec{V}_0 Y) - M(WY)]^2 + [(\vec{V}_0 Z) - M(WZ)]^2} \cong 1$$

$$\vec{e}_v = \frac{1}{L_p}\begin{bmatrix} (\vec{V}_0 X) - M(WX) \\ (\vec{V}_0 Y) - M(WY) \\ (\vec{V}_0 Z) - M(WZ) \end{bmatrix} = \begin{bmatrix} (VX) \\ (VY) \\ (VZ) \end{bmatrix} = \begin{bmatrix} 0.7051 \\ 0.7091 \\ -1.41\times10^{-3} \end{bmatrix} \qquad \text{(B.44)}$$

The unit vector $\vec{e}_u = \vec{e}_v \times \vec{e}_w$ is given as follows:

$$\vec{e}_u = \begin{bmatrix} (UX) \\ (UY) \\ (UZ) \end{bmatrix} = \begin{bmatrix} (VY)(WZ) - (VZ)(WY) \\ (VZ)(WX) - (VX)(WZ) \\ (VX)(WY) - (VY)(WX) \end{bmatrix} = \begin{bmatrix} 0.7091 \\ -0.7051 \\ 1.41\times10^{-3} \end{bmatrix} \qquad \text{(B.45)}$$

From Equations B.43 through B.45, the rotation matrix of segment 1 is

$$[R]_1 = \begin{bmatrix} 0.7091 & -0.7051 & 1.41\times10^{-3} \\ 0.7051 & 0.7091 & -1.41\times10^{-3} \\ 0 & 2\times10^{-3} & 1 \end{bmatrix} \qquad \text{(B.46)}$$

Similarly, the rotation matrix of segment 2 can be calculated as follows:

$$[R]_2 = \begin{bmatrix} 0.7091 & -0.7051 & -1.41 \times 10^{-3} \\ 0.7051 & 0.7091 & 1.41 \times 10^{-3} \\ 0 & -2 \times 10^{-3} & 1 \end{bmatrix} \qquad (B.47)$$

$[R]_1$ and $[R]_2$ are used to formulate the segment stiffness matrix, $[\overline{sk}]_1$ and $[\overline{sk}]_2$, corresponding to (X_R, Y_R, Z_R), respectively, based on Equation B.29.

Appendix C: Unbalanced Forces of a Finite Segment

At load step t, let P_c, M_{Uc}, and M_{Vc} represent the calculated average segment forces of a segment. P_c, M_{Uc}, and M_{Vc} can be calculated as follows:

$$P_c = \frac{P_{ca} - P_{cb}}{2} \tag{C.1}$$

$$M_{Uc} = \frac{M_{Uca} - M_{Ucb}}{2} \tag{C.2}$$

$$M_{Vc} = \frac{M_{Vca} - M_{Vcb}}{2} \tag{C.3}$$

in which P_{ca}, M_{Uca}, and M_{Vca} represent the calculated segment local forces at the a end of the segment in the local coordinate system (U, V, W), from $\{F^t_{LOC,SEG\,j}\}_{12\times1}$ in Appendix B. P_{cb}, M_{Ucb}, and M_{Vcb} are calculated force vector at the b end of the segment. The segment cross-sectional resultant force vector can be calculated using the following equations:

$$P = \sum_{i=1}^{N} \sigma_{zi} A_i \tag{C.4}$$

$$M_U = \sum_{i=1}^{N} V_i'' \sigma_{zi} A_i \tag{C.5}$$

$$M_V = \sum_{i=1}^{N} U_i'' \sigma_{zi} A_i \tag{C.6}$$

Note that U_i'' and V_i'' are the location of the ith cross-sectional element, corresponding to the (U'', V'') axes as shown in Figure B.4. The (U'', V'') axes are parallel to the instantaneous principal axes (U, V). The origin of the (U'', V'') axes is the same as that of the (U_0, V_0) axes and is usually chosen as the geometric centroid (i.e., point C in Figure B.4) of the elastic cross section. Normally, P, M_U, and M_V will deviate from P_c, M_{Uc}, and M_{Vc} in the analysis. In order to adjust P to P_c (i.e., $\Delta P = P_c - P = 0$), an iteration process is required to adjust the normal strain increment, $\Delta \varepsilon_c^j$, at the centroid of segment's cross section (see Equation 4.24). Currently, INSTRUCT does not adjust $\Delta \varphi_u^j$ and $\Delta \varphi_v^j$ during the iteration (see Equations 4.25 and 4.26). Once ΔP

converges to zero, the corresponding M_U and M_V are calculated from Equations C.5 and C.6, respectively. The unbalanced moment, $\{U\}_{bending}$, between average moments (M_{Uc}, M_{Vc}) and segment cross-sectional moments (M_U, M_V) is

$$\{U\}_{bending} = \begin{Bmatrix} M_{Uc} - M_U \\ M_{Vc} - M_V \end{Bmatrix} \tag{C.7}$$

Therefore, the moments at end a are adjusted approximately by

$$\begin{Bmatrix} M_{Ua} \\ M_{Va} \end{Bmatrix} = \begin{Bmatrix} M_{Uca} \\ M_{Vca} \end{Bmatrix} \mp \{U\}_{bending} = \begin{Bmatrix} M_{Uca} \\ M_{Vca} \end{Bmatrix} \mp \begin{Bmatrix} M_{Uc} - M_U \\ M_{Vc} - M_V \end{Bmatrix} \tag{C.8}$$

Similarly, the moments at end b are adjusted approximately by

$$\begin{Bmatrix} M_{Ub} \\ M_{Vb} \end{Bmatrix} = \begin{Bmatrix} M_{Ucb} \\ M_{Vcb} \end{Bmatrix} \mp \{U\}_{bending} = \begin{Bmatrix} M_{Ucb} \\ M_{Vcb} \end{Bmatrix} \mp \begin{Bmatrix} M_{Uc} - M_U \\ M_{Vc} - M_V \end{Bmatrix} \tag{C.9}$$

As mentioned previously, the disadvantage of using this method is that the plastification at each end of the segment is not actually calculated, and a uniform plastification distribution along the segment based on the average curvature increment is assumed. Because of this assumption, the unbalanced force calculation at each end of a segment is also approximated. By comparing with experimental column test results, this approximate approach provides good results if a column element is divided into more than five segments. The program only calculates the unbalanced forces for segments with single curvatures. For a segment with double curvature, the unbalanced forces will not be calculated by the program. In order to reduce numerical instability (i.e., zigzagged stress reversals at some segment's cross-sectional elements due to the unbalanced force adjustment), it is recommended that (1) more segments be used on each column, so the curvature distribution along each segment is close to a uniform distribution and (2) smaller incremental load steps be used. If a numerical instability still exists, try to use the simple Euler incremental approach (see Appendix D) with small incremental steps and without consideration of unbalanced forces (i.e., choose the UNBAL = .FALSE. option in the SOL04 input data block).

The segment's unbalanced shear forces can be obtained from the unbalanced bending moments by force equilibrium.

Appendix D: Nonlinear Incremental Solution Algorithms

As mentioned previously, the loadings in the pushover analysis may consist of joint loads (force control), imposed displacements (displacement control), or a combination thereof. The loading is divided into increments and applied to the structure in steps. At the beginning of each load step, the tangent stiffness of the structure is determined, and the structure is assumed to behave linearly for the duration of load step. Unbalanced forces, when they exist, are calculated at the end of each load step and added to the incremental loads for the next step. The structural stiffness is updated at each step, if necessary.

There are several other incremental methods commonly used for nonlinear pushover analysis, for example, the simple Euler incremental method, Newton–Raphson method, arc length control method, and work control method. In the simple Euler incremental method, the unbalanced forces that exist in each load increment are ignored. In the Newton–Raphson method, iterations are used to eliminate the unbalanced forces that exist at each load step. Schematic representations of the simple Euler incremental method, the current method used in INSTRUCT as described in the previous paragraph, and the Newton–Raphson method are shown in Figure D.1. In the figure, the subscript number in the Newton–Raphson method represents the iteration number. It can be seen that the incremental scheme used in the current method combines the pure Euler incremental method with a "single" equilibrium correction without going through an iteration at each load step.

A drawback of both the current method and the Newton–Raphson method is that they fail at the limit point, the peak point of the load–deflection curve. At the limit point, the solution will diverge as shown in Figure D.2. In order to trace the descending branch of the load–deflection curve, the current stiffness parameter, S_p, is used here to detect the limit point. In the current method, a "single" equilibrium correction at each incremental step is applied to trace the ascending branch of the load–deflection curve. However, once the limit point is near, the simple Euler incremental method is used to trace the curve past the limit point. The current method is then resumed to trace the descending branch of the curve. This process is shown in Figure D.3.

The current stiffness parameter has the form

$$S_p = \left(\frac{\left\| \Delta F^t \right\|}{\left\| \Delta F^1 \right\|} \right)^2 \frac{\{\Delta F^1\}^T \{\Delta \delta^1\}}{\{\Delta F^t\}^T \{\Delta \delta^t\}} \tag{D.1}$$

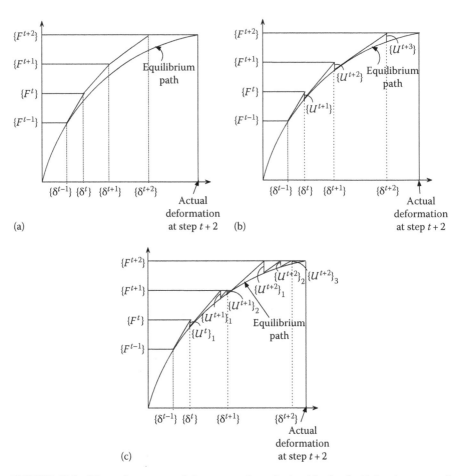

FIGURE D.1 Three force control incremental methods: (a) simple Euler incremental method, (b) current method used in **INSTRUCT**, and (c) Newton–Raphson method.

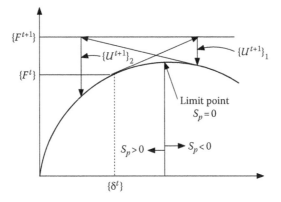

FIGURE D.2 Divergence of solution in the Newton–Raphson method.

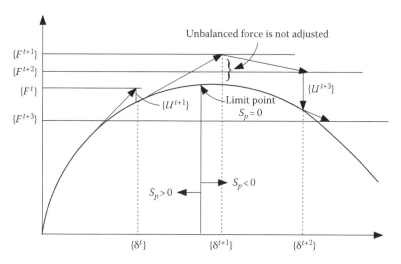

FIGURE D.3 Incremental method using current stiffness parameter.

where

Index 1 represents the initial increment step 1

t represents the incremental step t

$\|\Delta F^t\|$ is the norm of the increment load vector $\{\Delta F^t\}$

The unstable behavior (i.e., the descending branch of the curve) is characterized by a value of S_p less than zero. The value becomes zero at the limit point. The value of S_p is equal to one at the initial step 1 and decreases as the structure gets softer. In the program, when the absolute value of S_p is less than a user-defined value, simple Euler incremental steps are performed, and the unbalanced forces will not be added to the incremental loads for the next steps until the absolute value of S_p is greater than or equal to the user defined limit value. When S_p changes from a positive to a negative value at a step t, it is necessary to reanalyze step t by decreasing loads (i.e., using negative $\{\Delta F^t\}$ instead of positive $\{\Delta F^t\}$). The program will not automatically change from a positive incremental load step to a negative incremental load step for the reanalysis of step t. Therefore, a user needs to redefine the load step at step t and the subsequent steps and rerun the program. The norm of the incremental load vector, $\|\Delta F^t\|$, can be calculated as follows:

$$\|\Delta F^t\| = \left[\frac{1}{N} \sum_{i=1}^{N} \left| \frac{\Delta F_i^t}{F_{i,ref}^t} \right|^2 \right]^{1/2} \tag{D.2}$$

where

N is the total number of global degrees of freedom (including restrained degrees of freedom)

$F_{i,ref}^t$ is taken as the largest force component of the corresponding category (i.e., translational force or rotational moment) at step t

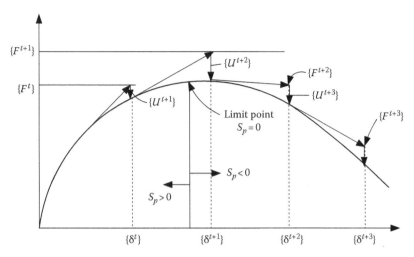

FIGURE D.4 Displacement control.

If an imposed displacement (i.e., displacement control) is used in the program, the unbalanced forces are considered at all displacement incremental steps. There is no limit point divergence problem, as equilibriums are carried out at displacement steps rather than at load steps. A schematic representation of displacement control is shown in Figure D.4.

Appendix E: Plastic Curvature Capacities and Neutral Axis Depth in Columns

The FHWA publication (FHWA, 2006) entitled *Seismic Retrofitting Manual for Highway Structures* provides a closed-form formula to estimate the plastic curvature capacities of concrete columns subjected to different failure modes. Note that the plastic curvature capacities based on these formulas are approximate. To obtain more accurate plastic curvature capacities, the FSFS method should be used to perform the moment–curvature analysis, as described in Chapter 4. The FHWA approximate formulas are shown in Table E.1.

TABLE E.1
Plastic Curvature Capacities

Column Failure Mode	Plastic Curvature Capacity (ϕ_p)	Equation
Compression failure, unconfined concrete	$\phi_p = \dfrac{\varepsilon_{cu}}{c} - \phi_n$	(E.1)
Compression failure, confined concrete	$\phi_p = \dfrac{\varepsilon_{cu}}{(c - d'')} - \phi_n$	(E.2)
Buckling of longitudinal bars	$\phi_p = \dfrac{\varepsilon_b}{(c - d')} - \phi_n$	(E.3)
Fracture of longitudinal reinforcement	$\phi_p = \dfrac{\varepsilon_{s\,max}}{(d - c)} - \phi_n$	(E.4)
Low-cycle fatigue of longitudinal reinforcement	$\phi_p = \dfrac{2\varepsilon_{ap}}{(d - d')} = \dfrac{2\varepsilon_{ap}}{D'}$	(E.5)
Lap-splice failure	$\phi_p = (\mu_{lap\phi} + 7)\phi_n$	(E.6)

where
  ϕ_n is the curvature corresponding to the nominal moment, M_n
  c is the depth from the extreme compression fiber of the cover concrete to the neutral axis, which can be estimated by the plastic section analysis as described later
  d'' is the distance from the extreme compression fiber of the cover concrete to the centerline of the perimeter hoop (thus, $c - d''$ is the depth of confined concrete under compression)
  ε_{cu} is the ultimate compression strain of the core concrete, given as

$$\varepsilon_{cu} = 0.004 + \frac{1.4\rho_s f_{yh}\varepsilon_{su}}{f_{cc}'} \quad \text{for confined concrete} \tag{E.7}$$

where
ε_{su} is the strain at the maximum stress of the transverse reinforcement
f_{yh} is the yield stress of the transverse steel
ρ_s is the volumetric ratio of transverse steel
f_{cc}' is the confined concrete strength

or

$$\varepsilon_{cu} = 0.004 \quad \text{for unconfined concrete} \tag{E.8}$$

(note that $\varepsilon_{cu}=0.005$ is used in the FHWA publication [2006] for seismic retrofit analysis)
$\varepsilon_b=2f_y/E_s$ is the buckling strain in the longitudinal reinforcing steel
ε_{smax} is the fracture tensile strain of the longitudinal steel. ε_{smax} should be limited to a value less than or equal to 0.1
ε_{ap} is the plastic strain amplitude, as given by

$$\varepsilon_{ap} = 0.08(2N_f)^{-0.5} \tag{E.9}$$

in which N_f is the effective number of equal-amplitude cycles of loading that lead to fracture, which can be approximated by

$$N_f = 3.5(T_n)^{-1/3} \tag{E.10}$$

provided that $2 \le N_f \le 10$, and T_n is the natural period of vibration of the bridge.
$\mu_{lap\phi}$ is the curvature ductility at the initial breakdown of bond in the lap-splice zone $\mu_{lap\phi}=0$ if $M_s < M_n$ where M_n is the nominal moment strength and M_s is the reduced moment strength given by

$$M_s = M_n\left(\frac{l_{lap}}{l_s}\right) \tag{E.11}$$

where
l_{lap} is the actual length of splice
l_s is the theoretical lap-splice length determined from

$$l_s = 0.04\frac{f_y}{\sqrt{f_c'}}d_b \text{ (in.)} \tag{E.12}$$

$\mu_{lap\phi}$ is the curvature ductility at which the concrete extreme fiber compression strain reaches 0.002, when $M_n < M_s < M_u$, in which M_u is the ultimate moment strength.

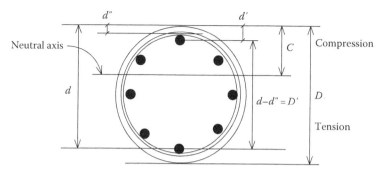

FIGURE E.1   Definition of c, D, D', d, d', and d''.

A user could calculate the least (controlled) plastic curvature capacity, ϕ_p, based on the equations in Table E.1. The plastic rotation capacity is then calculated by the PHL or CMR method as described in Chapter 4. The input parameter of plastic rotational capacity in INSTRUCT is PRMAX for "HINGE" and "IA_BILN" material types. Similarly, the input parameter of plastic curvature capacity is PCMAX for "R/CONCRETE1" and "MOMCURVA1" material types. INSTRUCT can perform a moment–curvature analysis using the FSFS method to obtain the plastic curvature capacity corresponding to the above failure modes.

In order to calculate the plastic curvature capacities corresponding to some of the failure modes shown in Table E.1, it is necessary to find the neutral axis depth, c, for a given value of strain at the extreme concrete compression fiber.

For rectangular sections, the neutral axis depth ratio is approximately given by

$$\frac{c}{D} = \frac{(P_e/f_c'A_g) + (\gamma\rho_t f_y/f_c')/(1 - 2d'/D)}{\alpha\beta + (2\gamma\rho_t f_y/f_c')/(1 - 2d'/D)} \tag{E.13}$$

For circular sections, the neutral axis depth ratio is approximately calculated by trial and error using the following equation:

$$\frac{c}{D} = \frac{1}{\beta}\left[\frac{(P_e/f_c'A_g) + 0.5\rho_t(f_y/f_c')(1 - 2c/D)/((1 - 2d'/D))}{1.32\alpha}\right]^{0.725} \tag{E.14}$$

where
  c is the depth to neutral axis (see Figure E.1)
  D is the overall depth of section
  P_e is the axial load on the section
  f_c' is the specified concrete strength
  f_y is the specified yield strength of the longitudinal reinforcement
  A_g is the gross cross-sectional area
  d' is the depth from the extreme compression fiber to the center of the compression reinforcement

ρ_t is the volumetric ratio of the longitudinal reinforcement

α, β are the concrete stress block parameters as defined below

γ is the reinforcing steel configuration factor

$\gamma = 0.5$ for square sections with steel placed symmetrically around the perimeter

$\gamma = 0.0$ for rectangular beam sections with steel lumped at the outer (top and bottom) faces

$\gamma = 0.0$ for wall section bending about the weak (out-of-plane) axis

$\gamma = 1.0$ for wall sections bending about the strong (in-plane) axis

α is the ratio of average concrete stress in the compression zone to confined concrete strength

$\alpha = 0.85 + 0.12(K - 1)^{0.4}$

K is the strength enhancement factor due to the confining action of the transverse reinforcement

$K = f'_{cc}/f'_c$, for circular and rectangular sections

f'_{cc} is the confined concrete strength

β is the depth of stress block

$\beta = 0.85 + 0.13(K - 1)^{0.6}$

For circular sections, the confined strength parameter (K) is given by Mander et al. (1988):

$$K = 2.254\sqrt{1 + 7.94\frac{f'_l}{f'_c}} - 2\frac{f'_l}{f'_c} - 1.254 \qquad (E.15)$$

where

$f'_l = (1/2)K_e\rho_sf_{yh}$ is the lateral stress supplied by the transverse reinforcement at yield

$\rho_s = 4A_{bh}/D''s$ is the volumetric ratio of spirals or circular hoops to the core concrete

A_{bh} is the cross-sectional area of the hoop or spiral bar

$K_e = (1 - \chi s/D'')/(1 - \rho_{cc})$ is the confinement effectiveness coefficient for spirals and hoop steel

ρ_{cc} is the ratio of area of longitudinal reinforcement to area of core of section

χ is the coefficient with values of 0.5 and 1.0 for spirals and hoops, respectively

s is the spacing of spirals or hoops

D'' is the diameter of transverse hoop or spiral (measured to the centerline of the hoop)

For rectangular sections, the confined strength parameter (K) is obtained from Figure E.2, which uses the x- and y-confining stresses (f'_{lx} and f'_{ly}, respectively) to derive K. Stresses f'_{lx} and f'_{ly} are defined as follows:

$f'_{lx} = K_e\rho_xf_{yh}$ is the lateral confining stress in the x-direction

$f'_{ly} = K_e\rho_yf_{yh}$ is the lateral confining stress in the y-direction

$\rho_x = A_{sx}/h''_ys$ is the volumetric ratio of transverse hoops or ties to the core concrete in x-direction

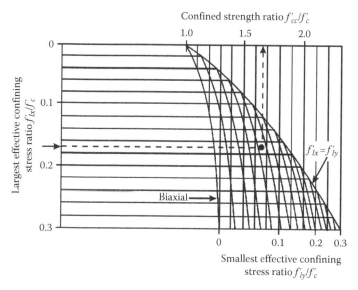

FIGURE E.2 Confined strength ratio (K) for reinforced concrete members (Paulay and Priestley, 1992).

$\rho_y = A_{sy}/h''_x s$ is the volumetric ratio of transverse hoops or ties to the core concrete in y-direction

A_{sx} is the total area of transverse reinforcement parallel to x-axis

A_{sy} is the total area of transverse reinforcement parallel to y-axis

K_e is the confinement effectiveness coefficient for rectangular sections with hoops or ties

$K_e = 0.75$ for rectangular columns

$K_e = 0.6$ for rectangular wall sections

f_{yh} is the yield stress of the transverse hoops

h''_x is the width of column paraller to x-direction (measured to the centerline of hoops or ties, see Figure 3.9)

h''_y is the width of column paraller to y-direction (measured to the centerline of hoops or ties)

Appendix F: Elastic and Inelastic Time History Analysis

The dynamic response of a multiple-degree-of-freedom system subjected to an earthquake excitation can be obtained by solving the following motion equation:

$$[M]\{\ddot{x}(t)\}+[C]\{\dot{x}(t)\}+[K]\{x(t)\} = \{F(t)\} \tag{F.1}$$

in which $\{x(t)\}$, $\{\dot{x}(t)\}$, and $\{\ddot{x}(t)\}$ are the structural displacement, velocity, and acceleration vectors, respectively, relative to the ground motion; $[M]$, $[C]$, and $[K]$ are the structural mass, damping, and stiffness matrix, respectively. $\{F(t)\}$ is the effective earthquake force vector expressed as follows:

$$\{F(t)\} = -[M]\{I_n\}\ddot{x}_G \tag{F.2}$$

where
   $\ddot{x}_G$ is the earthquake acceleration record expressed in terms of gravity, G
   $\{I_n\}$ is the system influence coefficient vector representing the structural displacement vector due to a unit ground movement

For example, $\{I_n\}$ of a four-degree-of-freedom structure shown in Figure F.1 is $\{1,0,1,0\}^T$.

Ground motion during an earthquake is measured by a strong motion accelerograph, which records the acceleration of the ground at a particular site. A typical accelerogram (i.e., $\ddot{x}_G$), a record of the north–south (N–S) component of the El Centro earthquake of May 18, 1940, is shown in Figure F.2.

The solution of Equation F.1 can be obtained by numerical integration techniques. The two best-known numerical integration techniques, Newmark and Wilson-θ methods are introduced here.

F.1   NEWMARK INTEGRATION METHOD

The Newmark integration method assumes that during an incremental time step, Δt, the acceleration varies linearly as shown in Figure F.3. The average acceleration from t to $t+\Delta t$ is $\{\ddot{x}\}_{avg}=(1/2)(\{\ddot{x}(t)\} + \{\ddot{x}(t+\Delta t)\})$. Thus, the velocity vector at $t+\Delta t$ can be expressed as follows:

$$\{\dot{x}(t + \Delta t)\} = \{\dot{x}(t)\} + \Delta t\{\ddot{x}\}_{avg} = \{\dot{x}(t)\} + \frac{\Delta t}{2}(\{\ddot{x}(t)\} + \{\ddot{x}(t + \Delta t)\}) \tag{F.3}$$

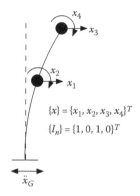

FIGURE F.1   Multiple-degree-of-freedom system.

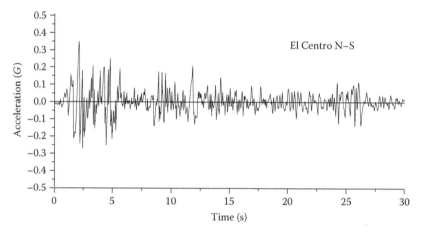

FIGURE F.2   Accelerogram for the N–S component of El Centro, the Imperial Valley Earthquake of May 18, 1940.

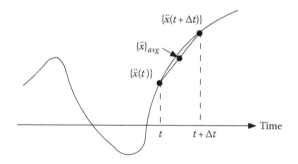

FIGURE F.3   Linear variation of acceleration.

The displacement vector at $t+\Delta t$ can be obtained from (F.3) as $\{x(t+\Delta t)\} = \{x(t)\} + \Delta t \{\dot{x}\}_{avg}$ in which $\{\dot{x}\}_{avg} = (1/2)(\{\dot{x}(t)\} + \{\dot{x}(t+\Delta t)\})$ or

$$\{x(t+\Delta t)\} = \{x(t)\} + \{\dot{x}(t)\}\Delta t + \frac{(\Delta t)^2}{4}(\{\ddot{x}(t)\} + \{\ddot{x}(t+\Delta t)\}) \tag{F.4}$$

Equations F.3 and F.4 represent the Newmark trapezoidal rule or the average acceleration method. The general Newmark integration may be expressed as follows:

$$\{\dot{x}(t+\Delta t)\} = \{\dot{x}(t)\} + [(1-\delta)\{\ddot{x}(t)\} + \delta\{\ddot{x}(t+\Delta t)\}]\Delta t \tag{F.5}$$

$$\{x(t+\Delta t)\} = \{x(t)\} + \{\dot{x}(t)\}\Delta t + \left[\left(\frac{1}{2} - \alpha\right)(\{\ddot{x}(t)\} + \alpha\{\ddot{x}(t+\Delta t)\}\right](\Delta t)^2 \tag{F.6}$$

where α and δ are parameters that can be determined to optimize integration accuracy and stability. When $\delta = 1/2$ and $\alpha = 1/4$, Equations F.5 and F.6 correspond to the average acceleration method. When $\delta = 1/2$ and $\alpha = 1/6$, Equations F.5 and F.6 are then associated with the linear acceleration method.

From Equation F.6,

$$\{\ddot{x}(t+\Delta t)\} = \frac{1}{\alpha\Delta t^2}\left[\{x(t+\Delta t)\} - \{x(t)\} - \Delta t\{\dot{x}(t)\} - \Delta t^2\left(\frac{1}{2} - \alpha\right)\{\ddot{x}(t)\}\right] \tag{F.7}$$

Substituting Equation F.7 into Equation F.5 leads to

$$\{\dot{x}(t+\Delta t)\} = \{\dot{x}(t)\} + \left[(1-\delta)\{\ddot{x}(t)\} + \delta\left\{\frac{1}{\alpha\Delta t^2}[\{x(t+\Delta t)\} - \{x(t)\}\right.\right.$$
$$\left.\left. - \Delta t\{\dot{x}(t)\} - \Delta t^2\left(\frac{1}{2} - \alpha\right)\{\ddot{x}(t)\}\right]\right\}\right] \tag{F.8}$$

Employing Equations F.7 and F.8 in Equation F.1 at $t+\Delta t$, and letting $a_0 = 1/\alpha\Delta t^2$, $a_1 = \delta/\alpha\Delta t$, $a_2 = 1/\alpha\Delta t$, $a_3 = 1/(2\alpha - 1)$, $a_4 = \delta/(\alpha - 1)$, and $a_5 = (\Delta t/2)(\delta/(\alpha - 2))$, Equation F.1 becomes

$$(a_0[M] + a_1[C] + [K])\{x(t+\Delta t)\}$$
$$= \{F(t+\Delta t)\} + [M](a_0\{x(t)\}$$
$$+ a_2\{\dot{x}(t)\} + a_3\{\ddot{x}(t)\}) + [C](a_1\{x(t)\} + a_4\{\dot{x}(t)\} + a_5\{\ddot{x}(t)\}) \tag{F.9a}$$

or

$$[\bar{K}]\{x(t+\Delta t)\} = [\bar{F}] \tag{F.9b}$$

from which $\{x(t+\Delta t)\}$ can be obtained, because all of the response parameters at time t are known. Substituting $\{x(t+\Delta t)\}$ in Equation F.7 leads to

$$\{\ddot{x}(t+\Delta t)\} = a_0[\{x(t+\Delta t)\}-\{x(t)\}]-a_2\{\dot{x}(t)\}-a_3\{\ddot{x}(t)\} \tag{F.10}$$

Employing $\{\ddot{x}(t+\Delta t)\}$ from Equation F.10 in Equation F.5, we have

$$\{\dot{x}(t+\Delta t)\} = \{\dot{x}(t)\}+a_6\{\ddot{x}(t)\}+a_7\{\ddot{x}(t+\Delta t)\} \tag{F.11}$$

where $a_6 = \Delta t(1-\delta)$ and $a_7 = \delta\Delta t$. When $\delta = 1/2$ and $\alpha = 1/6$, the Newmark integration method becomes the linear acceleration method; Equations F.9a, F.10, and F.11 are then expressed as Equations F.12a, F.13, and F.14, respectively

$$\left(\frac{6}{\Delta t^2}[M]+\frac{3}{\Delta t}[C]+[K]\right)\{x(t+\Delta t)\} = \{F(t+\Delta t)\}-[M]\{A\}-[C]\{B\} \tag{F.12a}$$

or

$$[\bar{K}]\{x(t+\Delta t)\} = \{\bar{F}(t+\Delta t)\} \tag{F.12b}$$

$$\{\ddot{x}(t+\Delta t)\} = \frac{6}{\Delta t^2}\left[\{x(t+\Delta t)\}-\{x(t)\}-\frac{6}{\Delta t}\{\dot{x}(t)\}-2\{\ddot{x}(t)\}\right] = \frac{6}{\Delta t^2}\{x(t+\Delta t)\}+\{A\}$$

$$\tag{F.13}$$

$$\{\dot{x}(t+\Delta t)\} = \{\dot{x}(t)\}+\frac{\Delta t}{2}\left[\{\ddot{x}(t)\}+\frac{6}{\Delta t^2}\{x(t+\Delta t)\}-\frac{6}{\Delta t^2}\{x(t)\}-\frac{6}{\Delta t}\{\dot{x}(t)\}-2\{\ddot{x}(t)\}\right]$$

$$= \frac{3}{\Delta t}\{x(t+\Delta t)\}+\{B\} \tag{F.14}$$

in which

$$\{A\} = -\frac{6}{\Delta t^2}\{x(t)\}-\frac{6}{\Delta t}\{\dot{x}(t)\}-2\{\ddot{x}(t)\} \tag{F.15}$$

and

$$\{B\} = -2\{\dot{x}(t)\}-\frac{\Delta t}{2}\{\ddot{x}(t)\}-\frac{3}{\Delta t}\{x(t)\} \tag{F.16}$$

Equations F.12a, F.13, and F.14 are typically used in elastic time history analysis. For inelastic (i.e., nonlinear) time history analysis, at each time increment, Δt, the stiffness matrix $[K]$ may be changed according to the hysteresis models of individual members. Thus, the linear acceleration method in an incremental form is desired, and described as follows.

Let $\{\Delta x\} = \{x(t+\Delta t)\} - \{x(t)\}$, $\{\Delta \dot{x}\} = \{\dot{x}(t+\Delta t)\} - \{\dot{x}(t)\}$, and $\{\Delta \ddot{x}\} = \{\ddot{x}(t+\Delta t)\} - \{\ddot{x}(t)\}$, Equation F.12a can be written as follows:

$$\left(\frac{6}{\Delta t^2}[M] + \frac{3}{\Delta t}[C] + [K]\right)\{\Delta x\} = (\{F(t+\Delta t)\} - \{F(t)\}) - [M]\{A\} - [C]\{B\} \quad \text{(F.17)}$$

Therefore, $\{\Delta x\}$ can be obtained from Equation F.17. From Equation F.13

$$\{\Delta \ddot{x}\} = \{\ddot{x}(t+\Delta t)\} - \{\ddot{x}(t)\} = \frac{6}{\Delta t^2}\{x(t+\Delta t)\} + \{A\} - \{\ddot{x}(t)\} = \frac{6}{\Delta t^2}\{x(t+\Delta t)\}$$

$$-\frac{6}{\Delta t^2}\{x(t)\} - \frac{6}{\Delta t}\{\dot{x}(t)\} - 2\{\ddot{x}(t)\} - \{\ddot{x}(t)\} = \frac{6}{\Delta t^2}\{\Delta x\} + \{\bar{A}\} \quad \text{(F.18)}$$

in which

$$\{\bar{A}\} = -\frac{6}{\Delta t}\{\dot{x}(t)\} - 3\{\ddot{x}(t)\} \quad \text{(F.19)}$$

From Equation F.14

$$\{\Delta \dot{x}\} = \{\dot{x}(t+\Delta t)\} - \{\dot{x}(t)\} = \frac{3}{\Delta t}\{x(t+\Delta t)\} + \{B\} - \{\dot{x}(t)\}$$

$$= \frac{3}{\Delta t}\{x(t+\Delta t)\} - 2\{\dot{x}(t)\} - \frac{\Delta t}{2}\{\ddot{x}(t)\} - \frac{3}{\Delta t}\{x(t)\} - \{\dot{x}(t)\} = \frac{3}{\Delta t}\{\Delta x\} + \{\bar{B}\}$$

$$\text{(F.20)}$$

in which

$$\{\bar{B}\} = -3\{\dot{x}(t)\} - \frac{\Delta t}{2}\{\ddot{x}(t)\} \quad \text{(F.21)}$$

The displacement, velocity, and acceleration vectors are then determined from

$$\{x(t+\Delta t)\} = \{x(t)\} + \{\Delta x\} \quad \text{(F.22)}$$

$$\{\dot{x}(t+\Delta t)\} = \{\dot{x}(t)\} + \{\Delta \dot{x}\} \quad \text{(F.23)}$$

$$\{\ddot{x}(t+\Delta t)\} = \{\ddot{x}(t)\} + \{\Delta \ddot{x}\} \quad \text{(F.24)}$$

It is noted that the specified incremental time interval, Δt, can affect the results of the calculation. In order to avoid numerical divergence, $\Delta t \leq 0.55T$ is recommended for the Newmark method, where T represents the fundamental period of the structure. The Newmark integration procedure is summarized as follows:

Step 1: Obtain time history record, $\ddot{x}_G$, with total excitation from $t=0$ s to t_{final} s. Determine the incremental time interval, Δt.

Step 2: Perform time history analysis at time t_i, where $t_i = t_{i-1} + \Delta t$, to calculate $\{\Delta x\}$, $\{\Delta \dot{x}\}$, and $\{\Delta \ddot{x}\}$ from Equations F.17, F.20, and F.18, respectively.

Step 3: Obtain $\{x(t_i)\}$, $\{\dot{x}(t_i)\}$, and $\{\ddot{x}(t_i)\}$ from Equations F.22 through F.24, respectively, where $t_i = t_{i-1} + \Delta t$ and $\{x(t_{i-1} + \Delta t)\} = \{x(t_{i-1})\} + \{\Delta x\}$, etc.

Step 4: If $t_i < t_{final}$, go to Step 2. If $t_i = t_{final}$, the numerical integration is completed.

F.2   WILSON-θ METHOD

The Wilson-θ method is an extension of the linear acceleration method in which a linear variation of the acceleration from time t to $t + \Delta t$ is assumed. In the Wilson-θ method, acceleration is assumed to be linear from time t to $t + \theta \Delta t$, with $\theta \geq 1$ ($\theta = 1$ is the linear acceleration method), as shown in Figure F.4.

Let $\{\Delta x\} = \{x(t + \Delta t)\} - \{x(t)\}$, $\{\Delta \dot{x}\} = \{\dot{x}(t + \Delta t)\} - \{\dot{x}(t)\}$, and $\{\Delta \ddot{x}\} = \{\ddot{x}(t + \Delta t)\} - \{\ddot{x}(t)\}$, the incremental velocity vector from t to $t + \tau$ can be expressed as follows:

$$\{\Delta \dot{x}_\tau\} = \{\dot{x}(t + \tau)\} - \{\dot{x}(t)\} = \frac{1}{2}\left[\{\ddot{x}(t)\} + \{\ddot{x}(t + \tau)\}\right]\tau$$

$$= \frac{1}{2}\left[\{\ddot{x}(t)\} + \{\ddot{x}(t)\} + (\{\ddot{x}(t + \Delta t)\} - \{\ddot{x}(t)\})\frac{\tau}{\Delta t}\right]\tau$$

$$= \{\ddot{x}(t)\}\tau + \frac{\tau^2}{2\Delta t}[\{\ddot{x}(t + \Delta t)\} - \{\ddot{x}(t)\}] = \{\ddot{x}(t)\}\tau + \frac{\tau^2}{2\Delta t}\{\Delta \ddot{x}\} \qquad (F.25)$$

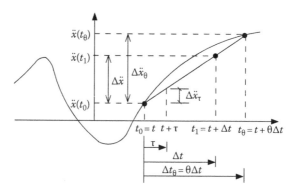

FIGURE F.4   Wilson-θ integration method.

Integrating Equation F.25, we have

$$\int_0^\tau \{\dot{x}(t+\tau)\}d\tau - \int_0^\tau \{\dot{x}(t)\}d\tau = \int_0^\tau \{\ddot{x}(t)\}\tau\,d\tau + \int_0^\tau \frac{\tau^2}{2\Delta t}\{\Delta\ddot{x}\}d\tau \qquad (F.26)$$

$$\rightarrow \{x(t+\tau)\} - \{x(t)\} - (\tau-0)\{\dot{x}(t)\} = \frac{\tau^2}{2}\{\ddot{x}(t)\} + \frac{\tau^3}{6\Delta t}\{\Delta\ddot{x}\}$$

$$\rightarrow \{x(t+\tau)\} = \{x(t)\} + \tau\{\dot{x}(t)\} + \frac{\tau^2}{2}\{\ddot{x}(t)\} + \frac{\tau^3}{6\Delta t}\{\Delta\ddot{x}\}$$

or

$$\{\Delta x_\tau\} = \tau\{\dot{x}(t)\} + \frac{\tau^2}{2}\{\ddot{x}(t)\} + \frac{\tau^3}{6\Delta t}\{\Delta\ddot{x}\} \qquad (F.27)$$

Since the Wilson-θ method assumes a linear variation of acceleration from t to $t+\theta\Delta t$, Equations F.25 and F.27 can also be expressed as Equations F.28 and F.29, respectively, with $t \le \tau \le t+\theta\Delta t$:

$$\{\Delta\dot{x}_\tau\} = \{\dot{x}(t+\tau)\} - \{\dot{x}(t)\} = \{\ddot{x}(t)\}\tau + \frac{\tau^2}{2\theta\Delta t}\{\Delta\ddot{x}_\theta\} \qquad (F.28)$$

$$\{\Delta x_\tau\} = \{x(t+\tau)\} - \{x(t)\} = \tau\{\dot{x}(t)\} + \frac{\tau^2}{2}\{\ddot{x}(t)\} + \frac{\tau^3}{6\theta\Delta t}\{\Delta\ddot{x}_\theta\} \qquad (F.29)$$

in which $\{\Delta\ddot{x}_\theta\} = \{\ddot{x}(t+\theta\Delta t)\} - \{\ddot{x}(t)\}$.

When $\tau = \theta\Delta t$, let $\Delta t_\theta = \theta\Delta t$ and $t_\theta = t+\theta\Delta t$, and Equations F.28 and F.29 are reduced to

$$\{\Delta\dot{x}_\theta\} = \{\dot{x}(t+\theta\Delta t)\} - \{\dot{x}(t)\} = \{\ddot{x}(t)\}\Delta t_\theta + \frac{1}{2}\{\Delta\ddot{x}_\theta\}\Delta t_\theta \qquad (F.30)$$

$$\{\Delta x_\theta\} = \{\dot{x}(t)\}\Delta t_\theta + \frac{1}{2}\{\ddot{x}(t)\}(\Delta t_\theta)^2 + \frac{1}{6}\{\Delta\ddot{x}_\theta\}(\Delta t_\theta)^2 \qquad (F.31)$$

From Equation F.31

$$\{\Delta\ddot{x}_\theta\} = \frac{6}{(\Delta t_\theta)^2}\left[\{\Delta x_\theta\} - \{\dot{x}(t)\}\Delta t_\theta - \frac{1}{2}\{\ddot{x}(t)\}(\Delta t_\theta)^2\right]$$

$$= \frac{6}{(\Delta t_\theta)^2}\{\Delta x_\theta\} - \frac{6}{\Delta t_\theta}\{\dot{x}(t)\} - 3\{\ddot{x}(t)\} \qquad (F.32)$$

Substituting Equation F.32 into Equation F.30 results in

$$\{\Delta \dot{x}_\theta\} = \{\ddot{x}(t)\}\Delta t_\theta + \frac{3}{\Delta t_\theta}\{\Delta x_\theta\} - 3\{\dot{x}(t)\} - \frac{3\Delta t_\theta}{2}\{\ddot{x}(t)\} = \frac{3}{\Delta t_\theta}\{\Delta x_\theta\} - 3\{\dot{x}(t)\} - \frac{\Delta t_\theta}{2}\{\ddot{x}(t)\}$$

(F.33)

Based on Equation F.1, the incremental equation of motion can be expressed as follows:

$$[M]\{\Delta \ddot{x}_\theta\} + [C]\{\Delta \dot{x}_\theta\} + [K]\{\Delta x_\theta\} = \{\Delta F_\theta\}$$

(F.34)

in which

$$\{\Delta F_\theta\} = \theta\{\Delta F\} = \theta[\{F(t + \Delta t)\} - \{F(t)\}]$$

Substituting Equations F.32 into Equation F.33 and Equation F.34 leads to

$$[\bar{K}]\{\Delta x_\theta\} = \{\Delta \bar{F}\}$$

(F.35)

where

$$[\bar{K}] = [K] + \frac{6}{(\Delta t_\theta)^2}[M] + \frac{3}{\Delta t_\theta}[C]$$

(F.36)

and

$$\{\Delta \bar{F}\} = \{\Delta F_\theta\} + [M]\{Q\} + [C]\{R\}$$

(F.37)

in which

$$\{\Delta F_\theta\} = \theta\{\Delta F\}$$

(F.38)

$$\{Q\} = \frac{6}{\theta \Delta t}\{\dot{x}(t)\} + 3\{\ddot{x}(t)\}$$

(F.39)

$$\{R\} = 3\{\dot{x}(t)\} + \frac{\Delta t_\theta}{2}\{\ddot{x}(t)\}$$

(F.40)

Equation F.35 is solved for $\{\Delta x_\theta\}$ as follows:

$$\{\Delta x_\theta\} = [\bar{K}]^{-1}\{\Delta \bar{F}\}$$

(F.41)

Substituting $\{\Delta x_\theta\}$ into Equation F.32, $\{\Delta \ddot{x}_\theta\}$ is obtained, and $\{\Delta \ddot{x}\}$ is determined by the following formula:

$$\{\Delta \ddot{x}\} = \frac{1}{\theta}\{\Delta \ddot{x}_\theta\} \tag{F.42}$$

The incremental velocity vector, $\{\Delta \dot{x}\}$, and displacement vector, $\{\Delta x\}$, are obtained at $\tau = \Delta t$ from Equations F.28 and F.29, respectively. The total displacement, velocity, and acceleration vectors are determined from

$$\{x(t + \Delta t)\} = \{x(t)\} + \{\Delta x\} \tag{F.43}$$

$$\{\dot{x}(t + \Delta t)\} = \{\dot{x}(t)\} + \{\Delta \dot{x}\} \tag{F.44}$$

$$\{\ddot{x}(t + \Delta t)\} = \{\ddot{x}(t)\} + \{\Delta \ddot{x}\} \tag{F.45}$$

Similar to the Newmark method, $\theta \geq 1.4$ is recommended to avoid numerical divergence.

For the inelastic time history analysis, at each time increment, Δt, the stiffness matrix $[K]$ is a tangent stiffness matrix, which may be changed in accordance with the hysteretic models of individual members. The Wilson-θ integration procedure is summarized as follows:

Step 1: Obtain time history record, $\ddot{x}_G$, with total excitation from $t=0$ s to t_{final} s. Determine the incremental time interval, Δt.

Step 2: Perform time history analysis at time t_i, where $t_i = t_{i-1} + \Delta t$, to obtain $\{\Delta x_\theta\}$ from Equation F.41. Substitute $\{\Delta x_\theta\}$ into Equation F.32 to obtain $\{\Delta \ddot{x}_\theta\}$, and then $\{\Delta \ddot{x}\} = (1/\theta)\{\Delta \ddot{x}_\theta\}$, per Equation F.42.

Step 3: Once $\{\Delta \ddot{x}\}$ is known, calculate $\{\Delta \dot{x}\}$ and $\{\Delta x\}$ from Equations F.28 and F.29, respectively.

Step 4: Obtain $\{x(t_i)\}$, $\{\dot{x}(t_i)\}$, and $\{\ddot{x}(t_i)\}$ from Equations F.43 through F.45, respectively, where $t_i = t_{i-1} + \Delta t$ and $\{x(t_{i-1} + \Delta t)\} = \{x(t_{i-1})\} + \{\Delta x\}$, etc.

Step 5: If $t_i < t_{final}$, go to Step 2. If $t_i = t_{final}$, the numerical integration is complete.

F.3   PROPORTIONAL DAMPING MATRIX

To form the damping matrix in Equation F.1, it is assumed that the damping matrix is linearly related to the mass and stiffness matrices:

$$[C] = \alpha[M] + \beta[K] \tag{F.46}$$

where α and β are constants. For a multiple-degree-of-freedom structure, Equation F.1 can be decoupled by the orthogonality relationship, yielding the damping term for mode i as follows:

$$2\rho_i\omega_i = \alpha + \beta\omega_i^2 \qquad (F.47)$$

where
   ρ_i is the damping ratio for mode i
   ω_i is the natural frequency for mode i

One method of determining the constants α and β is by estimating the damping ratio and natural frequencies of two modes i and j, then solving

$$\begin{bmatrix} 2\rho_i\omega_i \\ 2\rho_j\omega_j \end{bmatrix} = \begin{bmatrix} 1 & \omega_i^2 \\ 1 & \omega_j^2 \end{bmatrix}\begin{bmatrix} \alpha \\ \beta \end{bmatrix} \qquad (F.48)$$

Thus

$$\alpha = \frac{2\omega_i\omega_j(\rho_i\omega_j - \rho_j\omega_i)}{\omega_j^2 - \omega_i^2}, \quad \beta = \frac{2(\rho_j\omega_j - \rho_i\omega_i)}{\omega_j^2 - \omega_i^2} \qquad (F.49)$$

When $\rho_i = \rho_j = \rho$, Equation F.49 becomes

$$\alpha = \omega_i\omega_j\beta, \quad \beta = \frac{2\rho}{\omega_j + \omega_i} \qquad (F.50)$$

For the inelastic time history analysis, it can be seen that the damping matrix $[C]$ is also updated when the tangent stiffness matrix $[K]$ is changed.

Appendix G: Elastic and Inelastic Response Spectra

In the seismic design of structures, the maximum structural response subjected to a design earthquake is of interest to practicing engineers. The maximum structural response could be the maximum relative displacement with respect to the ground motion displacement or the maximum absolute acceleration (i.e., inertia force) with respect to the ground at its rest condition prior to the earthquake.

As shown in Figure G.1, for a single-degree-of-freedom (sdof) structure subjected to earthquake excitation, the motion equation, based on force equilibrium (Clough and Penzien, 1975), is

$$m\ddot{x}^t(t) + c\dot{x}(t) + kx(t) = 0 \qquad (G.1)$$

where
   m, c, and k are mass, damping coefficient, and stiffness, respectively
   $\dot{x}$ and x are the relative velocity and relative displacement, respectively
   superscript t represents the total displacement

The total displacement x^t is

$$x^t(t) = x(t) + x_G(t) \qquad (G.2)$$

$x_G(t)$ is the ground motion displacement. Similar to Equation F.1, Equation G.1 can also be expressed in terms of relative displacement, given as

$$m\ddot{x}(t) + c\dot{x}(t) + kx(t) = -m\ddot{x}_G(t) \qquad (G.3)$$

in which $\ddot{x}_G$ is the earthquake acceleration record expressed in terms of gravity, G. Dividing Equation G.3 by the mass, m, leads to

$$\ddot{x}(t) + 2\rho\omega\dot{x}(t) + \omega^2 x(t) = -\ddot{x}_G(t) \qquad (G.4)$$

where ρ is the damping ratio (or so-called damping factor) expressed as follows:

$$\rho = \frac{c}{c_{cr}} = \frac{c}{2\sqrt{km}} = \frac{c}{2m\omega} \qquad (G.5)$$

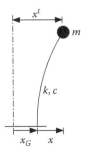

FIGURE G.1   sdof system subjected to ground motion.

where
   c_{cr} is the critical damping of the system
   ω is the angular frequency expressed as follows:

$$\omega = \sqrt{\frac{k}{m}} \ (\text{rad/s}) \qquad\qquad (G.6)$$

from which the structural natural frequency and the period can be calculated by using

$$f = \frac{\omega}{2\pi} \ (\text{cycle/s}) \qquad\qquad (G.7)$$

and

$$T = \frac{2\pi}{\omega} \ (\text{s/cycle}) \qquad\qquad (G.8)$$

G.1   ELASTIC RESPONSE SPECTRUM

A displacement response spectrum represents the maximum relative displacements of sdof oscillators with different periods (or frequencies) of vibration corresponding to a specified elastic damping ratio (typically 5%), subjected to ground motion, $\ddot{x}_G(t)$. The elastic displacement spectrum can be calculated using the step-by-step numerical integration method described in Appendix F to solve either Equation G.3 or G.4. Since the abscissas of the spectrum represent the structural periods of vibration, in practice, it is convenient to use Equation G.4 to generate the elastic response spectrum. The maximum displacement corresponding to each frequency, ω_i, can be obtained as follows:

$$R_d(\omega_i, \rho) = \max\left|x(t, \omega_i, \rho)\right|; \quad i = \text{the } i\text{th frequency or period} \qquad (G.9)$$

To generate the acceleration spectrum, the equation of motion in terms of total displacement should be used. By substituting Equation G.2 into Equation G.1, the equation of motion becomes

$$m\ddot{x}^t(t) + c\dot{x}^t(t) + kx^t(t) = c\dot{x}_G(t) + k\ddot{x}_G(t) \tag{G.10}$$

From this, the absolute acceleration spectrum can be calculated from Equation G.10 by the step-by-step numerical integration method as follows:

$$R_a(\omega_i,\rho) = \max\left|\ddot{x}^t(t,\omega_i,\rho)\right|; \qquad i = \text{the } i\text{th frequency or period} \tag{G.11}$$

From Equations G.1 and G.5, the total acceleration can be expressed as follows.

$$\ddot{x}^t(t) = -2\rho\omega\dot{x}(t) - \omega^2 x(t) \tag{G.12}$$

In practice, the damping terms in Equation G.12 can be neglected considering the damping force contribution to the equilibrium condition is small. Therefore, Equation G.12 can be simplified to

$$\ddot{x}^t(t) = -\omega^2 x(t) \tag{G.13}$$

The approximate calculation of the absolute acceleration spectrum can be formulated based on Equation G.13:

$$R_a(\omega_i,\rho) = \omega^2 \max\left|x(t,\omega_i,\rho)\right| = \omega^2 R_d \tag{G.14a}$$

Equation G.14a is called the pseudo-acceleration spectrum. Similarly, the pseudo-velocity spectrum is given as follows:

$$R_v(\omega_i,\rho) = \omega R_d \tag{G.14b}$$

Typical elastic spectra for the 1940 El Centro earthquake N–S component with 5% damping factor are shown in Figure G.2. The elastic displacement, velocity, and acceleration spectra were generated by the linear acceleration numerical integration method using Equation G.4, and they are in good agreement with those shown in other references (Naeim, 1989).

The response spectra generated from a specified earthquake such as those in Figure G.2 cannot be used for design, because the response of a structure due to this earthquake will be different from that due to another earthquake with similar magnitude, and the local peaks and valleys are specific to the earthquake record and may not represent general peak responses. For this reason, in practical applications, the response spectra from many earthquake records with common characteristics are averaged to develop the design spectrum with a smooth curve or several straight lines. Since the peak ground acceleration (PGA), velocity, and displacement for various earthquake records differ, the computed response spectra from these records cannot be averaged

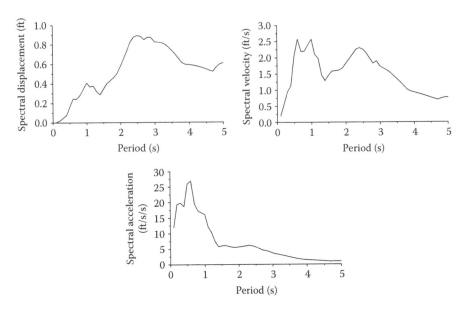

FIGURE G.2 Elastic response spectra (5% damping).

on an absolute basis. Therefore, various procedures are used to normalize response spectra before the averaging process is carried out. The most common normalization procedure is to normalize each spectrum to the corresponding peak ground motion. For example, a normalized design spectrum for a certain soil type in the AASHTO bridge design specifications (1992–2008) is shown in Figure 1.4. This spectrum is the average of many real earthquake spectra, developed by dividing their spectral ordinates by the corresponding PGAs. For bridge design, the design acceleration spectrum value, R_a in Equation G.14a, is equal to gC_{sm} in which g is the gravitational acceleration and C_{sm} is the elastic seismic coefficient are shown in Equation 1.8.

G.2 INELASTIC RESPONSE SPECTRUM

Structures subjected to severe earthquake ground motion experience deformation beyond the elastic range. The inelastic deformations depend on the hysteretic behavior (i.e., load–deformation characteristics) of the structures.

Similar to the elastic response spectrum, the inelastic response spectrum can be generated by the inelastic time history analysis described in Appendix F. An inelastic displacement response spectrum represents the maximum relative displacements of sdof oscillators with different periods (or frequencies) of vibration corresponding to a specified ductility level, subjected to ground motion, $\ddot{x}_G(t)$. To illustrate how to generate the inelastic response spectrum, a simple elastoplastic hysteresis model shown in Figure G.3 is used here.

In the figure, x_y and x_m represent the yield displacement and the maximum displacement of an sdof structure subjected to ground motion, $\ddot{x}_G(t)$. The ductility is expressed as follows:

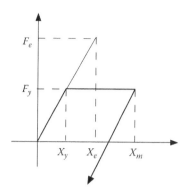

FIGURE G.3 Elastoplastic model for sdof system.

$$\mu = \left| \frac{x_m}{x_y} \right| \tag{G.15}$$

The step-by-step procedure below illustrates how to generate an inelastic spectrum with a target ductility level of μ_{target} and an elastic damping ratio ρ.

Step 1: Define a range of structural periods $(T_1 < T \le T_n)$, the incremental period ΔT, and an assumed structural mass m.

Step 2: Select the ith period T_i, $T_i = T_{i-1} + \Delta T$; $i = 1, n$; $T_0 = 0\,\text{s}$.

Step 3: Perform an elastic time history analysis for T_i to find the corresponding elastic strength demand F_e.

Step 4: Assume the structural yield strength, F_y, as a fraction of F_e (i.e., $F_y = ratio \times F_e$). The *ratio* is increased from 0.001 to 1. For each *ratio* increment, calculate the yield displacement $x_y = F_y/k$, in which $k = \omega_i^2 m$ and $\omega_i = 2\pi/T_i$, per Equations G.6 and G.8, respectively.

Step 5: Using F_y, x_y, and k perform a nonlinear time history analysis using the damping coefficient of $c = 2\rho\sqrt{km}$ in the following equation:

$$m\ddot{x}(t) + c\dot{x}(t) + kx(t) = -m\ddot{x}_G(t) \tag{G.16}$$

At each incremental time step, check the structural demand force $F(t)$. If $F(t) < F_y$, the elastic condition controls; if $F(t) \ge F_y$, the post-yield condition controls. Adjust the unbalanced force so that $F(t) = F_y$. Since the elastoplastic model is used for this example, use $k = 0$ and $\omega_i = \sqrt{k/m} = 0$ for the next incremental time step. It is noted that, for a structure with the hysteresis model other than the elastoplastic model, the tangent post-yield stiffness, k, should be used in accordance with the hysteresis model, and $F(t)$ adjusted accordingly. From the time history analysis, find the maximum displacement, $x_m = \max|x(t)|$.

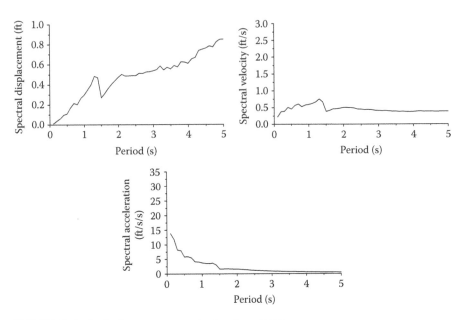

FIGURE G.4   Inelastic response spectra ($\mu=4$, $\rho=0.05$).

Step 6: Calculate ductility $\mu=|x_m/x_y|$. Compare μ with μ_{target}. If $|\mu-\mu_{target}|$ is more than a specified tolerance, increase values of *ratio* and go to step 4. If $|\mu-\mu_{target}|$ is less than a specified tolerance, $x_m=\max|x(t)|$ corresponding to μ_{target} is obtained. Similarly, the absolute maximum acceleration is $\ddot{x}_m=\max|\ddot{x}^t(t)|$, where $\ddot{x}^t(t)$ is calculated from Equation G.1. Per Equation G.13, for simplicity, the pseudoacceleration of $\ddot{x}_m=(T_i/2\pi)^2|x_y|$ can be used instead of using absolute maximum acceleration and pseudovelocity as given as $\dot{x}_m=T_i/2\pi|x_y|$.

Step 7: If $T_i=T_n$, the inelastic displacement spectrum is completed. If $T_i<T_n$, go to step 2 and select the next period T_i, and repeat steps 3–6.

Inelastic spectra for the 1940 El Centro earthquake N–S component generated using ductility $\mu=4$ are shown in Figure G.4. These spectra were generated by the Wilson-θ method using Equation G.3. The mass and the initial damping ratio are 20 kip-s²/in. and 0.05, respectively.

G.3   FORCE-REDUCTION *R*-FACTOR SPECTRUM

As described in Section 1.2.3, the force-reduction factor R is the ratio of the elastic strength demand to the inelastic strength demand of a structure subjected to the ground motion, $\ddot{x}_G(t)$. Therefore, the force-reduction R-factor spectrum represents the ratio of the elastic strength demand to the inelastic (or yield) strength demand, corresponding to a specified ductility demand, within a range of periods of vibration. By performing the elastic and inelastic response spectrum analyses described in the precious sections, both elastic and inelastic strength demands can be obtained (i.e., F_e and F_y). A typical $F_e(\mu=1)$ and $F_y(\mu=4)$ spectrum for the 1940 El Centro earthquake

N–S component is shown in Figure G.5, and the corresponding force-reduction R-factor spectrum (i.e., F_e/F_y) is shown in Figure G.6.

Figure G.6 clearly shows that the R-factor is a function of structural type and period. The R-factor is less than the ductility demand $\mu = 4$ in the short period range ($T < 0.5$ s), while between $0.5 < T < 5$, the R-factor varies significantly. This observation is only based on one earthquake acceleration record. However, similar to the design response spectrum, a design force-reduction R-factor spectrum should be generated based on a large number of ground acceleration time history records, soil conditions at site, initial damping, and the hysteretic behavior of structures.

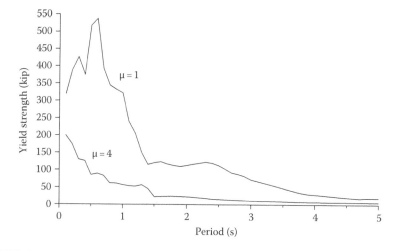

FIGURE G.5   Elastic strength demand ($\mu = 1$) and inelastic strength demand ($\mu = 4$) spectra for the 1940 El Centro earthquake N–S component (initial damping $\rho = 0.05$).

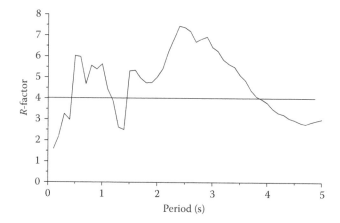

FIGURE G.6   Force-reduction R-factor spectrum for the 1940 El Centro earthquake N–S component (initial damping $\rho = 0.05$).

G.4 ELASTIC DISPLACEMENT SPECTRUM WITH EQUIVALENT VISCOUS DAMPING FOR DDBD

As described in Chapter 2, the calculation of the inelastic displacement demand is essential to the performance-based seismic design of highway bridges. In the direct displacement-based design (DDBD) procedures, the inelastic displacement demand and the corresponding equivalent viscous damping, ξ_{eq}, are calculated, so that the effective period, T_{eff}, of the substitute sdof system can be obtained from the elastic design displacement spectrum. A typical elastic design displacement spectrum is shown in Figure 2.2.

The concept of equivalent viscous damping was first proposed by Jacobsen (1930). He developed the equivalent viscous damping coefficient, c_{eq}, of a linear system, which can be used to estimate the maximum nonlinear displacement of a nonlinear system with high power of velocity of motion, subjected to harmonic motion (i.e., replace the equation of motion of a nonlinear system, $m\ddot{x} + c_n(\dot{x})^n + kx = Q\sin\omega t$, with the linear system, $m\ddot{x} + c_{eq}\dot{x} + kx = Q\sin\omega t$). By equating the work dissipated by the nonlinear system to the work dissipated by the linear system, the equivalent viscous damping coefficient, c_{eq}, can be obtained.

Jacobsen's concept was adopted by many researchers to obtain the equivalent viscous damping, ξ_{eq}, of an elastic substitute system to estimate the peak displacement response of an inelastic hysteretic system. For example, equating the energy dissipated in one cycle by an sdof bilinear hysteretic system (see Figure G.7) under the steady-state harmonic motion between the positive and negative maximum displacements to the viscous damping energy dissipated by the associated elastic substitute system undergoing the same displacements, the equivalent viscous damping, ξ_{eq}, can be expressed as (ATC-40, 1996)

$$\xi_{eq} = \frac{2(\mu-1)(1-r)}{\pi\mu(1+r\mu-r)} \tag{G.17}$$

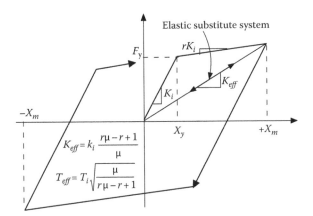

FIGURE G.7 Idealized equivalent viscous damping (bilinear sdof system).

in which r is the postyield stiffness ratio as shown in Figure G.7. It can be seen that the equivalent viscous damping, ξ_{eq}, is strongly dependent on the ductility demand, $\mu = x_m/x_y$, of the inelastic system. Since the actual earthquake motion is not a steady-state harmonic motion, the magnitudes of many small hysteresis loops due to earthquake are considerately lower than the maximum displacement, x_m. Many modified $\xi_{eq}-\mu$ models have been developed (Gulkan and Sozen, 1974; Iwan and Gates, 1979; ATC-40, 1996; Kwan and Billington, 2003; Dwairi et al., 2007; Priestley et al., 2007), based on a large number of ground acceleration time history records. From the $\xi_{eq}-\mu$ relationship, an inelastic displacement spectrum such as the one in Figure G.4 can be replaced by an elastic displacement spectrum with equivalent ξ_{eq}.

To demonstrate that the elastic displacement spectrum corresponding to ξ_{eq} can be used to estimate the maximum displacement of an sdof inelastic system, the following three $\xi_{eq}-\mu$ models are used to generate elastic displacement spectra representing the maximum inelastic displacement of an elastoplastic hysteretic system using the 1940 El Centro earthquake N–S component. These spectra will then be compared with the inelastic displacement spectrum in Figure G.4.

1. Model No. 1 (Dwairi et al., 2007):

$$\xi_{eq} = \xi_{elas} + \xi_{hyst} = 5 + C\left(\frac{\mu - 1}{\pi\mu}\right)\% \qquad (G.18)$$

in which $C = 85 + 60(1 - T_{eff})$ if $T_{eff} < 1$ s, and $C = 85$ if $T_{eff} \geq 1$ s. The equivalent viscous damping is the sum of elastic damping, ξ_{elas}, and hysteretic damping, ξ_{hyst}. $\xi_{elas} = 5\%$ is typically used here for concrete structure. T_{eff} is the effective period of the substitute elastic sdof system.

2. Model No. 2 (Priestley et al., 2007):

$$\xi_{eq} = \xi_{elas} + \xi_{hyst} = 5 + 67\left(\frac{\mu - 1}{\pi\mu}\right)\% \qquad (G.19)$$

3. Model No. 3 (ATC-40, 1996):

$$\xi_{eq} = \xi_{elas} + \kappa\xi_{hyst} = 5 + \kappa\left(200\left(\frac{\mu - 1}{\pi\mu}\right)\right)\% \qquad (G.20)$$

$\kappa = 1.13 - 0.51(\mu - 1/\mu)$, if $\xi_{hyst} > 16.25$; $\kappa = 1.0$, if $\xi_{hyst} \leq 16.25$.

Substituting $\mu = 4$ into Equations G.18, $\xi_{eq} = 25.3\%$ for $T_{eff} \geq 1$ s, and $38.2\% \leq \xi_{eq} \leq 26.7\%$ for 0.1 s $\leq T_{eff} \leq 0.9$ s. Similarly, substituting $\mu = 4$ into (G.19) and (G.20) leads to $\xi_{eq} = 21\%$ and 40.7%, respectively. The elastic displacement spectra corresponding to the above calculated ξ_{eq} s, generated from the elastic time history analysis (see Appendix F for elastic time history analysis), are shown in Figure G.8.

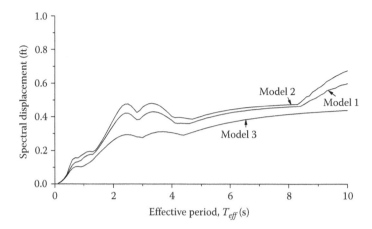

FIGURE G.8 Elastic displacement spectra with equivalent viscous dampings.

Since these spectra represent the maximum responses of a substitute elastic sdof system, the period on the abscissa of Figure G.8 is the effective period, T_{eff}, not the initial elastic period, T_i. In order to compare Figure G.8 with the inelastic displacement spectrum in Figure G.4, which has the elastic period, T_i, on the abscissa, the T_{eff} in Figure G.8 needs to be shifted to T_i. From Figure G.7 for the elastoplastic system (i.e., $r=0$)

$$T_i = \frac{T_{eff}}{\sqrt{\mu = 4}} = \frac{T_{eff}}{2} \tag{G.21}$$

Figure G.9 shows the comparison between the inelastic displacement spectrum and the elastic displacement spectra with equivalent viscous dampings calculated from the $\xi_{eq}-\mu$ models after shifting the period per Equation G.21. It can be seen that

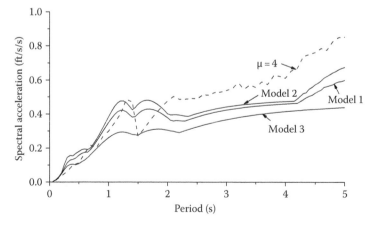

FIGURE G.9 Elastic and inelastic displacement spectra.

Models 1 and 2 provide a good estimate of inelastic response when T_i is less than 1.5 s. For $T_i > 1.5$ s, the estimates of all of the models are less than the inelastic displacement values from the inelastic displacement spectrum. As shown in the figure, Model No. 3 significantly underestimates the inelastic response due to overestimating the equivalent viscous damping. Note that models 1 and 2 were developed based on results from extensive time history analysis, using many ground motion records. Based on just one ground motion record, comparing the elastic displacement spectra with ξ_{eq} s with an inelastic displacement spectrum is not adequate. However, the main purpose of the above comparison is to demonstrate that the elastic displacement spectrum with appropriate ξ_{eq} can be used to estimate maximum inelastic displacement.

Instead of using elastic displacement design spectra with associated equivalent viscous damping, it is possible to develop inelastic displacement design spectral curves in terms of ductility demands for the DDBD. However, the disadvantage of using inelastic displacement design spectra is that the equivalent ductility demand of the substitute sdof system is not easy to obtain, due to the difficulty of estimating the equivalent yield displacement of the substitute sdof system, especially for bridges with non-regular geometry or nonuniform distribution of weight and stiffness.

Appendix H: Response Spectrum Analysis of Multiple-dof System

As described in Chapter 1, response spectrum analysis has been accepted by AASHTO for the seismic design of highway bridges since 1992. Most bridge engineers are familiar with this method. This appendix is mainly for engineers or students who are not knowledgeable in response spectrum analysis. To understand response spectrum analysis, fundamental structural dynamics concepts, such as free vibration, force vibration, and natural frequencies and mode shapes, are first introduced, followed by an introduction to response spectrum analysis.

H.1 DAMPED FREE VIBRATION SYSTEM

For a single-degree-of-freedom (sdof) system in free vibration, the equation of motion is

$$m\ddot{x}(t) + c\dot{x}(t) + kx(t) = 0 \tag{H.1}$$

where
    m, c, and k are mass, damping coefficient, and stiffness of the structure, respectively
    $\ddot{x}$, $\dot{x}$, and x are the relative acceleration, velocity, and displacement, respectively

Dividing Equation H.1 by the mass m produces

$$\ddot{x}(t) + 2\rho\omega\dot{x}(t) + \omega^2 x(t) = 0 \tag{H.2}$$

where
    ρ is the damping factor
    ω is the angular frequency

The general solution of Equation H.2 is

$$x(t) = e^{-\rho\omega t}(C_1 \cos \omega t + C_2 \sin \omega t) \tag{H.3}$$

The constants C_1 and C_2 can be determined from the initial conditions at $t=0$ s. If the initial displacement and velocity at $t=0$ s are x_0 and $\dot{x}_0$, substituting x_0 and $\dot{x}_0$ into Equation H.3 and its first derivative with respect to time, respectively, we have

$$C_1 = x_0 \tag{H.4a}$$

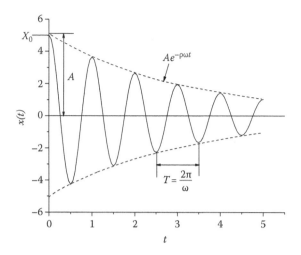

FIGURE H.1 Damped free vibration.

$$C_2 = \frac{\dot{x}_0 + \rho\omega x_0}{\omega} \tag{H.4b}$$

Then, Equation H.3 becomes

$$x(t) = e^{-\rho\omega t}\left(x_0 \cos\omega t + \frac{\dot{x}_0 + \rho\omega x_0}{\omega}\sin\omega t\right) \tag{H.5}$$

or

$$x(t) = Ae^{-\rho\omega t}\cos(\omega t - \alpha_d) \tag{H.6}$$

in which $A = \sqrt{C_1^2 + C_2^2}$ and $\alpha_d = \tan^{-1} C_2/C_1$. In Equation H.6, the factor $e^{-\rho\omega t}$ decreases with time, and so damping effects will gradually reduce the magnitude of vibration as demonstrated in Figure H.1. The damped vibration in Figure H.1 is based on $x_0 = 5$, $\dot{x}_0 = 0$, $\rho = 0.05$, and structural period $T = 1$ s.

H.2 DAMPED VIBRATION WITH DYNAMIC FORCING FUNCTION

For the sdof system shown in Figure G.1, subjected to a dynamic forcing function, $P(t)$, the equation of motion is

$$m\ddot{x}(t) + c\dot{x}(t) + kx(t) = P(t) \tag{H.7}$$

The general forcing function $P(t)$ is shown in Figure H.2 and consists of a series of impulse forces, $P\,dt'$, where t' varies from 0 to t s, and t is the structural response to

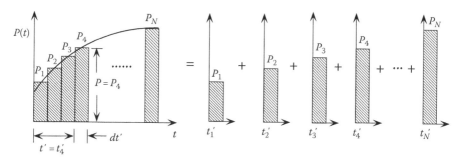

FIGURE H.2   General forcing function.

be calculated. Since the impulse force is equal to the change in momentum, the following equation is obtained for the velocity change due to each impulse force:

$$m\,d\dot{x} = P\,dt' \quad \text{or} \quad d\dot{x} = \frac{P\,dt'}{m} \tag{H.8}$$

Following the application of an impulse load at time t', the sdof structure is in free vibration. The displacement increment given in Equation H.5 becomes

$$dx = e^{-\rho\omega(t-t')}\left(\frac{P\,dt'}{m\omega}\sin\omega(t-t') \right) \tag{H.9}$$

Because all the impulse forces between $t'=0$ and $t'=t$ have such an effect, the total structural response due to all the impulse forces (i.e., $P(t)$ as a whole) can be obtained by integrating Equation H.9 as follows:

$$x(t) = \frac{1}{\omega}\int_0^t \frac{P(t')}{m}e^{-\rho\omega(t-t')}\sin\omega(t-t')dt' \tag{H.10}$$

Equation H.10 is called Duhamel's integral, which can be used to calculate the structural response due to any forcing function $P(t)$. For an sdof structure subjected to an earthquake excitation, the earthquake induced force is $m\ddot{x}_G$. Thus, Equation H.10 is expressed as follows:

$$x(t) = \frac{1}{\omega}\int_0^t \ddot{x}_G(t')e^{-\rho\omega(t-t')}\sin\omega(t-t')dt' \tag{H.11}$$

Since ground acceleration, $\ddot{x}_G$, is not a simple periodic function, it is very difficult to integrate. Normally, instead of using Duhamel's integral of Equation H.11, the structural response is calculated by the step-by-step numerical integration methods described in Appendix F.

H.3   STRUCTURAL NATURAL FREQUENCIES AND MODE SHAPES

For a structure with m total free dofs, the dimensions of the mass and stiffness matrices, $[M]_{m \times m}$ and $[K]_{m \times m}$, are $m \times m$. The characteristics of the structure's vibration are governed by the natural frequencies and corresponding mode shapes of the structure. The total number of governing modes is n and usually n is less than or equal to m. The ith natural frequency, ω_i, and the mode shape, $\{\Phi\}_i$, can be obtained by the following eigenvalue equation:

$$([K] - \omega_i^2[M])\{\Phi\}_i = \{0\}, \quad i = 1, 2, \ldots, n \tag{H.12}$$

To find the nontrivial solution of Equation H.12, set the determinant of $[K] - \omega_i^2[M]$ equal to zero:

$$\left| [K] - \omega_i^2[M] \right| = 0 \tag{H.13}$$

The expansion of Equation H.13 is a polynomial with order n. The n roots of the polynomial equation are eigenvalues of the natural frequencies ω_1, ω_2, ..., and ω_n. Substituting each of the natural frequency into Equation H.12 leads to n eigenvectors defining mode shapes $\{\Phi\}_1$, $\{\Phi\}_2$, ..., and $\{\Phi\}_n$. Normalizing each of the mode shapes $\{\Phi\}_i$, so that the largest positive or negative value of the term in the mode shape vector is equal to one, yields the normalized mode shapes called normal modes, denoted as $\{X\}_1$, $\{X\}_2$, ..., and $\{X\}_n$ here.

H.3.1   Orthogonality of Normal Modes

Let $\{X\}_u$ and $\{X\}_v$ be two normal modes corresponding to the natural frequencies of ω_u and ω_v, respectively. From Equation H.12

$$([K] - \omega_u^2[M])\{X\}_u = \{0\}$$

or

$$\omega_u^2[M]\{X\}_u = [K]\{X\}_u \tag{H.14}$$

Similarly

$$\omega_v^2[M]\{X\}_v = [K]\{X\}_v \tag{H.15}$$

Postmultiplying the transpose of Equation H.14 by $\{X\}_v$ yields

$$\omega_u^2\{X\}_u^T[M]^T\{X\}_v = \{X\}_u^T[K]^T\{X\}_v \tag{H.16}$$

Premultiplying Equation H.15 by $\{X\}_u^T$ yields

$$\omega_v^2\{X\}_u^T[M]\{X\}_v = \{X\}_u^T[K]\{X\}_v \tag{H.17}$$

Since $[M]$ and $[K]$ are symmetric, $[M]^T=[M]$ and $[K]^T=[K]$. Subtracting Equation H.17 from Equation H.16 gives

$$(\omega_u^2 - \omega_v^2)\{X\}_u^T[M]\{X\}_v = 0 \tag{H.18}$$

Since $\omega_u \neq \omega_v$, Equation H.18 is satisfied only if

$$\{X\}_u^T[M]\{X\}_v = 0 \quad \text{for } u \neq v \tag{H.19}$$

which is the orthogonality condition for the uth and vth normal modes with respect to the mass matrix $[M]$. From Equations H.17 and H.19

$$\{X\}_u^T[K]\{X\}_v = 0 \quad \text{for } u \neq v \tag{H.20}$$

which is the orthogonality condition with respect to stiffness matrix $[K]$. Also, from Equation F.46, since the proportional damping matrix $[C]$ is linearly related to the mass and stiffness matrices, the orthogonality condition also applies to the damping matrix $[C]$:

$$\{X\}_u^T[C]\{X\}_v = 0 \quad \text{for } u \neq v \tag{H.21}$$

H.4 MULTIPLE-MODE RESPONSE SPECTRUM ANALYSIS

The equation of motion of a multiple-dof system subjected to an earthquake excitation is shown in Equation F.1 and reproduced here:

$$[M]\{\ddot{x}(t)\} + [C]\{\dot{x}(t)\} + [K]\{x(t)\} = \{F(t)\} = -[M]\{I_n\}\ddot{x}_G \tag{H.22}$$

Let

$$\{x(t)\} = [X]\{x'(t)\} = \sum_{n=1}^{N}\{X\}_n x_n'(t), \quad N = \text{number of modes considered} \tag{H.23}$$

where
    $[X]$ is the normal mode matrix
    $\{x'(t)\}$ is the generalized response vector
    $\{X\}_n$ is the nth normal mode vector
    $x_n'(t)$ is the generalized model response corresponding to the nth mode

Substituting Equation H.23 into Equation H.22 yields

$$[M][X]\{\ddot{x}'(t)\} + [C][X]\{\dot{x}'(t)\} + [K][X]\{x'(t)\} = \{F(t)\} \qquad \text{(H.24)}$$

Multiplying the above equation by the transpose of any normal mode vector, $\{X\}_n$, corresponding to the nth mode, gives

$$\{X\}_n^T[M][X]\{\ddot{x}'(t)\} + \{X\}_n^T[C][X]\{\dot{x}'(t)\} + \{X\}_n^T[K][X]\{x'(t)\} = \{X\}_n^T\{F(t)\} \quad \text{(H.25)}$$

Using the orthogonality conditions of normal modes in Equations H.19 through H.21, Equation H.25 can be decoupled into the following generalized form:

$$M_n\ddot{x}_n'(t) + C_n\dot{x}_n'(t) + K_n x'(t) = P_n(t), \quad n = 1 \text{ to } N \qquad \text{(H.26)}$$

in which the generalized properties for the nth mode are given as follows:

$$M_n = \{X\}_n^T[M]\{X\}_n = \text{generalized mass} \qquad \text{(H.27)}$$

$$C_n = \{X\}_n^T[C]\{X\}_n = 2\rho_n\omega_n M_n = \text{generalized damping} \qquad \text{(H.28)}$$

$$K_n = \{X\}_n^T[K]\{X\}_n = \omega_n^2 M_n = \text{generalized stiffness} \qquad \text{(H.29)}$$

$$P_n(t) = \{X\}_n^T\{F(t)\} = \text{generalized loading} \qquad \text{(H.30)}$$

From the above equations, Equation H.26 can be further simplified to

$$\ddot{x}_n'(t) + 2\rho_n\omega_n\dot{x}_n'(t) + \omega_n^2 x_n'(t) = \frac{P_n(t)}{M_n} = \frac{\{X\}_n^T[M]\{I_n\}}{M_n}\ddot{x}_G(t), \quad n = 1 \text{ to } N \qquad \text{(H.31)}$$

As described previously, the response of a damped vibration can be obtained by Equation H.11. Similar to Equation H.11, the response of the nth mode in Equation H.31 at time t can be expressed as follows:

$$x_n'(t) = \frac{\{X\}_n^T[M]\{I_n\}}{M_n\omega_n} \int_0^t \ddot{x}_G(t')e^{-\rho\omega(t-t')}\sin\omega_n(t-t')dt' = \frac{\{X\}_n^T[M]\{I_n\}}{M_n}S_n(t) \quad \text{(H.32)}$$

in which

$$S_n(t) = \frac{1}{\omega_n}\int_0^t \ddot{x}_G(t')e^{-\rho\omega(t-t')}\sin\omega_n(t-t')dt' \qquad \text{(H.33)}$$

$S_n(t)$ is same as Duhamel's integral shown in Equation H.11, which represents the displacement response of an sdof system at time t, with natural frequency ω_n subjected to ground motion $\ddot{x}_G$. As described in Appendix G, the displacement-response spectrum, $R_d(\omega_n, \rho_n)$, represents the maximum relative displacements of sdof oscillators with different periods (or frequencies) of vibration, subjected to ground motion, $\ddot{x}_G(t)$. Hence,

$$R_d(\omega_n, \rho_n) = \max |S_n(t)| \tag{H.34}$$

From Equations H.32 and H.34, the maximum displacement of the generalized model response corresponding to the nth mode is

$$x'_n = x'_n(t) \, |_{\max} = \frac{\{X\}_n^T [M] \{I_n\}}{M_n} R_d(\omega_n, \rho_n) \tag{H.35}$$

Since the pseudo-acceleration spectrum $R_a(\omega_n, \rho_n)$ is equal to $\omega_n^2 R_d(\omega_n, \rho_n)$ (see Equation G.14), Equation H.35 can also be expressed as follows:

$$x'_n = \frac{\{X\}_n^T [M] \{I_n\}}{M_n \omega_n^2} R_a(\omega_n, \rho_n) = \gamma_n R_a(\omega_n, \rho_n) \tag{H.36}$$

The maximum response of a multiple-dof system corresponding to the nth mode, $\{x\}_n$, can be expressed as follows:

$$\{x\}_n = \{X\}_n x'_n = \{X\}_n \gamma_n R_a(\omega_n, \rho_n) \tag{H.37a}$$

where
γ_n is called the participation factor for the nth mode, and

$$\{x\}_n = \{x_n^1, x_n^2, \ldots, x_n^k, \ldots, x_n^m\} \tag{H.37b}$$

The superscript m represents the mth dof of the structural system. Using Equation H.37a, the maximum modal response $\{x\}_n$ is obtained for each mode. The next question to arise is how should these model maxima be combined for the best estimate of maximum total response? The response expression in Equation H.23 provides accurate results only as long as $\{x(t)\}$ is evaluated concurrently with time. However, in response spectrum analysis, time is removed from the equation. The maximum response values for individual modes cannot possibly occur at the same time. Therefore, a combination of modal maximum such as

$$\{x\} = \{x^1, x^2, \ldots, x^k, \ldots, x^m\}^T = \sum_{n=1}^{N} \{x\}_n \tag{H.38}$$

is too conservative for design applications. To resolve this, two modal combination methods are typically accepted, as giving a more reasonable estimate of maximum structural response. These two methods are described as follows.

H.4.1 SRSS Model Combination Method

The square-root-of-the-sum-of-the-squares (SRSS) method can be expressed as follows:

$$x^k = \sqrt{\sum_{n=1}^{N} (x_n^k)^2}, \quad N = \text{number of modes considered} \qquad (\text{H.39})$$

in which x^k is the maximum response of the dof k. SRSS provides good approximation of the response for frequencies distinctly separated from neighboring modes.

H.4.2 Complete-Quadratic-Combination Model Combination Method

In general, the complete-quadratic-combination (CQC) method (Wilso et al., 1981) offers a significant improvement in estimating structural response. The CQC combination is expressed as follows:

$$x^k = \sqrt{\sum_{i=1}^{N} \sum_{j=1}^{N} x_i^k \alpha_{ij} x_j^k}, \quad i, j = 1, N \qquad (\text{H.40})$$

where
   α_{ij} is the cross-correlation coefficient, indicating the cross correlation between modes i and j
   α_{ij} is a function of frequency and damping ratio of a structure and can be expressed as follows:

$$\alpha_{ij} = \frac{8\rho^2(1+q)q^{3/2}}{(1-q)^2 + 4\rho^2 q(1+q)^2} \qquad (\text{H.41})$$

with

$$q = \frac{\omega_j}{\omega_i} \qquad (\text{H.42})$$

H.4.3 Combination of Structural Responses due to Multiple-Component Ground Motions

In the design and analysis of structures subjected to seismic loading, multicomponent ground motions should be considered. The AASHTO bridge design specifications described in Chapter 1 require the consideration of two horizontal orthogonal seismic components (X and Y components). AASHTO allows using the same design-response spectrum for both orthogonal seismic components. The structural response obtained from Equation H.39 or H.40 due to seismic component X is then combined with that due to seismic component Y, by the following 30% rule:

$$(x^k)_{X\text{-}component} + 0.3(x^k)_{Y\text{-}component} \tag{H.43}$$

and

$$(x^k)_{Y\text{-}component} + 0.3(x^k)_{X\text{-}component} \tag{H.44}$$

with the larger of the two used for design. In some cases, the 30% rule underestimates the structural response. Past research indicated that when using the identical design-response spectrum for each of the seismic component, the SRSS combination rule in Equation H.45 provides more realistic results (Menun and Kiureghian, 1998) than the 30% rule:

$$\sqrt{(x^k)^2{}_{X\text{-}component} + (x^k)^2{}_{Y\text{-}component}} \tag{H.45}$$

If the design-response spectra for two seismic components are not identical and have different magnitudes, the CQC3 rule proposed by Menun and Kiureghian (1998) is recommended. The CQC3 provides a more general procedure than either the 30% rule or the SRSS rule for multicomponent ground motion combinations. It not only estimates maximum structural response due to the two seismic horizontal components but the seismic vertical component as well. It also accounts for the correlation between individual seismic components.

Appendix I: Polynomial Curve Fitting

Given m data points, $P_1(x_1, y_1)$, $P_2(x_2, y_2), \ldots, P_m(x_m, y_m)$, a defined guess function shown in Figure I.1 is expressed as follows:

$$g(x) = \gamma_1 \phi_1(x) + \gamma_2 \phi_2(x) + \cdots + \gamma_n \phi_n(x) = \sum_{j=1}^{n} \gamma_j \phi_j(x) \tag{I.1}$$

where
$\phi_j(x)$ is the jth specified function
γ_j is the jth parameter to be determined

If we define $\phi_j(x) = x^{j-1}$, Equation I.1 becomes a $(n-1)$-order polynomial, given by

$$g(x) = \gamma_1 + \gamma_2 x + \cdots + \gamma_n x^{n-1} = \sum_{j=1}^{n} \gamma_j x^{j-1} \tag{I.2}$$

In order to find the guess function $g(x)$, which lies near the data points P_1, P_2, ..., P_m, define a least square function $E(g)$ as follows:

$$E(g) = \sum_{k=1}^{m} \left[g(x_k) - y_k \right]^2 = \sum_{k=1}^{m} \left[\gamma_1 \phi_1(x_k) + \gamma_2 \phi_2(x_k) + \cdots + \gamma_n \phi_n(x_k) - y_k \right]^2 \tag{I.3}$$

Minimizing Equation I.3 to best fit the points P_1, P_2, ..., P_m

$$0 = \frac{\partial E(g)}{\partial \gamma_i} = \sum_{k=1}^{m} 2 \left[g(x_k) - y_k \right] \frac{\partial}{\partial \gamma_i} \left[g(x_k) - y_k \right], \quad i = 1, \ldots, n \tag{I.4}$$

Since $\phi_1(x_k)$, $\phi_2(x_k), \ldots, \phi_n(x_k)$ and y_k are constants, and $\partial \gamma_j / \partial \gamma_i = 0$ for $j \neq i$, from Equation I.3, where $\partial [g(k_k) - y_k] / \partial \gamma_i = \phi_i(x_k)$, Equation I.4 becomes

$$\sum_{k=1}^{m} \left[\sum_{j=1}^{n} \gamma_j \phi_j(x_k) - y_k \right] \phi_i(x_k) = 0, \quad i = 1, \ldots, n \tag{I.5}$$

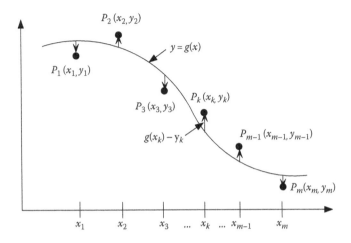

FIGURE I.1 Fitting $g(x_k)$ to $P_1, P_2, ..., P_m$.

or

$$\sum_{k=1}^{m}\sum_{j=1}^{n}\gamma_j\phi_j(x_k)\,\phi_i(x_k) = \sum_{k=1}^{m}y_k\phi_i(x_k), \quad i = 1,...,n \tag{I.6}$$

Interchanging the order of j and k summations gives

$$\sum_{j=1}^{n}\gamma_j\left(\sum_{k=1}^{m}\phi_i(x_k)\,\phi_j(x_k)\right) = \sum_{k=1}^{m}y_k\phi_i(x_k), \quad i = 1,...,n \tag{I.7}$$

Equation I.7 can be expressed as (Maron, 1982)

$$\begin{bmatrix} \sum\phi_1(x_k)\phi_1(x_k) & \sum\phi_1(x_k)\phi_2(x_k) & \cdots & \sum\phi_1(x_k)\phi_n(x_k) \\ \sum\phi_2(x_k)\phi_1(x_k) & \sum\phi_2(x_k)\phi_2(x_k) & \cdots & \sum\phi_2(x_k)\phi_n(x_k) \\ \vdots & \vdots & \cdots & \vdots \\ \sum\phi_n(x_k)\phi_1(x_k) & \sum\phi_n(x_k)\phi_2(x_k) & \cdots & \sum\phi_n(x_k)\phi_n(x_k) \end{bmatrix}\begin{Bmatrix} \gamma_1 \\ \gamma_2 \\ \vdots \\ \gamma_n \end{Bmatrix} = \begin{Bmatrix} \sum\phi_1(x_k)y_k \\ \sum\phi_2(x_k)y_k \\ \vdots \\ \sum\phi_n(x_k)y_k \end{Bmatrix} \tag{I.8}$$

In which $\sum$ denotes $\sum_{k=1}^{m}$ (i.e., summation of all the data points). Substituting Equation I.2 into Equation I.8 leads to

$$
\begin{bmatrix}
m & \sum x_k & \sum x_k^2 & \cdots & \sum x_k^{n-1} \\
\sum x_k & \sum x_k^2 & \sum x_k^3 & \cdots & \sum x_k^n \\
\vdots & \vdots & \vdots & \cdots & \vdots \\
\sum x_k^{n-1} & \sum x_k^n & \sum x_k^{n+1} & \cdots & \sum x_k^{2n-2}
\end{bmatrix}
\begin{bmatrix}
\gamma_1 \\ \gamma_2 \\ \vdots \\ \gamma_n
\end{bmatrix}
=
\left\{
\begin{matrix}
\sum y_k \\ \sum x_k y_k \\ \vdots \\ \sum x_k^{n-1} y_k
\end{matrix}
\right\}
\qquad (I.9)
$$

For a third-order polynomial (i.e., $n=4$), Equation I.9 becomes

$$
\begin{bmatrix}
m & \sum x_k & \sum x_k^2 & \sum x_k^3 \\
\sum x_k & \sum x_k^2 & \sum x_k^3 & \sum x_k^4 \\
\sum x_k^2 & \sum x_k^3 & \sum x_k^4 & \sum x_k^5 \\
\sum x_k^3 & \sum x_k^4 & \sum x_k^5 & \sum x_k^6
\end{bmatrix}
\begin{bmatrix}
\gamma_1 \\ \gamma_2 \\ \gamma_3 \\ \gamma_4
\end{bmatrix}
=
\left\{
\begin{matrix}
\sum y_k \\ \sum x_k y_k \\ \sum x_k^2 y_k \\ \sum x_k^3 y_k
\end{matrix}
\right\}
\qquad (I.10)
$$

or

$$
[A]_{n\times n}\{\gamma\}_{n\times 1} = \{B\}_{n\times 1} \qquad (I.11)
$$

Therefore, $\{\gamma\}_{n\times 1}$ can be solved as follows:

$$
\{\gamma\}_{n\times 1} = [A]_{n\times n}^{-1}\{B\}_{n\times 1} \qquad (I.12)
$$

Example I.1

From the moment–curvature analysis of a column section, seven axial load–nominal moment ($P - M_n$) data points are shown in Table I.1. Find the coefficients γ_1, γ_2, γ_3, and γ_4 of the third-order polynomial interaction curve.

Solution

Let the x-axis represents axial load, P, and the y-axis represents the nominal moment, M_n. Since the total number of data points is 7, $m=7$. From Equations I.10 and I.11

$$
[A] =
\begin{bmatrix}
a_{11} & a_{12} & a_{13} & a_{14} \\
a_{21} & a_{22} & a_{23} & a_{24} \\
a_{31} & a_{32} & a_{33} & a_{34} \\
a_{41} & a_{42} & a_{43} & a_{44}
\end{bmatrix}
$$

TABLE I.1
Data Points, $P(x,y)$

Axial Load, P (kip)	Moment, M_n (k-ft)
0	200
210	350
440	414
560	373
740	296
922	184.5
1143	0

$$a_{11} = 7$$

$$a_{12} = \sum_{k=1}^{7} x_k = 0 + 210 + 440 + 560 + 740 + 922 + 1143 = 4015$$

$$a_{13} = \sum_{k=1}^{7} x_k^2 = 0 + (210)^2 + (440)^2 + (560)^2 + (740)^2 + (922)^2 + (1143)^2 = 0.3255E + 7$$

$$a_{22} = a_{13} = a_{31}$$

$$a_{14} = \sum_{k=1}^{7} x_k^3 = 0 + (210)^3 + (440)^3 + (560)^3 + (740)^3 + (922)^3 + (1143)^3 = 0.2952E + 10$$

$$a_{24} = \sum_{k=1}^{7} x_k^4 = 0.2867E + 13$$

$$a_{33} = a_{24} = a_{42}$$

$$a_{34} = \sum_{k=1}^{7} x_k^5 = 0.2911E + 16$$

$$a_{44} = \sum_{k=1}^{7} x_k^6 = 0.3047E + 19$$

$$[B] = \begin{Bmatrix} \sum y_k \\ \sum x_k y_k \\ \sum x_k^2 y_k \\ \sum x_k^3 y_k \end{Bmatrix} = \begin{Bmatrix} b_1 \\ b_2 \\ b_3 \\ b_4 \end{Bmatrix}$$

$$b_1 = \sum_{k=1}^{7} y_k = 200 + 350 + 414 + 373 + 296 + 184.5 + 0 = 1817.5$$

$$b_2 = \sum_{k=1}^{7} x_k y_k = 0 \times 200 + 210 \times 350 + 440 \times 414 + 560 \times 373 + 740$$
$$\times 296 + 922 \times 184.5 + 1143 \times 0 = 853,689$$

$$b_3 = \sum_{k=1}^{7} x_k^2 y_k = (210)^2 \times 350 + (440)^2 \times 414 + (560)^2 \times 373$$
$$+ (740)^2 \times 296 + (922)^2 \times 184.5 = 0.5315E + 9$$

$$b_4 = \sum_{k=1}^{7} x_k^3 y_k = (210)^3 \times 350 + (440)^3 \times 414 + (560)^3 \times 373 + (740)^3$$
$$\times 296 + (922)^3 \times 184.5 = 0.3686E + 12$$

From Equation I.12

$$\{\gamma\} = [A]^{-1}\{B\} = \begin{Bmatrix} 198.154 \\ 1.0685 \\ -0.1568E - 2 \\ 0.4215E - 6 \end{Bmatrix}$$

Therefore, the polynomial curve is

$$M_n = 198.154 + 1.0685P - 0.001568P^2 + (0.4215 \times 10^{-6})P^3$$

The axial load–nominal moment interaction curve generated by INSTRUCT is shown in Figure I.2.

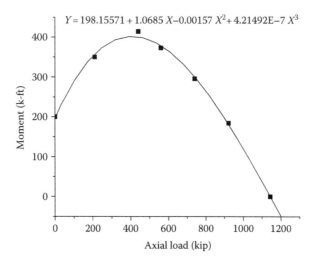

$Y = 198.15571 + 1.0685\,X - 0.00157\,X^2 + 4.21492\text{E}{-}7\,X^3$

FIGURE I.2 Axial load–nominal moment interaction curve fit.

Appendix J: Plate Element Stiffness Matrix

Plate elements can be used to model bridge collision walls, bridge decks, or building floors. The closed form of the plate element stiffness matrix can be derived from the standard finite-element procedures of using assumed shape functions and the principle of virtual work (Weaver and Johnston, 1987). The shape functions relate generic displacements to nodal displacements. The principle of virtual work states that the virtual work of external actions on an element is equal to the virtual strain energy of internal stresses of the element. This appendix provides the plate element closed-form stiffness matrix derived from the above-mentioned finite-element procedure, so that the user can directly use it for the structural analysis.

A plate element and its degrees of freedom are shown in Figure 5.9 and reproduced here as Figure J.1. The element has 20 degrees of freedom, which include consideration of in-plane and out-of-plane deformations.

The element force vector, displacement vector, and stiffness matrix are as follows:

$$\{F_e\} = \{F_1, F_2, F_3, \ldots, F_{20}\}^T \tag{J.1}$$

$$\{\delta_e\} = \{\delta_1, \delta_2, \delta_3, \ldots, \delta_{20}\}^T \tag{J.2}$$

$$\{F_e\} = [k_e]\{\delta_e\} \tag{J.3}$$

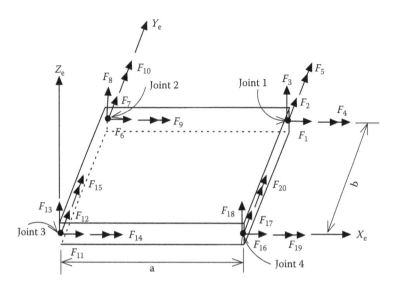

FIGURE J.1 Plate element.

$$
[k_e] =
$$

$$(J.4)$$

where

$$T_1 = \Psi\left[\frac{a}{9b} + \frac{b(1-\mu)}{45a}\right], \qquad T_2 = \Psi\left[\frac{b}{9a} + \frac{a(1-\mu)}{45b}\right]$$

$$T_3 = \Psi\left[\frac{a}{18b} - \frac{b(1-\mu)}{45a}\right], \qquad T_4 = \Psi\left[\frac{b}{18a} - \frac{a(1-\mu)}{45b}\right]$$

$$T_5 = \Psi\left[\frac{a}{18b} - \frac{b(1-\mu)}{180a}\right], \qquad T_6 = \Psi\left[\frac{b}{18a} - \frac{a(1-\mu)}{180b}\right]$$

$$T_7 = \Psi\left[\frac{a}{96b} + \frac{b(1-\mu)}{180a}\right], \qquad T_8 = \Psi\left[\frac{b}{36a} + \frac{a(1-\mu)}{180b}\right]$$

$$T_9 = \Psi\left[\frac{a}{6b^2} + \frac{(1+4\mu)}{60a}\right], \qquad T_{10} = \Psi\left[\frac{b}{6a^2} + \frac{(1+4\mu)}{60b}\right]$$

$$T_{11} = \Psi\left[\frac{a}{12b^2} - \frac{(1+4\mu)}{60a}\right], \qquad T_{12} = \Psi\left[\frac{b}{12a^2} - \frac{(1+4\mu)}{60b}\right]$$

$$T_{13} = \Psi\left[\frac{a}{12b^2} - \frac{(1-\mu)}{60a}\right], \qquad T_{14} = \Psi\left[\frac{b}{12a^2} - \frac{(1-\mu)}{60b}\right]$$

$$T_{15} = \Psi\left[\frac{a}{6b^2} + \frac{(1-\mu)}{60a}\right], \qquad T_{16} = \Psi\left[\frac{b}{6a^2} + \frac{(1-\mu)}{60b}\right]$$

$$T_{17} = \Psi\left[-\frac{b}{3a^3} + \frac{a}{6b^3} - \frac{(14-4\mu)}{60ab}\right], \qquad T_{18} = \Psi\left[-\frac{a}{3b^3} + \frac{b}{6a^3} - \frac{(14-4\mu)}{60ab}\right]$$

$$T_{19} = 0, \qquad T_{20} = 0$$

$$T_{21} = \Psi\left[\frac{\mu}{12}\right], \qquad T_{22} = 0$$

$$T_{23} = \Psi\left[\frac{b}{3a^3} + \frac{a}{3b^3} + \frac{(14-4\mu)}{60ab}\right], \qquad T_{24} = \Psi\left[-\frac{b}{6a^3} - \frac{a}{6b^3} + \frac{(14-4\mu)}{60ab}\right]$$

$$T_{25} = \Omega\left[\frac{4b}{a} + \frac{2a(1-\mu)}{b}\right], \qquad T_{26} = \Omega\left[-\frac{4b}{a} + \frac{a(1-\mu)}{b}\right]$$

$$T_{27} = \Omega\left[-\frac{2b}{a} - \frac{a(1-\mu)}{b}\right], \qquad T_{28} = \Omega\left[\frac{2b}{a} - \frac{2a(1-\mu)}{b}\right]$$

$$T_{29} = \Omega\left[\frac{3(1+\mu)}{2}\right], \quad T_{30} = \Omega\left[\frac{3(3\mu-1)}{2}\right]$$

$$T_{31} = \Omega\left[\frac{3(1-3\mu)}{2}\right], \quad T_{32} = \Omega\left[\frac{4a}{b} + \frac{2b(1-\mu)}{a}\right]$$

$$T_{33} = \Omega\left[\frac{2a}{b} - \frac{2b(1-\mu)}{a}\right], \quad T_{34} = \Omega\left[-\frac{2a}{b} - \frac{b(1-\mu)}{a}\right]$$

$$T_{35} = \Omega\left[-\frac{4a}{b} + \frac{b(1-\mu)}{a}\right], \quad \Psi = \frac{Et^3}{1-\mu^2}, \quad \text{and} \quad \Omega = \frac{Et}{12(1-\mu^2)}$$

where

E and μ are the elastic modulus and Poisson's ratio of the material, respectively

t is the thickness of the plate element.

References

American Association of State Highway Officials (AASHO), 1969, *Standard Specifications for Highway Bridges*, 10th edn., Washington, DC.

American Association of State Highway Transportation Officials (AASHTO), 1977, *Standard Specifications for Highway Bridges*, 12th edn., Washington, DC.

American Association of State Highway Transportation Officials (AASHTO), 1996, *Standard Specifications for Highway Bridges*, 16th edn., Washington, DC.

American Association of State Highway Transportation Officials (AASHTO), 2007, *LRFD Bridge Design Specifications*, 4th edn., With 2008 Interim, Washington, DC.

American Association of State Highway Transportation Officials (AASHTO), 2009, *Guide Specifications for LRFD Seismic Bridge Design*, Washington, DC.

Applied Technology Council, 1996, *Seismic Evaluation and Retrofit of Concrete Buildings*, Vol. I, California Seismic Safety Commission, Sacramento, CA, ATC 40.

ASCE, 1989, *Manual of Steel Construction*, 9th edn., Chicago, IL.

ATC, 1981, Seismic Design Guidelines for Highway Bridges, ATC-6 Report, Applied Technology Council, Redwood City, CA.

ATC-32, 1994, *Revised Caltrans Bridge Design Specifications*, Applied Technology Council, Redwood City, CA.

ATC and MCEER, 2003, Recommended LRFD Guidelines for the Seismic Design of Highway Bridges, Part I: Specifications and Part II: Commentary and Appendixes, ATC-49 Reports (NCHRP 12–49), Applied Technology Council, Redwood City, CA.

Black, R.G., Wenger, W.A.B., and Popov, E.P., 1980, Inelastic Buckling of Steel Structures under Cyclic Load Reversals, EERC, University of California, Berkeley, CA, Report No. UCB/EERC-80140.

Bracci, J.M., Kunnath, S.K., and Reinhorn, A.M., 1997, Seismic performance and retrofit evaluation of reinforced concrete structures, *Journal of Structural Engineering*, ASCE, 123(1), 3–10.

BSSC, 1998, NEHRP Recommended Provisions for Seismic Regulations for New Buildings and Other Structures: Part II Commentary, Report FEMA 302 and 303, Building Seismic Safety Council, Washington, DC.

Caltrans Seismic Design Criteria, 1999, Version 1.0, California Department of Transportation, Sacramento, CA.

Chen, W.F. and Atsuta, T., 1977, *Theory of Beam-Columns*, McGraw-Hill Book Co., New York, Vol. 2, pp. 504–527.

Chen, W.F. and Lui, E.M., 1991, *Stability Design of Steel Frames*, CRC Press, Inc., Boca Raton, FL.

Cheng, C.T., 1997, New paradigms for the seismic design and retrofit of bridges, PhD dissertation, Department of Civil Engineering, State University of New York, Buffalo, NY.

Cheng, F., 2000, *Matrix Analysis of Structural Dynamics*, Marcel Dekker, Inc., New York, pp. 534–539.

Cheng, F. and Ger, J., 1992, *Inelastic Response and Collapse Behavior of Steel Building Structures Subjected to Multi-Component Earthquake Excitations*, Civil Engineering Study Structural Series 92-30, University of Missouri-Rolla, Rolla, MO.

Cheng, F., Ger, J., Li, D., and Yang, J.S., 1996a, *INRESB-3D-SUPII User's Manual: General Purpose Program for Inelastic Analysis of RC and Steel Building Systems for 3D Static and Dynamic Loads and Seismic Excitations*, NSF Report, U.S. Department of Commerce, National Technical Information Service, Springfield, VA, NTIS No. PB97-123624.

Cheng, F., Ger, J., Li, D., and Yang, J.S., 1996b, *INRESB-3D-SUPII Program Listing: General Purpose Program for Inelastic Analysis of RC and Steel Building Systems for 3D Static and Dynamic Loads and Seismic Excitations*, NSF Report, U.S. Department of Commerce, National Technical Information Service, Springfield, VA, NTIS No. PB97-123616.

Cheng, F.Y. and Mertz, G., 1989a, *User's Manual for INRESB-3D-SUP, A Computer Program for Inelastic Analysis of 3-Dimentional Reinforced-Concrete and Steel Buildings*, Civil Engineering Study Structural Series 89-31, University of Missouri-Rolla, Rolla, MO.

Cheng, F.Y. and Mertz, G., 1989b, *Inelastic Seismic Response of Reinforced-Concrete Low-Rise Shear Walls and Building Structures*, Civil Engineering Study Structural Series 89-30, University of Missouri-Rolla, Rolla, MO.

Cheng, F.Y., Lu, L.W., and Ger, J.F., 1992, Observations on behavior of tall steel building under earthquake excitations, *Proceedings of Structural Stability Research Council*, Pittsburgh, PA, pp. 15–26.

Chopra, A.K., 2005, *Earthquake Dynamics of Structures, a Primer*, 2nd edn., Earthquake Engineering Research Institute, Oakland, CA, MNO-11.

Clough, R. and Penzien, J., 1975, *Dynamics of Structures*, McGraw-Hill, Inc., New York.

Dutta A. and Mander, J., 1998, Capacity Design and Fatigue Analysis of Confined Concrete Columns, MCEER Report MCEER-98-0007.

Dwairi, H. and Kowalsky, M.J., 2006, Implementation of inelastic displacement patterns in direct displacement-based design of continuous bridge structures, *Earthquake Spectra*, 22(3), 631–662.

Dwairi, H., Kowalsky, M.J., and Nau, J.M., 2007, Equivalent damping in support of direct displacement-based design, *Journal of Earthquake Engineering*, 11(4).

FEMA-273, Building Seismic Safety Council (BSSC), 1997, *NEHRP Guidelines for the Seismic Rehabilitation of Buildings*, Federal Emergency Management Agency, Washington, DC.

FEMA-356, Building Seismic Safety Council (BSSC), 2000, *Prestandard and Commentary for the Seismic Rehabilitation of Buildings*, Federal Emergency Management Agency, Washington, DC.

FHWA, 1996, Seismic Design of Bridges, Design Example No. 4, Three-span continuous CIP concrete bridge, FHWA-SA-97-009.

FHWA, 2006, *Seismic Retrofitting Manual for Highway Structures: Part 1—Bridges*, FHWA-HRT-06-032.

Frankel, A., Mueller, C., Barnhard, T., Perkins, D., Leyendecker, E., Dickman, N., Hanson, S., and Hooper, M., 1996, National Seismic Hazard Maps: Documentation, Open-File Report 96-32, U.S. Geological Survey, Reston, VA.

Ger, J.F. and Cheng, F.Y., 1992, Collapse assessment of a tall steel building damaged by 1985 Mexico earthquake, *Proceedings of the Tenth World Conference on Earthquake Engineering*, Madrid, Spain, July 1992, pp. 51–56.

Ger, J.F. and Cheng, F.Y., 1993, Post-buckling and hysteresis models of open-web girders, *Journal of Structural Engineering Division*, ASCE, 119, March, 831–851.

Ger, J.F., Cheng, F.Y., and Lu, L.W., 1993, Collapse behavior of Pino Suarez building during 1985 Mexico earthquake, *Journal of Structural Engineering Division*, ASCE, 119, March, 852–870.

Ghosn, M. and Moses, F., 1998, Redundancy in Highway Bridge Superstructures, NCHRP Report 406, TRB, National Research Council, Washington, DC.

Goel, R.K., 2005, Evaluation of modal and FEMA pushover procedures using strong-motion records of buildings, *Earthquake Spectra*, 21(30), 653–684.

Goel, R.K. and Chopra, A.K., 2004, Evaluation of Modal and FEMA pushover analyses: SAC Buildings, *Earthquake Spectra*, 20(1), 225–254.

Gulkan, P. and Sozen, M.A., 1974, Inelastic responses of reinforced concrete structures to earthquake motions, *ACI Journal*, 71(6), 604–610.

Gupta, B. and Kunnath, S.K., 2000, Adaptive spectra-based pushover procedure for seismic evaluation of structures, *Earthquake Spectra*, 16(2), 367–391.

ICBO, 1997, Structural engineering design provision, Uniform Building Code, Vol. 2, *International Conference of Building Officials*, Birmingham, AL.

International Code Council, Inc. (ICC), 2000, International Building Code: Building Officials and Code Administrators International, Inc., *International Conference of Building Officials, and Southern Building Code Congress International, Inc.*, Birmingham, AL.

Iwan, W.D. and Gates, N.C., 1979, Estimating earthquake response of simple hysteretic structures, *Journal of the Engineering Mechanics Division*, ASCE, 105(EM3), 391–405.

Jacobsen, L.S., 1930, Steady forced vibrations as influenced by damping, *ASME Transactions*, APM 52-15, 51, 169–181.

Jain, A.K., Goel, S.C., and Hanson, R.D., 1980, Hysteretic cycles of axially loaded steel members, *Journal of the Structural Division*, ASCE, 106, 1777–1795.

Kwan, W.P. and Billington, S.L., 2003, Influence of hysteretic behavior on equivalent period and damping of structural systems, *Journal of the Structural Engineering*, ASCE, 129(5), 576–585.

Liu, W.D., Ghosn, M. and Moses, F., 2001, Redundancy in Highway Bridge Substructures, NCHRP Report 458, TRB, National Research Council, Washington, DC.

Mander, J.B., Priestley, J.N., and Park, R., 1988, Theoretical stress-strain model for confined concrete, *Journal of Structural Engineering*, ASCE, 114(8), 1804–1826.

Maron, M.J., 1982, *Numerical Analysis, a Practical Approach*, Macmillan Publishing Co., Inc., New York.

MCEER/ATC Joint Venture, 2003, "Design Examples, Recommended LRFD guidelines for the seismic design of highway bridges", NCHRP 12-49 project, MCEER Report No. MCEER-03-SP09.

McGuire, W., Gallagher, R., and Ziemian, R., 2000, *Matrix Structural Analysis, with MASTAN2*, 2nd edn., John Wiley & Sons, Inc., New York.

Menun, C. and Kiureghian, A.D., 1998, A replacement for the 30%, 40%, and SRSS rules for multicomponents seismic analysis, *Earthquake Spectra*, 14(1), 153–163.

Miranda, E. and Bertero, V., 1994, Evaluation of strength reduction factors for earthquake-resistant design, *Earthquake Spectra*, 10(2), 357–379.

Moehle, J. et al., 1995, Highway bridges and traffic management, *Earthquake Spectra*, 11(S2), 287–372.

Naeim, F., 1989, *The Seismic Design Handbook*, Van Nostrand Reinhold, New York.

Ohtaki, T., Benzoni, G., and Priestley, N., 1996, Seismic Performance of a Full Scale Bridge Column—as Built and as Repaired, Report No. SSRP-96/07, University of California, San Diego, CA.

Park, Y. and Ang, A.H.S., 1985, Seismic damage analysis of reinforced concrete buildings, *Journal of Structural Division*, ASCE, 111(4), 740–757.

Paulay, T. and Priestley, M.J.N., 1992, *Seismic Design of Reinforced Concrete and Masonry Buildings*, John Wiley & Sons, Inc., New York.

Popov, E.P. and Black, R.G., 1981, Steel struts under severe cyclic loadings, *Journal of Structural Engineering Division*, ASCE, 107(ST9), 1857–1881.

Priestley, M.J.N., Calvi, G.M., and Kowalsky, M.J., 2007, *Displacement-Based Seismic Design of Structures*, IUSS Press, Pavia, Italy.

Priestley, J.N., Seible, F., and Calvi, G.M., 1996, *Seismic Design and Retrofit of Bridges*, Wiley, New York.

Ramberg, W. and Osgood, W.R., 1943, Description of stress–strain curves by three parameters, Technical Note 902, National Advisory Committee for Aeronautics, Washington, DC.

Seed, H.B., Ugas, C., and Lysmer, J., 1976, Site dependent spectra for earthquake resistant design, *Bulletin of Seismological Society of America*, 66(1), 1323–1342.

Seible, F., Priestley, N., Latham, C., and Silva, P., 1994, Full-Scale Bridge Column/Superstructure Connection Tests under Simulated Longitudinal Seismic Loads, Report No. SSRP-94/14, University of California, San Diego, CA.

SEQMC, Demo version 1.00.06, 1998, Moment-curvature analysis package for symmetric sections, SC Solutions, Santa Clara, CA.

South Carolina Department of Transportation (SCDT), 2001, *Seismic Design Specifications for Highway Bridges*.

Stone, W.C. and Cheok, G.S., 1989, *Inelastic Behavior of Full-Scale Bridge Columns Subjected to Cyclic Loading*, NIST Building Science Series 166, U.S. Government Printing Office, Washington, DC.

Suarez, V. and Kowalsky, M., 2006, Implementation of displacement based design for highway bridges, *Fifth National Seismic Conference on Bridges and Highways*, San Francisco, CA, September 18–20, 2006.

Takeda, T., Sozen, M.A., and Nielsen, N.N., 1970, Reinforced concrete response to simulated earthquakes, *Journal of the Structural Division*, ASCE, 96(ST12), 2557–2573.

Taylor, A.W., Kuo, C., Wellenius, K., and Chung, D., 1997, A Summary of Cyclic Lateral Load Tests on Rectangular Reinforced Concrete Columns, NIST, Report No. NISTIR 5984.

Weaver, W. and Johnston, P., 1987, *Structural Dynamics by Finite Elements*, Prentice-Hall, Inc., Englewood Cliffs, NJ.

Wilson, E.L., Kiureghian, A.D. and Bayo, E.P., 1981, A replacement for the SRSS method in seismic analysis, *Journal of Earthquake Engineering and Structural Dynamics*, 9, 187–194.

Zahn, F.A., Park, R., and Priestley, M.J.N., 1986, Design of Reinforced Concrete Bridge Columns for Strength and Ductility, Report 86-7, Department of Civil Engineering, University of Canterbury, Christchurch, New Zealand.

Zahrai, M.S. and Bruneau, M., 1999, Ductile end-diaphragms for seismic retrofit of slab-on-girder steel bridges, *Journal of Structural Engineering*, ASCE, 125(1), 71–80.

Index